T0234523

the essential DAVID**BOHM**

There are few scientists of the twentieth century whose life's work has created more excitement and controversy than that of physicist David Bohm (1917–1992). Exploring the philosophical implications of both physics and consciousness, Bohm's penchant for questioning scientific and social orthodoxy was the expression of a rare and maverick intelligence.

For Bohm, the world of matter and the experience of consciousness were complementary aspects of a more fundamental process he called the *implicate order*. Without a working sensibility of what an implicate order might be, our conceptions of the various threads of Bohm's work – whether in quantum theory or social dialogue – remain incomplete. But with an enhanced understanding of such an order, the wholeness of Bohm's work becomes apparent and accessible.

For the first time in a single volume, *The Essential David Bohm* offers a comprehensive overview of Bohm's original works from a non-technical perspective. Including three chapters of previously unpublished material, each reading has been selected to highlight some aspects of the implicate order process, and to provide an introduction to one of the most provocative thinkers of our time.

the essential
DAVID **BOHM**
edited by **lee nichol**

With a Reminiscence by H.H. The Dalai Lama

Routledge
Taylor & Francis Group
LONDON AND NEW YORK

First published 2003
by Routledge
2 Park Square, Milton Park, Abingdon, Oxon OX14 4RN

Simultaneously published in the USA and Canada
by Routledge
270 Madison Avenue, New York, NY 10016

Reprinted 2004, 2006, 2007 (twice)

Routledge is an imprint of the Taylor & Francis Group, an informa business

© 2003 Sarah Bohm; Lee Nichol for selection, editorial matter
and Chapter 8; Yitzhak Woolfson for Chapter 7

Typeset in Celeste by RefineCatch Limited, Bungay, Suffolk

British Library Cataloguing in Publication Data
A catalogue record for this book is available from the British Library

Library of Congress Cataloging in Publication Data
Bohm, David.
    [Selections. 2002]
    The essential David Bohm / edited by Lee Nichol.
        p. cm.
    Includes bibliographical references and index.
        1. Science–Philosophy. 2. Creative ability in science. 3. Order (Philosophy) I. Nichol,
Lee, 1955– II. Title.
    QC175 .B6652 2002
    501–dc21   2002031683

ISBN 10: 0–415–26173–2 (hbk)
ISBN 10: 0–415–26174–0 (pbk)

ISBN 13: 978–0–415–26173–9 (hbk)
ISBN 13: 978–0–415–26174–6 (pbk)

It was a hard thing to undo this knot.
The rainbow shines, but only in the thought
Of him that looks. Yet not in that alone,
For who makes rainbows by invention?

And many standing round a waterfall
See one bow each, yet not the same to all,
But each a hand's breadth further than the next.
The sun on falling waters writes the text
Which yet is in the eye or in the thought.
It was a hard thing to undo this knot.

—Gerard Manley Hopkins

CONTENTS

ACKNOWLEDGEMENTS

The editor would like to express his gratitude to the following individuals, each of whom uniquely contributed to the completion of this volume: His Holiness the Dalai Lama, Saral Bohm, Neil Nelson, Sonia Nelson, Yitzhak Woolfson, Leroy Little Bear, Amethyst First Rider, Michael Krohnen, Bruce Robertson, Jean Robertson, Sylvia Gretchen, Ron Purser, John Erskine, Steve McCarl, Bill Nichol, Brenda Goings, Fred Calhoun, Bruce Garber, Arthur Braverman, Richard Kortum, Peter Gold, Richard Burg, Jack Petranker, Leslie Bradburn, Tony Bruce, Muna Khogali, Nann Henderson, Kevin Shluker and Aja Bulla-Richards.

Special thanks are due to Basil Hiley, David Peat, Renée Weber, Paavo Pylkkänen, Don Factor, Mark Edwards, Arleta Griffor, and Jane Fleming, without whose efforts over many years the body of work from which this volume is distilled would have been greatly diminished.

The untitled poem by Gerard Manley Hopkins (1844–1889) on page v is included to honor the vision and perseverance of David Bohm.

The editor and publisher wish to thank the following for kind permission to reprint:

Chapter 4: reprinted from *Dialogues with Scientists and Sages: The Search for Unity* by Renée Weber, Routledge and Kegan Paul, London © Renée Weber, 1986.

Chapter 9: reprinted from *Physics and the Ultimate Significance of Time*, ed. David Ray Griffin by permission of the State University of New York Press, New York © 1986, State University of New York. All rights reserved.

THE DALAI LAMA

Reminiscence About David Bohm

David Bohm was a scientist well known for his work in quantum physics. I met him on several occasions and my personal impression of him as a human being was that although he was a famous scientist, he was neither arrogant nor proud. On the contrary, he was humble and easy to talk to. True to his scientific training he was unbiased and open-minded. Consequently, although he was mostly focused on his own field of scientific research, he was also interested in consciousness because of its implications with regard to the theories of quantum physics. Therefore, he also showed interest in some of the ancient Indian schools of philosophy.

He seems to have known Krishnamurti quite well and had stimulating interactions with him. When we met, he would ask about the Buddhist concept of interdependence and the corresponding theory of emptiness. For my part I had the opportunity to listen to him explain quantum physics on a number of occasions and I made some effort to understand these difficult ideas. Because he was the one I could question in detail about quantum mechanics, I consider him to have been one of my scientific 'gurus'.

November 17, 2001

EDITOR'S INTRODUCTION

There are few scientists of the twentieth century whose life's work has generated more excitement and controversy than that of physicist David Bohm. Marginalized by his colleagues in the physics élite for challenging the standard interpretation of quantum theory, forced into *de facto* expatriation for resisting McCarthyism, admired and derided alike for his interest in the phenomenology of consciousness, Bohm's penchant for questioning scientific and social orthodoxy was the natural expression of a rare and maverick intelligence.

Bohm, who in the 1940s refined the theory of plasma as the fourth state of matter, and in the 1950s recast the Einstein–Podolsky–Rosen paradox and shared in the discovery of the Aharonov–Bohm effect, was widely considered one of the most talented and promising physicists of his generation. But his primary work from the 1950s to the 1990s – the ongoing development of his "causal interpretation" of quantum mechanics as an alternative to the standard Copenhagen interpretation – was met with surprising hostility by the majority of the world physics community.

Due largely to a 1994 *Scientific American* cover story[1] and F. David Peat's *Infinite Potential – The Life and Times of David Bohm* (1997),[2] the means by which Bohm's alternative quantum theory had been effectively suppressed came to light, and the general outlines of this alternative were finally presented to a substantial reading public. This theory, developed in collaboration with Prof. Basil Hiley and known in its mature form as the "ontological interpretation" of quantum mechanics,[3] is now widely viewed as a serious critique of the Copenhagen interpretation, and proffers a revisioning of quantum theory in which objective reality is restored and undivided wholeness is fundamental.

Concurrent with the development of the ontological interpretation – and ultimately as an integral part of it – was the evolution of Bohm's philosophy of consciousness. At once ancient and modern, this philosophy employs concepts of contemporary psychology and cognitive science, yet is deeply resonant with perspectives as diverse as those of Native America, Buddhism, and Vedanta. And while there exists no school of Bohmian philosophy of consciousness, Bohm's insistence that the subtle unity of the "observer and the observed" be brought out of the laboratory and into the domain of immediate experience has contributed significantly to the mainstreaming of cognitive philosophy and consciousness studies.

For many observers there has been an easy temptation to think of these lines of inquiry as two separate and unrelated threads. To many physicists, Bohm squandered precious time and resources in exploring the nature of mind. For those non-scientists who were intuitively drawn to Bohm's philosophy of consciousness, particularly as it manifested in his work with social dialogue, Bohm's physics was the stuff of academia – respectable arcana perhaps, but forever out of reach, and without much concrete relevance.

For Bohm himself, however, these seemingly disparate lines of inquiry were manifestations of an integral primary vision he eventually came to call the *implicate order*. This "order" was a propositional template for plotting the emergence and dynamics of both matter and consciousness. Hinted at by topical analogies such as the hologram, Bohm's vision of the implicate order posited a dynamic "structure-process" from which our coordinate-based experience of three-dimensional space and time is but a derivative, temporary projection. Within that coordinate-based "explicate" order, we may have any number and variety of experiences, yet never come to consider that the very coordinate frame for all those experiences is itself a projection, another experience arising from a deeper "implicate" order.

Consequently, grasping what Bohm was alluding to with the implicate order can be an elusive undertaking, and as it likely was for him, ever a work in progress. Yet without a working sensibility of what an implicate order might be, our conceptions of the various aspects of Bohm's work will necessarily remain limited and incomplete. With such a sensibility in hand, however, not only does the wholeness of Bohm's work become readily apparent, but each facet – be it scientific, sociological, or contemplative – is now illuminated as if from within, infused with new and richer meaning.[4]

It is thus the intention of the present volume to provide the reader with the essential materials for acquiring such an enhanced understanding of one of the most radical and important thinkers of the twentieth century. Offering for the first time between two covers a comprehensive overview of David Bohm's original works from a non-technical perspective, each reading has been selected to contextualize or highlight some feature of the implicate order process.

In conceiving the implicate order, Bohm was originally concerned with the nature of order *per se*. Both scientifically and socially, he felt that we are largely unaware of the degree to which inherited orders, or paradigms, dominate our perception and thought. In our current time, it is the order of *mechanism* that prevails. The essence of a mechanistic view is that all natural phenomena exist in a strictly external relationship to one another, and exhibit precise and discernible chains of cause and effect. In such a world of externalized entities, the theory goes, it is only a matter of time before the "ultimate particle" – the basic stuff of which the universe is made – will be discovered and explained.

However, in both theory and experiment this mechanistic order has been undermined by a series of developments in twentieth-century physics. Bohm points to four features of the "new" physics that, taken together, require the formulation of a larger, more encompassing order. The first significant challenge to mechanism came from Einstein, who claimed there were deep contradictions in the very notion of an independently existing particle. He proposed that what we normally think of as a particle is actually a temporary localized *pulse* emerging from a larger field, very much as a vortex temporarily forms from the dynamic flowing of a stream. In Bohm's words, "the field structure associated with two pulses will merge and flow together in one unbroken whole."[5] In this way, the mechanistic notion of the inherent separability of material structures was initially questioned. Yet Einstein ultimately took recourse in a mechanistic view by asserting that the fields themselves were separate and external from one another.

It was to be three subsequent discoveries in quantum mechanics that would more profoundly challenge a mechanistic world view. The first of these is that the movement of quantum entities (photons, electrons, etc.) does not possess the classical (and common sense) attribute of steady, continuous development from one state or position to another. Rather, the movement of quantum entities is *discontinuous*. Action or movement is constituted of indivisible *quanta* – discrete pulses or packets of energy – which "jump" from one energy state to the next, without passing through intermediary energy states. To Bohm, this quantum discontinuity implies a universe woven together in a dynamic, tapestry-like configuration: "If all actions are in the form of discrete quanta, the interactions between different entities (e.g., electrons) constitute a single structure of indivisible links, so that the entire universe has to be thought of as an unbroken whole."[6]

The second quantum discovery is that matter may behave like a solid particle in one context, like a wave in another context, and like both together in yet a third context. This context-dependent aspect of matter violates the mechanistic axiom that *an entity is what it is*, until it irreversibly transforms through disintegration. That an entity (e.g., an electron) can manifest as a wave in one context and a particle in another suggests, according to Bohm, a kind of information exchange that is more akin to transformations in organisms than to the interacting parts of machines. From this perspective, the "particle" would seem to be gathering information about its environment and responding according to the *meaning* of the information. This capacity for transformation – the famous "wave–particle duality" – is vividly illustrated in the double slit experiment.[7]

The third quantum discovery, and perhaps the strangest, is the well-demonstrated non-local connection between particles (e.g., electrons) – referred to as "entanglement" in current terminology. When under certain conditions electrons combine to form a coupled pair, and are then again separated by significant distances, instantaneous influence between the particles is discovered, for which there is no "local" or mechanical explanation. One of the possibilities that arises from this observation is that the speed of light has been violated; at the very least it becomes clear that local causality cannot fully account for the phenomena of quantum mechanics. Thus, either the theory of quantum mechanics is wrong – very unlikely, based on a century of unqualified experimental validation – or some form of non-locality permeates reality.

For Bohm, a reality structure founded on discontinuous quantized movement, context-dependent form, and non-local action requires the envisioning of a new order that both incorporates and goes beyond the order of mechanism. However, it was Bohm's position that this requirement is understood by the majority of the physics community only via highly abstract mathematical algorithms, with little or no concern for what these rather shocking assaults on common sense might *mean*. Indeed, the prevailing quantum orthodoxy maintains that any search for meaning or structure behind experimental phenomena is itself pointless and without meaning. Thus, the three quantum discoveries – with their potentially vast implications – are shoe-horned into a

---

tacit metaphysical commitment to mechanism, in spite of the fact that such a commitment is philosophically inconsistent with experimental fact. And as go the assumptions of the scientific community, felt Bohm, so go the assumptions of the culture at large. In this way the philosophy of mechanism continues to trickle into and structure the perception, imagination, and creativity of contemporary society.

Bohm's intention in outlining a philosophy of implicate order is therefore twofold. On the one hand he is pointing to the quite precise wholeness of the natural world suggested in the intersection of relativity and quantum theory. For Bohm, this wholeness is actual, it matters, and it warrants metaphysical investigation.[8] The implicate order therefore becomes a means of overtly modeling aspects of wholeness implied in that intersection, of providing a framework for approaching the contradictory aspects of that intersection, and of urging the community of physicists away from its investment in outmoded world views.

On the other hand, the structure of reality implied by twentieth-century physics is seen by Bohm *to be enfolded within and therefore available to human experience.* Wholeness is thus understood to be more than a theoretical construct – it is a meaning-field, a living totality that includes us. Here we enter new terrain in which we are active participants, not simply detached observers. For Bohm, the shift from an abstract observational perspective to the embodied knowledge of a participatory lifeworld was of the highest importance and carried the full force of necessity. Yet to the extent we remain invested in a non-participatory, mechanistic outlook, we deny ourselves even a glimpse of such new possibilities. And currently, the assumptions of mechanism continue to be tacitly embraced by each of us, made only stronger by the further assumption *that we have no metaphysical assumptions* – that we are more or less seeing reality "as it is."

In the social domain, these nested assumptions virtually assure conflict and fragmentation, because they tie us to existential identifications as absolutes rather than as functional categories. Self, family, nation, and ideology become restrictions and distortions, rather than membranes which connect us to larger wholes. When defending and sustaining these identifications become paramount, we are faced with a deep irony in which our shared creative potential – mind itself – has turned against us.

It was with this understanding in hand that Bohm ventured into the realm of experience, both individual and collective. It is not necessary to wait for the community of scientists to transcend a mechanistic view, felt Bohm. We can begin to inquire now, in the course of daily life, as to the efficacy and limits of a mechanistic orientation. And much may be at stake. As Paavo Pylkkänen suggests in his introduction to the *Bohm–Biederman Correspondence*,

> We live in an age where science and technology increasingly need to serve the interests of economic growth and immediate practical applications. The adverse effects of science and technology are well known, but it seems to be impossible to control and foresee developments such as those in information technology. For example, how will the current developments in this technology affect human consciousness? Will consciousness have to mechanize itself in order for the new information technology to function efficiently?[9]

It is issues of precisely this magnitude that led Bohm to repeatedly breach the artificial boundary between science and experience, and to question not only the structure of objectifiable reality, but also the structure of human subjectivity. He came to a position as old as civilization itself, though much out of favor in our time: mind and matter cannot be understood as two. But dedicated as he was to exploring the implications of such a view, Bohm repeatedly warned against assuming the final truth of any perspective, including his own. It was his view that, like the universe itself, no theory is ever complete. Each advance in knowledge may serve as a temporary resting point, but ultimately should function as a portal to yet further comprehension and insight. In this light, the implicate order model and *The Essential David Bohm* are perhaps best approached in a spirit of leisurely exploration and free play, and with a willingness to test the soundness of our world views, including those we hold most dear.

Lee Nichol
Albuquerque, New Mexico,
January 2002

Notes

1   Albert, David Z., "Bohm's Alternative to Quantum Mechanics," *Scientific American*, May 1994.
2   Peat, F. David, *Infinite Potential: The Life and Times of David Bohm*, Addison-Wesley, Reading, Mass. (1997).
3   See Bohm, D. and Hiley, B. J., *The Undivided Universe: An Ontological Interpretation of Quantum Theory*, Routledge, London (1993).
4   In working with the material in this volume, the reader may find the following illustrations

useful. Viewing the Necker cube gives a very good intuitive sense of an implicate order, in this case a perceptual implicate order. Generally speaking, it is only possible to perceive one of the two possible cubes at a time – while one is enfolded in perception (an implicate order), the other is unfolded in perception (an explicate order). This phenomenon is significantly amplified when one views the hypercube – many possible geometric orders remain enfolded in perception (a complex implicate order), while only one is unfolded at any given moment (an explicate order).

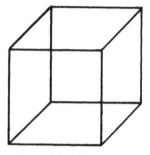

Figure I.1 A Necker cube

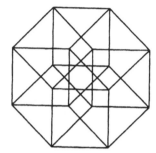

Figure I.2 A hypercube

5   Bohm, D., *Wholeness and the Implicate Order*, Routledge, London (1980), p. 174.

6   *Ibid*, p. 175.

7   For an explanation of the double slit experiment, see Chapter 6 of the present volume; also see Zukav, G., *The Dancing Wu Li Masters – An Overview of the New Physics*, William Morrow, New York (1979), pp. 82–9.

8   Bohm uses the term "metaphysics" in its original sense of making philosophical claims about the way *all things are*, not in its currently dismissive connotation which suggests fanciful speculation or mysticism. When "metaphysics" is considered in its original meaning, Bohm claims that all cultures and the individuals within them hold thoroughgoing metaphysical presuppositions. See Bohm, D., *On Creativity*, Routledge, London (1998), Ch. 4, for a fuller treatment of Bohm's views of metaphysics.

9   Bohm, D. and Biederman, C., *Bohm–Biederman Correspondence* (ed. Paavo Pylkkänen), London, Routledge (1999), p. xxi.

Part One – Universal Orders

1 THE QUALITATIVE INFINITY OF NATURE (1957)

In this essay Bohm argues for a philosophical perspective that transcends classical mechanism, while still utilizing a mechanistic framework when applicable. With such an approach we can sidestep unnecessary "either–or" philosophical choices, and thus maintain maximum flexibility of view with regard to a philosophy of science. Innocuous on its face, this perspective ultimately casts doubt on the widely-held view that the whole of nature can be reduced to a set of laws that are in principle finite and knowable.

To illustrate the basis for such a non-mechanical view, Bohm points to research into the structure of the atom. In spite of perennial claims that the ultimate structure of the atom is about to be found, the simple fact is that no such structure has yet been discovered. On the contrary, the depth and scale of atomic substructure give no indication whatsoever of being exhaustively determined, giving way to ever greater subtlety and theoretical complexity. An ultimate structure will not be found, says Bohm, because it does not exist. There is no ultimate or final structure to anything at all, including atoms. There are only relatively invariant dynamic structures that appear as "things" – electrons, starfish, trees – within limited contexts, inexorably transforming into new structures, forms, and processes when these contexts change. This theme of relative invariance recurs in subsequent chapters, and plays a central role in conceiving the implicate order.

Intersecting with the inexhaustible nature of form and structure is what Bohm refers to as the "background." This background is the totality of causes and conditions that contribute to the creation and dissolution of each

Extract from Chapter 5 of D. Bohm, *Causality and Chance in Modern Physics*, University of Pennsylvania Press, Philadelphia (1957) and Routledge, London (1957).

temporary "thing." Each thing not only has its own relatively stable sub-structure, but has a potentially unlimited context as well, much like a particle's derivative relation to its field. In Bohm's view, the complex relation of each thing to its background, together with the richness of its substructure, suggest that mechanistic accounts will always be superseded by new levels of complexity and order. If indeed elementary material "things" are thus potentially infinite in their nature, far more so will be the case for the whole of nature. Consequently, we must reformulate our conceptions of the laws of nature so that they account for context and contingency, rather than generalizing them in absolute fashion.

The pairing of substructure and background is an early formulation of the relation between implicate and explicate orders. While the investigation of material substructures is oriented toward increasingly subtle aspects of an explicate three-dimensional order, the notion of an infinite background lays the foundation for conceiving a multi-dimensional implicate order, from which all material structure – of any degree of subtlety – derives its relative existence and sustenance.

In outlining the dynamic between structure and background, Bohm introduces the notion of reciprocal relationship. In the mechanistic concept of interaction, the various components of a given system affect one another in strictly quantitative fashion, exemplified by Newton's laws of motion which specify equality of action and reaction, as well as the maintenance of the identity of the components. By contrast, a reciprocal relationship enables a qualitative relation between structure and background, in which each has the potential not only to "impact" the other, but to generate transformations in the nature of what each actually is.

From this perspective, each "thing," whether particle or starfish, derives its relative stability through the influence of background and substructure. But this influence is not a one-way affair – each particular thing in its turn may affect and influence its respective background and substructure. Thus qualitative changes may flow in both directions, with particles subtly affecting their fields (as well as the reverse) or starfish subtly affecting their ecosystems (as well as the reverse). More broadly considered, the notion of reciprocal relation allows for nested, mutual influence even between macroscopic processes and those at the atomic level, indicating the complexity of the pathways through which the qualitative infinity of nature may manifest.

1 INTRODUCTION

For several centuries there has existed a very strong tendency for one form or another of the philosophy of mechanism to be generally adopted among physicists. In the present chapter we shall criticize this philosophy, demonstrating the weaknesses in its basic assumptions, and then we shall go on to propose a different and broader point of view which we believe to correspond more nearly than does mechanism to the implications of scientific research in a wide range of fields. In addition to presenting this broader point of view in some detail, we shall also show how it permits a more satisfactory resolution of several important problems, scientific as well as philosophical, than is possible within the framework of a mechanistic philosophy.

2 SUMMARY OF THE ESSENTIAL CHARACTERISTICS OF A MECHANISTIC PHILOSOPHY

The essential characteristics of a mechanistic philosophy in the most general form that it has developed thus far in physics are the following:

The enormous diversity of things found in the world, both in common experience and in scientific research, can all be reduced completely and perfectly and unconditionally (i.e. without approximation and in every possible domain) to nothing more than the effects of some definite and limited general framework of laws. While it is admitted that the details of these laws may be subjected to changes in accordance with new experimental results that may be obtained in the future, its basic general features are regarded as absolute and final. This means that the fundamental entities that are supposed to exist, the kinds of qualities that define the modes of being of these entities, and the general kinds of relationships in terms of which the basic laws are to be expressed, are supposed to fit into some fixed and limited physical and mathematical scheme, which could in principle be subjected to a complete and exhaustive formulation, if indeed it is not supposed that this has already been done. At bottom, the only changes that are regarded as possible within this scheme are quantitative changes in the parameters or functions defining the state of the system (as precisely as the nature of the system permits this state to be defined[1]), while fundamental qualitative changes in the modes of being of the basic entities and in the forms in which the basic laws are to be expressed are not regarded as possible. Thus, the essence of the mechanistic position lies in its assumption of fixed basic qualities, which means that the laws themselves will finally reduce to purely quantitative relationships.

The philosophy of mechanism has undergone an extensive evolution in its specific form, all the while retaining the essential characteristics described above, in forms that tend, however, to become more and more complex and subtle with the further development of science.

3 CRITICISM OF THE PHILOSOPHY OF MECHANISM

We shall now review some of the most important criticisms that can be made against the philosophy of mechanism.

First of all, the historical development of physics has not confirmed the basic assumptions of this philosophy, but rather, has continually contradicted them. Thus, since the time of Newton, there have been introduced, not only the whole series of specific changes in the conceptual structure of physics,[2] but also the revolutionary changes in the whole general framework, brought about by the theory of relativity and the quantum theory. Moreover, physics is now faced with a crisis in which it is generally admitted that further changes will have to take place, which will probably be as revolutionary compared to relativity and the quantum theory as these theories are compared to classical physics.

Secondly, the mechanistic assumption of the absolute and final character of any feature of our theories is never necessary. For the possibility is always open that such a feature has only a relative and limited validity, and that the limits of its validity may be discovered in the future. Thus, Newton's laws of motion, regarded as absolute and final for over two hundred years, were eventually found to have a limited domain of validity, these limits having finally been expressed with the aid of the quantum theory and the theory of relativity. Indeed, the mechanistic thesis that certain features of our theories are absolute and final is an assumption that is not subject to any conceivable kind of experimental proof, so that it is, at best, purely philosophical in character.

Thirdly, the assumption of the absolute and final character of any feature of our theories contradicts the basic spirit of the scientific method itself, which requires that *every* feature be subjected to continual probing and testing, which may show up contradictions at any point where we come into a new domain or to a more accurate study of previously known domains than has hitherto been carried out. Indeed, the normal pattern that has developed without exception in every field of science studied thus far has been just the appearance of an endless series of such contradictions, each of which has led to a new theory permitting an improved and deeper understanding of the material under investigation. Thus, the full and consistent application of the scientific method makes sense only in a context in which we refrain from assuming the absolute and final character of any feature of any theory and in which we therefore do not accept a mechanistic philosophy.

Of course, the above arguments do not prove that a mechanistic philosophy is definitely wrong. For it is always conceivable that the trouble thus far has been that we have just not found the true absolute and final theory, and that this theory may be somewhere beyond the horizon of current scientific research. On the other hand, the historically demonstrated inadequacy of this philosophy up to the present, the fact that its basic assumptions cannot possibly be proved, and the fact that they are in disagreement with the whole spirit of the scientific method, would suggest to us that it may well be worth our while to consider points of view that go outside the limits of a mechanistic philosophy. It is with the development of a point of view having such an aim that we shall be concerned throughout the rest of this chapter.

4   A POINT OF VIEW THAT GOES BEYOND MECHANISM

The nucleus of a point of view that goes beyond mechanism and that is also in better accord than is mechanism with general scientific experience and with the needs of scientific research has been previously presented,[3] in connection with the extremely rich and diversified structure that has thus far actually been found in the laws of nature. The most essential feature characterizing this general structure is this: Any given set of qualities and properties of matter and categories of laws that are expressed in terms of these qualities and properties is in general applicable only within limited contexts, over limited ranges of conditions and to limited degrees of approximation, these limits being subject to better and better determination with the aid of further scientific research. Indeed, both the very character of the empirical data and the results of a more detailed logical analysis show that beyond the above limitations on the validity of any given theory, the possibility is always open that there may exist an unlimited variety of additional properties, qualities, entities, systems, levels, etc., to which apply correspondingly new kinds of laws of nature. Or, lumping all of the above diverse possibilities into the single category of "things," we see that a systematic and consistent analysis of what we can actually conclude from experimental and observational data leads us to the notion that nature may have in it an infinity of different kinds of things.

It is clear that this point of view carries us completely outside the scope of what can be considered a mechanistic philosophy. For, as we recall, the mechanistic point of view involves the assumption that the possible variety in the basic properties and qualities existing in nature is limited, so that one is permitted at most to consider quantitative infinities, which come from making some finite number of kinds of things bigger and bigger or more and more numerous. Moreover, it is also clear that the notion to which we have been led is quite distinct from that of a series of successive approximations that converge to some fixed and limited set of final laws. For there is evidently no reason why new qualities and properties and the corresponding new laws should always lead just to smaller and smaller corrections that converge in this simple and uniform way towards definite results. This may well be what happens in certain contexts and within a definite range of conditions. Nevertheless, there is no conceivable empirical justification for excluding the possibility that in different contexts or under changed conditions these new qualities, properties, and laws will lead to effects that are not small in relation to those following from previously known properties, qualities, and laws.

Thus, for example, while the laws of relativity and quantum theory do in fact lead under special conditions to small corrections to those of Newtonian mechanics, they lead more generally, as is well known, to qualitatively new results of enormous significance, results that are not contained in Newtonian mechanics at all.[4] The same possibility evidently necessarily exists with regard to any other new laws that may eventually be discovered. Therefore, the assumption that the laws of nature constitute an infinite series of smaller and smaller steps that approach what is in essence a mechanistic limit is just as arbitrary and unprovable as is the assumption of a finite set of laws permitting an exhaustive treatment of the whole of nature.

We see then, that, as far as the empirical data of science themselves are concerned, they cannot justify any *a priori* restrictions at all, either on the character or on the relative importance in different conditions and contexts of the inexhaustibly rich and diversified qualities and properties that may exist in nature. Such qualities and properties – which can always, as far as we are able to tell, lie hidden behind the errors and inadequacies of any given set of theories – may be disclosed later in an investigation carried out under new conditions, in new contexts, or to new degrees of approximation.

Thus far, we have been led by our analysis of the character of empirical data and of scientific theories only to a consideration of the *possibility* that nature may have in it an infinity of potentially or actually significant qualities (i.e. qualities which are of major importance or which can become of major importance under suitable conditions and in suitable contexts). It is now clear, however, that there are really only two possibilities with regard to this problem. Either the qualities having this kind of significance are limited in number, or else they are not. To suppose the former is essentially to fall back into one form or another of the mechanistic philosophy, to which, as we have seen, so many objections can be raised. If we wish to go outside the mechanistic philosophy, we therefore really have no choice but to consider the consequences of the assumption that the number of such significant qualities is not limited.

We have thus been led to see what is the first crucial step towards a point of view that goes beyond the mechanistic philosophy. On the other hand, at this stage of the analysis, this point of view presents itself as one of two possible alternatives: i.e. either mechanism or an infinity of potentially or actually significant qualities. Evidently we must choose one or the other. But on what basis can we make such a choice? In order to answer this question we point out that the notion of the qualitative infinity of nature becomes more than merely an alternative to the philosophy of mechanism, if we take into account the role of conditions, context, and degree of approximation in limiting the domain of applicability of any given theory. For, with this addition, it constitutes a broader point of view, in the sense that it contains within it all of those consequences of mechanism which represent a genuine contribution to the progress of scientific research, while it does not contain those which make no such contribution and which impede scientific research.

To see this, we first note that, with regard to any given domain of phenomena, the specific form of the assumption of the qualitative infinity of nature that has been suggested above does not contradict the notion that these phenomena can be treated in terms of some finite set of qualities and laws, and indeed, in terms of a number much smaller than the number of items of empirical data that may be available. It is evident that if this were not possible, then one of the most important achievements of scientific theories would be lost, for they would no longer permit the explanation and prediction of a large number of at first sight independent phenomena on the basis of relatively few general qualities, properties, laws, principles, etc. The recognition of this possibility and its practical exploitation in a wide range of fields was indeed the basic contribution that the mechanistic philosophy brought to science in the early phases of its development.

As we have seen, however, as long as we qualify our theories by specifying the context, conditions, and degrees of approximation to which they are valid, or at least by admitting that these limitations on their validity must eventually be discovered, then the notion of the qualitative infinity of nature leads one to treat any given domain of phenomena in exactly the same way as is done if one adopts a mechanistic point of view. It is only with regard to predictions in new domains, in new contexts, and to new degrees of approximation that the qualitative infinity of nature dictates an additional measure of caution, since it implies that eventually (but exactly where must be determined only empirically) any limited number of qualities, properties, and laws will prove to be inadequate. But, as we have seen, the very form of the data themselves, as well as a logical analysis of their meaning, dictates exactly the same measure of caution. We see, then, that none of the really well-founded conclusions that can be obtained with the aid of the assumption of a finite number of qualities in nature can possibly be lost if we assume instead that the number of such qualities is infinite, and at the same time recognize the role of contexts, conditions, and degrees of approximation.[5] All that we can lose is the illusion that we have good grounds for supposing that in principle we can, or eventually will be able, to predict everything that exists in the universe in every context and under all possible conditions.

Not only can nothing of real value for scientific work be lost if we adopt the notion of the qualitative infinity of nature in the specific form that has been described here, but on the contrary, much can be gained by doing this. For, first of all, we can thereby free scientific research from irrelevant restrictions which tend to result from (and which have in fact so often actually resulted from) the supposition that a particular set of general properties, qualities, and laws must be the correct ones to use in all possible contexts and conditions and to all possible degrees of approximation. Secondly, we are led to a concept of the nature of things which is in complete accord with the most basic and essential characteristic of the scientific method; i.e. the requirement of continual probing, criticizing and testing of every feature of every theory, no matter how fundamental that theory may seem to be. For this view explains the necessity for doing scientific research in just this way and in no other way, since, if there is no end to the qualities in nature, there can be no end to our need to probe and test all features of all of its laws. Finally, as we shall show throughout the rest of this chapter, the assumption of the qualitative infinity of nature leads to a much more satisfactory solution of a number of important problems, both scientific and philosophical, than is possible within the framework of a mechanist philosophy; and this in turn gives further evidence that it is a better point of view for the guidance of scientific research.

In conclusion, then, the notion of the qualitative infinity of nature permits us to retain all the positive achievements that were made possible by the development of mechanism. In addition, it enables us to go beyond mechanism by showing the limitations of the latter philosophy and by pointing towards new directions in which our concepts and theories may undergo further development. Naturally, we do not wish to propose here that the qualitative infinity of nature is a final doctrine, beyond which no further steps can ever be made. Indeed, as science progresses, it seems very likely that the qualitative infinity of nature will eventually be found to fit into some still more general point of view, which in turn retains its positive achievements, and which goes beyond them, much as the notion of the qualitative infinity of nature goes beyond mechanism. But, in this chapter, our purpose is merely to call attention to the many factors that suggest the need for this important step carrying us outside the limits of a mechanistic philosophy, and to show the numerous advantages that come from taking this step.

5   MORE DETAILED EXPOSITION OF THE MEANING OF QUALITATIVE INFINITY OF NATURE

In this section we shall bring out in more detail what is the general view of the world implied by the notion of the qualitative infinity of nature, and we shall show how this view agrees with the actual results of research that have been obtained thus far in the field of physics.

In order to make possible a discussion in relatively concrete terms we shall begin by considering a specific example: viz., the atomic theory of matter. Now, as is well known, the earliest forms of the atomic theory were based on the assumption that the fundamental qualities and properties defining the modes of being of the atoms were limited in number. On the other hand, deeper studies of the atom have disclosed more and more details of a moving substructure, which has within it a richness of properties and qualities that has never yet shown the slightest sign of being exhausted or of approaching exhaustion. Thus, there was found in the atoms a structure of electrons moving around a central nucleus consisting of neutrons and protons which themselves took part in further characteristic kinds of motions of their own. Within all of these motions appeared quantum-mechanical fluctuations of various kinds. Then came the discovery of a structure for the electrons and protons involving in some as yet poorly understood way the motions of unstable particles such as mesons and hyperons. Still later came the realization that because these latter particles can be "created," "destroyed," and transformed into each other, they too are very likely to have a further structure that is related to the motions of some still deeper-lying kinds of entities the nature of which is not yet known.

An essential characteristic of the rich and highly interconnected substructure of moving matter described above is that not only do the quantitative properties in it continually change but that the basic qualities that define its mode of being can also undergo fundamental transformations when conditions alter sufficiently. Thus, in electrical discharges, atoms can be excited and ionized, in which case they obtain many new physical and chemical properties. Under bombardment with very high-energy particles, the nuclei of the various chemical elements can be excited and transformed into new kinds of nuclei, with even more radical changes in their physical and chemical properties. Moreover, in nuclear processes, neutrons can be transformed into protons, either by the emission of neutrinos or of mesons; and of course, as we have seen, mesons are unstable, so that their very mode of existence implies the necessity for their transforming into basically different kinds of particles. Thus, further research into the structure of matter has not only shown what is, as far as we have been able to tell, an unlimited variety of qualities, processes, and relationships, but it has also demonstrated that all of these things are subject to fundamental transformations that depend on conditions.

Thus far, we have tended to emphasize the inexhaustible *depth* in the properties and qualities of matter. In other words, we have considered how experiments have shown the existence of level within level of smaller and smaller kinds of entities, each of which helps to constitute the substructure of the entities above it in size, and each of which helps to explain, at least approximately, by means of its motions how and why the qualities of the entities above it are what they are under certain conditions, as well as how and why they can change in fundamental ways when conditions change. But now we must take into account the fact that the basic qualities and properties of each kind of entity depend not only on their substructures but also on what is happening in their general background. In physics, research thus far has not tended to stress this feature of the laws of nature as much as it has emphasized the substructure. Nevertheless, the various fields (e.g., electromagnetic, gravitational, mesonic, etc.) that have been introduced into the conceptual structure of physics represent to some extent an explicit recognition of the importance of the background. For, as we have seen, these fields (whose mode of existence requires that they be defined over broad regions of space) enter into the definition of the basic characteristics of all the fundamental particles of current physics. Moreover, when such fields are highly excited, they too can give rise to qualitative transformations in the particles, while, vice versa, the particles have an important influence on the character of the fields. Indeed, previous discussion[6] of the quantum theory shows that fields and particles are closely linked in an even deeper way, in the sense that both are probably opposite sides of some still more general type of entity, the detailed character of which remains to be discovered.[7] Thus, the next step in physics may well show the inadequacy of the simple procedure of just going through level after level of smaller and smaller particles, connected perhaps by fields which interact with these particles. Instead, we may find that the background enters in a very fundamental way even into the definition of the conditions for the existence of the new kinds of basic entities to which we will eventually come, whatever they may turn out to be. Thus, we may be led to a theory in which appears a much closer integration of substructure and background into a well-knit whole than is characteristic of current theories.

We see from the above discussion that the qualitative infinity of nature is not equivalent to the idea expressed by the well-known rhyme:

> "Great fleas have little fleas
> Upon their backs to bite 'em;
> Little fleas have lesser fleas,
> And so *ad infinitum.*"

For, firstly, we are not supposing that the same pattern of things is necessarily repeated at all levels; and secondly, we are not even supposing that the general pattern of levels that has been so widely found in nature thus far must necessarily continue without limit. While we cannot decide this question from what is known at present, we have already suggested reasons why we may perhaps now be approaching a point at which the notion of levels will, at the very least, have to be enriched a great deal by the explicit inclusion of the effects of a background that is essential for the very existence of the entities in terms of which our theories are to be formulated. Moreover, it is evidently quite possible that as we penetrate further still, we will find that the character of the organization of things into levels will change so fundamentally that even the pattern of levels itself will eventually fade out and be replaced by something quite different. Hence, while the qualitative infinity of nature is consistent with an infinity of levels, it does not necessarily imply such an infinity. And, more generally, this notion does not require *a priori* the continuation of any special feature of the general pattern of things that has been found thus far, nor does it exclude *a priori* the possibility that any such feature may continue to be encountered, perhaps in new contexts and in new forms, no matter how far we may go. Such questions are left to be settled entirely by the results of future scientific research.

There is, however, one general statement that can be made at this point about the inexhaustible diversity of things that may exist in the universe; namely, that they must have some degree of autonomy and stability in their modes of being. Now, thus far, we have always found that such autonomy exists.[8] Indeed, if it did not exist, then we would not be able to apply the concept of a "thing" and there would then be no way even to formulate any laws of nature. For how can there be an object, entity, process, quality, property, system, level, or whatever other thing one cares to mention, unless such a thing has some degree of stability and autonomy in its mode of existence, which enables it to preserve its own identity for some time, and which enables it to be defined at least well enough to permit it to be distinguished from other things? If such relatively and approximately autonomous things did not exist, then laws would lose their essential significance . . .

In conclusion, then, actual scientific research has thus far shown the need to analyse nature in terms of a series of concepts that involve the recognition of the existence of more and more kinds of things; and the development of such new concepts has never yet shown any signs of coming to an end. Up to the present, the various kinds of things existing in nature have, at least as far as investigations in the field of physics are concerned, been found to be organized into levels. Each level enters into the substructure of the higher levels, while, vice versa, its characteristics depend on general conditions in a background determined in part in other levels both higher and lower, and in part in the same level. It is quite possible, of course, that further studies will disclose a still more general pattern of organization of things. In any case, it is clear that the results of scientific research to date strongly support the notion that nature is inexhaustible in the qualities and properties that it can have or develop. If the laws of nature are to be expressible in any kind of terms at all, however, it is necessary that the things into which it can be analysed shall have at least some degree of approximate and relative autonomy in their modes of being, which is maintained over some range of variation of the conditions in which they exist.

6 CHANCE AND NECESSARY CAUSAL INTERCONNECTIONS

With the aid of the general world view described in Section 5, we shall now proceed to show that the hypothesis of the qualitative infinity of nature provides a framework within which can fit quite naturally the concept of chance and necessary causal interconnections as two sides of every real natural process.

First of all, we point out that if there are an unlimited number of kinds of things in nature, no system of purely determinate law can ever attain a perfect validity. For every such system works only with a finite number of kinds of things, and thus necessarily leaves out of account an infinity of factors, both in the substructure of the basic entities entering into the system of law in question and in the general environment in which these entities exist. And since these factors possess some degree of autonomy, one may conclude[9] that the things that are left out of any such system of theory are in general undergoing some kind of a random fluctuation. Hence, the determinations of any purely causal theory are always subject to random disturbances, arising from chance fluctuations in entities, existing outside the context treated by the theory in question. It thus becomes clear why chance is an essential aspect of any real process and why any particular set of causal laws will provide only a partial and one-sided treatment of this process, which has to be corrected by taking chance into account.

Of course, it should not be supposed that every inadequacy or breakdown of causal laws must necessarily be due to the effects of chance fluctuations. Indeed, as happened in connection with the experiments leading to the theory of relativity (Michelson–Morley experiment, etc.), the failure of a given set of causal laws may represent just a simple and reproducible deviation between the predictions of these laws and the experimental results. A deviation of this kind implies only that the causal laws in question must be replaced by newer, more extensive, and more accurate causal laws (as indeed happened with Newtonian mechanics, which was replaced by the more general and more nearly correct relativistic mechanics). Quite often, however, experiments have disclosed not just simple and reproducible deviations from the predictions of a certain set of causal laws, but rather a breakdown of the entire scheme by which a specified set of properties are found to be related in a unique and necessary way in terms of a set of causal laws of a given general kind. Such a breakdown manifests itself in the appearance of chance fluctuations, not coming from anything in the context under investigation, but coming rather from qualitatively different kinds of factors existing in contexts that are new relative to the one under consideration.[10] In such a case, the original causal law is seen to be valid only to the extent that the chance fluctuations in question cancel out, while in any given application the law will have a certain characteristic minimum range of error. This range of error is an objective property of the law in question, a property that is determined by the magnitudes of the chance fluctuations arising outside the context under investigation.

Vice versa, however, the characteristic limitation on the domain of validity of any given causal law which results from the neglect of the effects of chance fluctuations is balanced by a corresponding limitation on the domain of validity of any given law of chance, which results from the neglect of systematic causal interconnections between different contexts. In many cases (e.g. throws of a die) these interconnections are so unimportant that they have never yet been significant in any real applications. Nevertheless, this need not always be so. Consider, for example, the case of insurance statistics. Here, one is able to make approximate predictions concerning the mean lifetime of an individual in a given group (e.g. one of definite age, height, weight, etc.) without the need to go into a detailed investigation of the multitudes of complex factors that contribute to the life or death of each individual in this group. This is possible only because the factors responsible for the death of any individual are extremely manifold and diverse, and because they tend to work more or less independently in such a way as to lead to regular statistical laws. But the assumptions underlying the use of these statistical laws are not always true. Thus, in the case of an epidemic or a war, the systematic interconnection between the cause of death of different individuals grows so strong that statistical predictions of any kind become practically impossible. To apply the laws of chance uncritically, by ignoring the possibility of corrections due to causal interconnections that may be unimportant in some conditions but crucially important in others, is therefore just as capable of leading to erroneous results as is the uncritical application of causal laws, in which one ignores the corrections that may be due to the effects of chance fluctuations . . .

Neither causal laws nor laws of chance can ever be perfectly correct, because each inevitably leaves out some aspect of what is happening in broader contexts. Under certain conditions, one of these kinds of laws or the other may be a better representation of the effects of the factors that are dominant and may therefore be the better approximation for these particular conditions. Nevertheless, with sufficient changes of conditions, either type of law may eventually cease to represent even what is essential in a given context and may have to be replaced by the other. Thus, we are led to regard these two kinds of laws as effectively furnishing different views of any given natural process, such that at times we may need one view or the other to catch what is essential, while at still other times, we may have to combine both views in an appropriate way. But we do not assume, as is generally done in a mechanistic philosophy, that the whole of nature can eventually be treated completely perfectly and unconditionally in terms of just one of these sides, so that the other will be seen to be inessential, a mere shadow, that makes no fundamental contribution to our representation of nature as a whole. Thus, the notion of the qualitative infinity of nature leads us to the necessity of considering the laws of nature both from the side of causality and from that of chance, as well as more generally from new directions that may go beyond these two limits.

7 RECIPROCAL RELATIONSHIPS AND THE APPROXIMATE AND RELATIVE CHARACTER OF THE AUTONOMY OF THE MODES OF BEING OF THINGS

The qualitative infinity of nature has an important bearing on the problem of the reciprocal relationships between things, and on the question of the extent to which the modes of being of different things have an approximate autonomy.

First of all, we note that the universal interconnection of things has long been so evident from empirical evidence that one can no longer even question it. However, in a mechanistic point of view, it is assumed that this interconnection can ultimately be reduced to nothing more than *interaction* between the fundamental entities which compose the system. By this we mean that in the mutual action of these entities on each other, there can only be quantitative changes in their properties, while fundamental qualitative changes in their modes of being cannot take place, provided that these entities are really the basic ones out of which the system is composed. Thus, in Newton's laws of motion there is equality of action and reaction of the elementary particles on each other, but this action and reaction is not supposed to affect the properties of the particles in a fundamental way.

On the other hand, in terms of the notion of the qualitative infinity of nature, one is led, as we have seen in previous sections, to the conclusion that every entity, however fundamental it may seem, is dependent for its existence on the maintenance of appropriate conditions in its infinite background and substructure. The conditions in the background and substructure, however, must themselves evidently be affected by their mutual interconnections with the entities under consideration. Indeed, as we have shown in many examples, this interconnection can, under appropriate conditions, grow so strong that it brings about qualitative changes in the modes of being of every kind of entity known thus far.[11] This type of interconnection we shall denote by the name of *reciprocal relationship*, to distinguish it from mere interaction.

The question now follows quite naturally, "If everything is in this very fundamental kind of reciprocal relationship with everything else, a relationship in which even the basic qualities and modes of being can be transformed, then how can we disentangle these relationships in such a way as to obtain an intelligible treatment of the laws governing the universe, or any part of it?" The answer is that all effects of reciprocal connections are not in general of equal importance. Of course we have the well-known fact that within suitable contexts many of the reciprocal connections produce no significant effects, so that they can be ignored. On the other hand, if we consider a significant reciprocal connection between two things, then we must in general take both directions of this connection into account. If both directions are of comparable importance, then we will still find it very difficult to disentangle the real relationships between things, because one thing affects the basic qualities and laws determining the mode of being of the other; and this effect is returned in a complex process.

Experience in a wide range of fields of science shows, however, that both directions of a reciprocal connection do not always have to have comparable significance. When they do not have equal significance, the problem is evidently simplified because the thing which has the major effect on the other is the dominant and controlling factor in the relationship. In this case we can study the laws and modes of being of the factors of major importance to a good degree of approximation, independently of the effects which may originate in the minor factor. A fundamental problem in scientific research is then to find what are the things[12] that in a given context, and in a given set of conditions, are able to influence other things without themselves being significantly changed in their basic qualities, properties, and laws. These are, then, the things that are, within the domain under consideration, autonomous in their essential characteristics to an adequate degree of approximation. When we have found such things, then we can make use of them for the prediction and control of the other things whose modes of being and basic characteristics are dependent on them. For example, in the case of the relationship between the large-scale level and the atomic level, we find that under conditions that are usually met and in most of the contexts that have thus far been treated in research in physics, the effect of the atomic motions on the laws of the large-scale level is much more important than the effects of the large-scale level on the laws of the atomic motions. Thus, it becomes possible by studying the laws of the atomic motions to make many kinds of approximate predictions concerning the laws and properties of things at the large-scale level and in this way to improve our understanding and control of the large-scale level.

On the other hand, our prediction of the properties of the large-scale level through those of the atomic level can never be perfect, if only because there is a small but nevertheless real reciprocal influence of the large-scale level on the laws of the atomic level. This is due to the electronic and nucleonic substructure of the atom, which can be significantly affected by suitable conditions at the large-scale level (e.g. very high temperatures). Moreover, as we saw, the same possibility can arise with regard to the substructure of every entity that is known in physics (e.g. electrons, protons, mesons, etc.), provided that the conditions at the large-scale level are changed appropriately. As a result, we are led to the conclusion that in its reciprocal connections with the things existing in any *given* lower level, the entities at the macroscopic level must have at least some relative autonomy in their modes of being, in the sense that these modes cannot be predicted perfectly from the specific lower level (or levels) in question. Even though the effects of this autonomy may be negligible in a wide range of conditions and contexts, it may nevertheless become very important in other conditions and contexts.

We see, then, that the existence of reciprocal relationships of things implies that each "thing" existing in nature makes some contribution to what the universe as a whole is, a contribution that cannot be reduced completely, perfectly, and unconditionally, to the effects of any specific set or sets of other things with which it is in reciprocal interconnection. And, vice versa, this also means evidently that no given thing can have a complete autonomy in its mode of being, since its basic characteristics must depend on its relationships with other things. The notion of a thing is thus seen to be an abstraction, in which it is *conceptually* separated from its infinite background and substructure. Actually, however, a thing does not and could not exist apart from the context from which it has thus been conceptually abstracted. And therefore the world is not made by putting together the various "things" in it, but, rather, these things are only approximately what we find on analysis in certain contexts and under suitable conditions.

To sum up, then, the notion of the infinity of nature leads us to regard each thing that is found in nature as some kind of abstraction and approximation. It is clear that we *must* utilize such abstractions and approximations if only because we cannot hope to deal directly with the qualitative and quantitative infinity of the universe. The task of science is, then, to find the right kind of things that should be abstracted from the world for the correct treatment of problems in various contexts and sets of conditions. The proof that any particular kinds of things are the right ones for a given context is then obtained by showing that they provide us with a good approximation to the essential features of reality in the context of interest. In other words, we require that theories formulated in terms of these abstractions lead to correct predictions, and to the control of natural processes in accordance with the plans that are made on the basis of these theories. When this does not happen, we must, of course, revise our abstractions until success is obtained in these efforts. Scientific research thus brings us through an unending series of such revisions in which we are led to conceptual abstractions of things that are relatively autonomous in progressively higher degrees of approximation, wider contexts, and broader sets of conditions.

8 THE PROCESS OF BECOMING

Thus far, we have been discussing the properties and qualities of things mainly in so far as they may be abstracted from the processes in which these things are always changing their properties and qualities and becoming other things. We shall now consider in more detail the characteristics of these processes which may be denoted by the general term of "motion." By "motion" we mean to include not only displacements of bodies through space, but also all possible changes and transformations of matter, internal and external, qualitative and quantitative, etc.

Both the existence and the necessity for the process of motion described above have now been demonstrated in innumerable ways in all the sciences. Thus, the study of astronomy shows that the planets, stars, nebulae, and galaxies all take part in a very large number of kinds of characteristic motions. These motions follow from the effects of the gravitational forces which would start bodies moving even if they were initially at rest, and because of the inertia, which keeps them in motion. And as a result of these motions, over periods of time of the order of billions of years, new stars, new planets, new nebulae, new galaxies, new galaxies of galaxies, etc., can come into existence, while the older organization of things passes out of existence. On the earth, the science of geology has shown that the apparently permanent features of the surface are always changing. Thus, as a result of the flow of water and the action of wind, existing rocks and mountains, and even continents, are continually being worn away while subterranean motions are continually leading to the formation of new ones.

The science of biology shows that life is a continual process of inexhaustible complexity in which various kinds of organisms come into being, live, and die. Indeed, every organism is maintained in existence by characteristic metabolic processes taking place within it, as well as by the motions necessary for it to obtain food and other materials from its environment. Over longer time, as a result of the effects of natural selection and other factors, the forms of life have had to evolve; and in this process, new species of organisms have come into existence while old species have died out. Over still longer periods of time, life itself has come into existence out of a basis of inanimate matter, very probably as a result of motions at the inorganic level of the kind suggested by Opharin;[13] and as conditions change it may later have to pass out of existence, perhaps to give way to something new, of which we can at present have no idea.

In chemistry, one sees that as a result of thermal agitation of the molecules and other causes, different chemical compounds must react to produce new kinds of compounds, while already existing kinds of compounds must be dissociated into simpler compounds. In physics we find, at the atomic level and below, a universal and ceaseless motion which follows as a necessary consequence of the laws appropriate to these levels, and which is discovered to be more violent the deeper we penetrate into it. Thus, we have atomic motions, electronic and nucleonic motions, field motions, quantum fluctuations, probable fluctuations in a sub-quantum mechanical level, etc. Moreover, as happens at the higher levels, not only do the quantitative properties of things change in these motions (e.g., position, velocity, etc., of the various particles, the strength of the various fields, etc.), but so also do the basic qualities defining the modes of being of the entities, such as molecules, atoms, nucleons, mesons, etc., with which we deal in this theory.

In sum, then, no feature of anything has as yet been found which does not undergo necessary and characteristic motions. In other words, such motions are not inessential disturbances superimposed from outside on an otherwise statically existing kind of matter. Rather, they are inherent and indispensable to what matter is, so that it would in general not even make sense to discuss matter apart from the motions which are necessary to define its mode of existence.

Now, the various motions taking place in matter have the further very important characteristic that, in general, they are not and cannot be smoothly co-ordinated to produce simple and regular results. Rather, they are often quite complex and poorly co-ordinated and contain within them a great many relatively independent and contradictory tendencies.

There are two general reasons why such contradictory tendencies must develop; first because there are always chance disturbances arising from essentially independent causes, and secondly, because the systematic processes that are necessary for the very existence of the things under discussion are, as a rule, contradictory in some of their long-run effects. We shall give here a few examples taken from the fields that were discussed in the previous paragraph. Thus, in the field of astronomy, we find that partly as a result of chance disturbances from other galaxies and partly as a result of the laws of motion under the gravitational forces originating in the same galaxy, stars have a very complicated and irregular distribution of velocities going in all sorts of directions, etc., with the result that some systems of stars are being disrupted, while new systems are formed. On the surface of the earth, storms, earthquakes, etc., which are of chance origin relative to the life of a given individual, may produce conditions in which this individual cannot continue to exist; while a similar result can be brought about by old age, which follows from the effects of the very metabolic processes that are necessary to maintain life.

Going on to the subject of physics, we see that both the effects of chance fluctuations and of the operation of systematic causal laws are continually leading to complicated and violent fluctuations in the various levels, which are not at all well co-ordinated with each other, and which quite often lead to contradictory tendencies in the motions. Indeed, these contradictory tendencies not only follow necessarily from the laws governing the motions, but must exist in order for many things to possess characteristic properties which help define what they are. For example, a gas would not have its typical properties if all the molecules had a strong tendency to move together in a co-ordinated way. More generally, the relative autonomy in the modes of being of different things implies a certain independence of these things, and this in turn implies that contradictions between these things can arise. For if things were co-ordinated in such a way that they could not come into contradiction with each other, they could not be really independent.

We conclude, then, that opposing and contradictory motions are the rule throughout the universe, and this is an essential aspect of the very mode of things.

Now it may be asked how it is possible for any kind of quality, property, entity, level, domain, etc., to have even an approximately autonomous existence, in the face of the fact that an infinity of relatively independent kinds of motions with contradictory tendencies are taking place in its environment and in its substructure. The answer is that the existence of any particular quality, property, entity, level, domain, etc., is made possible by a balancing of the processes that are tending to change it in various directions. Thus, in the simple case of a liquid, we have a balancing of the effects of the inter-molecular forces tending to hold the molecules together, and the random thermal motions tending to disrupt the entire system. In a galaxy, we have a balancing of the gravitational forces against the centrifugal tendencies due to rotation and the disruptive effects of the random components of the motions of stars. In atoms we have a similar balancing of the attractive forces of the nucleus against the disruptive effects due to quantum fluctuations in the electronic motions and the centrifugal tendencies due to rotations of the electrons around the nucleus. With living beings, we have a much more subtle and complex system of balancing processes. The full analysis of this process naturally cannot yet be made. But already we can see that the two essential directions of processes in living beings are those leading to growth and those leading to decay. If the growth processes go unchecked, then a typical possible result is the development of a cancer, which eventually destroys the organism. On the other hand, if the opposite processes go unchecked, then the organs will atrophy and wither away, and the organism will again eventually be destroyed. The maintenance of life then requires an approximate balancing of the destruction and decay of tissue by fresh growth.

Now it is clear that if qualities, properties, entities, domains, levels, etc., are maintained in existence by a balance of the processes tending to change them, then this balance can, in general, be only an approximate and conditional one. As a result, any given thing is subject to being changed with changing conditions, both by changes of conditions that are produced externally and by changes that may be necessary consequences of internal motions connected with the very mode of being of the thing in question. To illustrate this point, let us return to the problem of a liquid. As long as the temperature, pressure, etc., of the liquid are held constant, the balance of molecular processes that maintains the liquid state will be continued. But to think of an isolated specimen of a liquid is evidently an abstraction. Any real liquid exists in some kind of environment, which cannot fail to change with enough time. Thus, if the container is on the earth, it will be subject to changes of temperature, to storms and earthquakes that may destroy any temperature-stabilizing mechanism surrounding it, and over longer periods of time to geological processes that may have similar effects. Thus, it can safely be predicted that if, for example, we consider a period of a hundred million years, no particular specimen of a liquid will remain a liquid throughout the whole of this time. Analysing this problem further, we see that, as we consider broader contexts and longer periods of time, there will be more and more opportunities for conditions to change in such a way that any particular balance of processes is fundamentally altered. This is because it will be able to come into reciprocal relationships with more and more relatively autonomous entities, domains, systems, etc., the motions of which can come to influence the processes in question.

Indeed, if we go to the extreme of considering supergalactic regions of space and corresponding epochs of time, we see that there is a possibility for such a broad range of changes of conditions that every kind of entity, domain, system, or level will eventually be subject to fundamental changes, even to destruction or extinction, while new kinds of entities, domains, and levels will come into existence in their place. For example, there is currently under discussion a theory in which it is assumed that some five billion years ago or more the parts of the universe that are now visible to us were originally concentrated in a comparatively small space having an extremely high temperature, and a density so high that neither atoms nor nuclei, nor electrons, nor protons, nor neutrons as we now know them could have existed. (Matter would then have taken some other form about which we cannot have much idea at the present.) This particular section of the universe is then assumed to have exploded, and subsequently to have cooled down to give rise ultimately to electrons, protons, neutrons, atoms, dust, clouds, galaxies, stars, planets, etc., by means of a series of processes into which we need not go further here. The recession of the stars, suggested by the so-called red shift,[14] would then be a residual effect of the velocities imparted to matter in this explosion. Now, it is very important to emphasize how speculative and provisional large parts of this theory are.[15] Nevertheless, for our purposes here, it is interesting in that it gives an example of how widespread could be the effects of a breaking of the balance of opposing processes within the previously existing highly dense state of matter; for the resulting explosion would have given rise to everything that exists in the part of the universe that is now visible to us.

In any case, whatever may have been the at present practically unknown earlier phase of the process of evolution of this particular part of the universe, there exists by now a considerable amount of evidence suggesting that the galaxies, the stars, and the earth come from some quite different previously existing state of things. With regard to what happened on our planet after it came into existence, we have of course much better evidence coming from traces left in the rocks, fossils, etc. Then, coming to the consideration of the origin of life, we have the hypothesis of Opharin, which gives at least the general outlines of how living matter could have come into existence on the earth. Here we see the importance of the incomplete co-ordination and contradictory character of the various kinds of processes that took place on the earth at the time in question; for storms, ocean currents, air currents, etc., would have led to a chance mixing of various organic compounds until at last a substance appeared that began to reproduce itself at the expense of the surrounding organic material. As a result, the contradictory character of the motions at the inorganic level created the conditions in which a whole new level could come into existence, the level of living matter. And from here on, changes in the inanimate environment ceased to be the only causes of development. For a fundamental property of life is that the very processes that are necessary for its existence will change it. Thus, in the case of the individual living being, the balance of growth and decay is never perfect, so that in the earlier phases of its life, the organism grows, then it reaches approximate balance at maturity, and then the processes of decay begin to win out, leading to death. With regard to the various species of living beings considered collectively, these provide each other with a mutual environment, both through their competition and through their co-operation. Thus as a result of the very development of many kinds of living beings, the environment is changed in such a way that the balance of the processes maintaining the heredity of such species is altered, and the result is the well-known evolution of the species.

In sum, then, we see that the very nature of the world is such that it contains an enormous diversity of semi-autonomous and conflicting motions, trends, and processes. Thus, if we consider any particular thing, either the motions taking place externally to it or those taking place internally and which are inherent aspects of its mode of being, will eventually alter or destroy the balance of processes that is necessary to maintain that thing in existence in its present form and with its present characteristics. For this reason, any given thing or aspect of that thing must necessarily be subjected to fundamental modifications and eventually to destruction or decay, to be replaced by new kinds of things.

In conclusion, the notion of the qualitative infinity of nature leads us to regard the eternal but ever-changing process of motion and development described above as an inherent and essential aspect of what matter is. In this process there is no limit to the new kinds of things that can come into being, and no limit to the number of kinds of transformations, both qualitative and quantitative, that can occur. This process, in which exist infinitely varied types of natural laws, is just the process of *becoming*, first described by Heraclitus several thousand years ago (although, of course, by now we have a much more precise and accurate idea of the nature of this process than the ancient Greeks could have had).

9   ON THE ABSTRACT CHARACTER OF THE NOTION OF DEFINITE AND UNVARYING MODES OF BEING

It is clear from the preceding section that the empirical evidence available thus far shows that nothing has yet been discovered which has a mode of being that remains eternally defined in any given way. Rather, every element, however fundamental it may seem to be, has always been found under suitable conditions to change even in its basic qualities, and to become something else. Moreover, as we have also seen, the notion of the qualitative infinity of nature implies that every kind of thing not only can change in this very fundamental way but that, given enough time, conditions in its infinite background and substructure will alter so much that it must do so. Hence, the notion of something with an exhaustively specifiable and unvarying mode of being can be only an approximation and an abstraction from the infinite complexity of the changes taking place in the real process of becoming. Such an approximation and abstraction will be applicable only for periods of time short enough so that no significant changes can take place in the basic properties and qualities defining the modes of being of the things under consideration.

When we come to times that are long enough for the basic kinds of things entering into any specific theory to undergo fundamental qualitative changes, then what breaks down is the assumption that we can specify the modes of being of these things *precisely* and *exhaustively* in terms of the concepts that were applicable before this change took place. Indeed, the very fact that a thing is able to undergo a qualitative change is itself a property that is an essential part of the mode of being of the thing, and yet a property that is not contained in the original concept of it. For example, the fact that the liquid, water, turns into steam when heated and ice when cooled, is a basic property of the liquid in question, without which it could not be water as we know it. Nevertheless, the original concept of water as nothing more than a liquid evidently does not contain these possibilities, either explicitly or implicitly, as necessary properties of this liquid. Hence, this concept does not give a precise and exhaustive representation of all the properties of the liquid in question.

Now the way one usually deals with this problem is to regard the transformations between solid, liquid, and vapour that take place at certain temperatures as part of the qualities defining the mode of being of a single broader category of substance; viz., water. But now the same kind of problem arises again at a new level. For the laws governing the transformations of these qualities are, in turn, being regarded as part of an eternal and exhaustive specification of the properties of the substance, water. On the other hand, in reality this law is applicable and has meaning only under limited conditions. For example, it will no longer have relevance at temperatures and denotes of matter so high that there can be no such things as atoms, and therefore no such a substance as water. Thus, we are led to include water as a special state of a still broader category of things (e.g., systems of electrons, protons, neutrons, etc.) and the laws governing the transformation of water into other kinds of substances as a part of the mode of being of this still broader category. But if *all* things eventually undergo qualitative transformations, then the process described above will never end. Thus we conclude that the notion that all things can become other kinds of things implies that a complete and eternally applicable definition of any given thing is not possible in terms of any finite number of qualities and properties.

If, however, we now start from the opposite side, viz., from the notion of the qualitative infinity of nature, we are then immediately able to arrive at a type of definition of the mode of being of any given kind of thing that does not contradict the possibility of its becoming something else. For, as we have previously seen, the reciprocal relationships between all things then imply that no given thing can be *exactly* and *in all respects* the kind of thing that is defined by any specified conceptual abstraction. Instead, it is always *something more* than this and, at least in some respects, *something different*. Hence, if the thing becomes something else, no unresolvable contradiction is now necessarily implied. For it is in any case never exactly represented by our original concept of it. Logically speaking, what this point of view towards the meaning of our conceptual abstractions does is, therefore, to create room for the possibility of qualitative change, by leading us to recognize that those aspects of things that have been ignored may, under suitable conditions, cease to have negligible effects, and indeed may become so important that they can bring about fundamental changes in the basic properties of the things under consideration.

We may illustrate the above conclusions by returning to a more detailed discussion of the transformations between steam, liquid water, and ice. Thus, the macroscopic concept of a certain state of matter (e.g., gaseous, liquid, or solid) leaves out of account an enormous number of kinds of factors that are not and cannot be defined in the macroscopic domain alone. Among these are the motions of the molecules constituting the fluid quantum fluctuations, field fluctuations, nuclear motions, mesonic motions, motions in a possible sub-quantum mechanical level, and so on. In short, we may say that the real fluid is enormously richer in qualities and properties than is our macroscopic concept of it. It is richer, however, in just such a way that these additional characteristics may, in a wide variety of applications, be ignored in the macroscopic domain. Nevertheless, when we come to the problem of understanding why transformations between gas, liquid, and solid are possible, we can no longer completely ignore the additional properties of the real fluid . . .

Not only is the notion of unvarying and exhaustively specifiable modes of being of things an abstraction that fails for periods of time that are too long (because of the possibility of fundamental qualitative changes), but it also fails for times that are too short. This is because the characteristic properties and qualities of a thing depend in an essential way on processes that are taking place in the background and substructure of the thing in question. Thus, for example, the properties of an atom (e.g. spectral frequencies, chemical reactivity, etc.) arise and are determined mainly in the process of motion of the electrons in the orbit, which take a period of time of the order of 10^{-15} seconds. Over shorter periods of time, however, the properties of an atom as a whole are so poorly defined that it is not even appropriate to consider them as such. A better conception of what the atom is can then be obtained by regarding it as a collection of electrons in motion around the nucleus. But as we shorten the period of time still further, the same problem arises with regard to electrons, protons, neutrons, mesons, etc. And if we go to a larger scale, the reader will readily see that a similar behaviour is obtained (e.g., the existence of a living being is maintained by inner metabolic and nervous processes that are fast in comparison with the period in which it makes sense to define the basic characteristics of such a being). Indeed, the notion of the qualitative infinity of nature implies that such behaviour is inevitable. For, as we saw in the previous section, each kind of thing is maintained in existence by a balance of opposing processes in its infinite background and substructure, which are tending to change it in different ways. Thus, the properties of such a thing can be defined only over periods of time long enough so that the average of the effects of all these processes does not fluctuate significantly.

It is clear, then, that all our concepts are, in a great many ways, abstract representations of matter in the process of becoming. The choice of such abstractions is, however, limited by the requirement that they shall represent what is essential in a certain context to a suitable degree of approximation and under appropriate conditions . . .

We conclude, then, that we must finally reach a stage in every theory where we introduce the notion of something with unvarying and exhaustively specifiable modes of being, if only because we cannot possibly take into account all the inexhaustibly rich properties, qualities, and relationships that exist in the process of becoming. At this point, then, we are making an abstraction from the real process of becoming. Whether the abstraction is adequate or not depends on whether or not the specific phenomena that we are studying depend significantly on what we have left out. With the further progress of science, we are then led through a series of such abstractions, which furnish ever better representations of more and more aspects of matter in the concrete and real process of becoming.

Now, when we refer to the process of becoming by the word "concrete," we mean by this to call attention to the quality of being special, peculiar, and unique that one always finds to be characteristic of real things when one studies them in sufficient detail. For example, if we consider any concept (e.g. apples), then this concept contains nothing in it that would permit us to distinguish one apple from another. We may then indicate other qualities which make such a distinction possible (red apples, hard apples, sweet apples, etc.). Evidently, no finite number of such qualities can ever give a complete representation of any specific example of a real apple. Of course, by going deeper (e.g., by giving the physical and chemical state of each part of the apple) we could come closer to our goal. But this process could never end. For even the modes of being of the individual atoms, electrons, protons, etc., inside the apple are in turn determined by an infinity of complex processes in their substructures and backgrounds. Thus, we see that because every kind of thing is defined only through an inexhaustible set of qualities each having a certain degree of relative autonomy, such a thing can and indeed must be *unique*; i.e., not completely identical with any other thing in the universe, however similar the two things may be.[16]

Carrying the analysis further, we now note that because all of the infinity of factors determining what any given thing is are always changing with time, *no such a thing can even remain identical with itself* as time passes. In certain respects, this brings us to a deeper notion of the process of becoming than we had before. For at each instant of time, each thing has, when viewed from one side, an enormous (in fact infinite) number of aspects which are in common with those that it had a short time ago. Indeed, if this were not so, it would not be a thing; i.e., it would not preserve any kind of identity at all. On the other hand, when viewed from another side, it has an equally enormous (in fact infinite) number of aspects that are not those that it had a short time ago. For typical sorts of things with which we commonly deal, however, these latter aspects are not essential in the normal contexts and conditions with which we work. In new contexts (e.g., a sub-atomic or a super-galactic time scale) or under new conditions (e.g., very high temperatures), these aspects may, however, take on a crucial importance.

We are in this way led to the conclusion that the process of becoming will necessarily have, at each moment, certain aspects that are concrete and unique. In other words, each thing in each moment of its existence must have certain qualities which, in some respects, belong uniquely to that thing and to that moment. The notion of unvarying and exhaustively specifiable modes of being is then an abstraction obtained, in general, by considering what is common to the same thing at different moments, or to many similar things at the same moment. In doing this, we evidently ignore the differences between these things, which are just as essential a side of them as are their similarities. By abstracting in more detail from these differences, we are then led to see newer but subtler aspects in which these differences contain common or similar relationships that apply to all of these things. Thus, the uniqueness of each thing at each instant of time is reflected in our abstract concepts by the limitless richness and complexity of the concepts that one needs to obtain a better and better abstract representation of matter in the process of becoming, or, in other words, by the inexhaustibility of the qualities that are to be found in nature.

Notes

1 For example, in the usual interpretation of the quantum theory, the state of a system is subject, in general, only to a statistical determination.

2 The field concept, the concept of quantitative changes that lead to qualitative changes, the concepts of chance and statistical law.

3 See Bohm, D., *Causality and Chance in Modern Physics*, University of Pennsylvania Press, Philadelphia (1957), Ch. 1, Sec. 10; Ch. 2, Sec. 15.

4 E.g., the "rest energy" of matter, the stability of atoms, etc.

5 It is by recognizing that a finite and generally limited number of qualities, properties and laws may be adequate in given contexts, conditions, and degrees of approximation that we avoid the procedure of simply falling back into an arbitrary multiplication of qualities that was characteristic of the pre-mechanistic point of view, especially in the scholastic form of the Aristotelian philosophy that was prevalent in the Middle Ages.

6 See Bohm, *Causality and Chance in Modern Physics*, Chs 3 and 4.

7 This is suggested by the wave–particle duality in the general properties of matter, which implies, as we have seen, that we may have to deal with some new kind of thing that can, under suitable circumstances, act either like a localized particle or like an extended field.

8 This autonomy may have many origins; e.g., the falling of the propagation of influences of one thing on another with an increase of separation between them, the decay of such influence with the passage of time, electrical screening, the existence of thresholds, such that influences which are too weak to surpass these thresholds produce no significant effects; the fact that individual constituents of an object (such as atoms) are too small to have an appreciable effect on the object as a whole, while collectively there is considerable independence of motions of the constituents leading to the cancellation of chance fluctuations. Many other such sources of autonomy exist, and doubtless more will be discovered in the future.

9 See Bohm, *Causality and Chance in Modern Physics*, Ch. 1, Sec. 8.

10 This is, for example, what happens to classical physics. For a particle such as an electron follows the classical orbit only approximately, and in a more accurate treatment is found to undergo random fluctuations in its motions, arising outside the context of the classical level (see Bohm, *Causality and Chance in Modern Physics*, Chs 3 and 4).

11 See *ibid*, Ch. 2, Sec. 13.

12 Let us recall that we are here using the word "thing" in a very general sense, so that it represents *anything* (e.g., objects, entities, qualities, properties, systems, levels, etc.).

13 See Bohm, *Causality and Chance in Modern Physics*, Ch. 1, Sec. 8.

14 The "red shift" of the spectral lines of stars has been interpreted as a Doppler shift due to a recessional motion. If this interpretation is correct, then the stars are receding from each other with a velocity that is more or less proportional to their distances. The most distant stars visible would have speeds as high as 10,000 miles a second, and still more distant stars would presumably have still higher velocities. However, there are many possible explanations for the same phenomenon; e.g., perhaps the behaviour of light over long distances is slightly different from that predicted by Maxwell's equations, in such a way that the frequency of light diminishes as it is transmitted through space.

15 In the actually published forms of this theory, it is assumed that the *whole universe* (and not just a part of it) was originally concentrated into the small space referred to above. Even if we do not make this additional assumption, the theory is already quite speculative. But this additional assumption is based on Einstein's theory of general relativity, which has been proved to a rather low level of approximation only in weak gravitational fields for low concentrations of matter and over limited regions of space. A gigantic extrapolation is then made to gravitational fields of fantastic intensity, to unheard of concentrations of matter, and to a region of space that includes nothing less than the whole universe. While this extrapolation cannot be proved to be wrong at

present, it is in any case an example of extreme mechanism. If we divest the theory of these irrelevant and unfounded extrapolations, then the hypothesis is still, however, interesting to consider.

16   According to the Pauli exclusion principle, any two electrons are said to be "identical." This conclusion follows from the fact that within the framework of the current quantum theory there can be no property by which they could be distinguished. On the other hand, the conclusion that they are *completely* identical in *all* respects follows only if we accept the assumption of the usual interpretation of the quantum theory that the present general form of the theory will persist in every domain that will ever be investigated. If we do not make this assumption, then it is evidently always possible to suppose that distinctions between electrons can arise at deeper levels.

2 PHYSICS AND PERCEPTION (1967)

Here Bohm argues that discoveries in contemporary physics, particularly those of Einstein, have striking correlates in the nature of human perception as outlined by Piaget and others. Since the development of the special theory of relativity, Newton's formulations of space, time, and mass have come to be understood not as absolutes, but rather as relatively invariant (stable) structures that depend for this stability upon the observer's perspective and frame of reference. Similarly, the work of Piaget suggests a process of perception which at its root also depends on the discernment of relatively invariant entities and processes.

Bohm explores at length Piaget's thesis that the experiential world of the infant is rooted in fluid frames of reference that give shape and form to the immensely complex totalities of the natural world. It is only through a cumulative process of abstraction that the child, and consequently the adult, comes to perceive this inherently fluid, relative domain in terms of fixed, solid objects existing within fixed structures of space and time. Ironically then, the full-blown abstracting process of the adult yields a world more in line with the absolutes of Newton, while the rudimentary perceptual processes of the infant are more consistent with Einstein's relativistic discoveries.

But underlying the fixity of the adult's highly abstract "world maps," a rich world of indeterminate perception is potentially available. In this context Bohm continues his investigation of reciprocal movements, now in the form of a "circular reflex" which Piaget claims to be the basis for all perceptual learning. This circular reflex is a feedback loop in which the incoming

Extract from the Appendix of D. Bohm, *The Special Theory of Relativity*, Routledge, London ([1965], 1996).

reception of sensory data is always complemented by active, outgoing impulses. These outgoing impulses run continuous "tests" upon the sensory data, accommodate new information, and form abstracted images of predictable (relatively invariant) structures and movements. As one loop, these impulses are the basis for mapping a world and its contents into existence.

Bohm envisions a continuum of such two-way perceptual interaction that extends to all knowledge structures. At the simplest level it is how a child formulates the relative stability of body and objects. At the everyday level it is how we negotiate yet more abstract processes, such as inserting a key into a keyhole. At a further level of abstraction, it is how science progresses, as in the case of shifting from a view of absolute space and time to one of relative space and time.

The existence of this continuum – from rudimentary perception to highly abstract constructions of thought – leads Bohm to the unorthodox conclusion that at its root, science is not a process of accumulating knowledge of the world. Rather it is one of extending this two-way circular reflex via the use of sophisticated instruments and progressive degrees of abstraction. Thus, knowledge is seen by Bohm as a secondary aspect of scientific endeavor, the frozen by-product of a potentially open-ended perceptual movement.

Such a perspective raises questions about the efficacy of the mental maps we abstract from the raw data of the world, whether these maps be scientifically "objective" or personally "subjective." In many respects, much of our accumulated knowledge has a very good fit with the objective world, in that it allows us to successfully navigate that world with a minimum of mishaps. We do step out of the way of moving cars, and would suffer the consequences if we did not. The two-way perceptual process and its resultant knowledge is thus far from arbitrary. But based on this relatively successful "mapping," we tacitly assume that our maps are true replicas of the world – a perfect match rather than a relatively good fit.[1] Once we make this assumption, and our maps fuse with the territory, we all too easily develop a tacit conservatism which diminishes the exploratory potential of the perceptual act itself. But if we can be aware of this fusion and ascertain in what ways it is correct and what ways it is not, we have taken a significant step in restoring the primacy of perception.

Whereas the reciprocal relations outlined in Chapter 1 primarily describe the "external" world, the thread of this chapter goes a step further, linking the "internal" process of perception to analogous processes in the external world. In positing this linkage, Bohm is anticipating his later formulation of soma-significance (Chapter 5), in which mind and matter are understood as reciprocal, evolving aspects of a more fundamental implicate order.

1   INTRODUCTION

In Einstein's theory of relativity, the notions of space, time, mass, etc., are no longer regarded as representing absolutes, existing in themselves as permanent substances or entities. Rather, the whole of physics is conceived as dealing with the discovery of what is *relatively invariant* in the ever-changing movements that are to be observed in the world, as well as in the changes of points of view, frames of reference, different perspectives, etc., that can be adopted in such observations. Of course, the laws of Newton and Galileo had already incorporated a number of relativistic notions of this kind (e.g., relativity of the centre of coordinates, of the orientation, and speed of the frame of reference). But in them the basic concepts of space, time, mass, etc., were still treated as absolutes. Einstein's contribution was to extend these relativistic notions to encompass the laws, not only of mechanics, but also those of electrodynamics and optics, in the special theory, and of gravitation in the general theory. In doing this he was led to make the revolutionary step to which we have referred, i.e., of ceasing to regard the properties of space, time, mass, etc., as absolutes, instead treating these as invariant features of the relationships of observed sets of objects and events to frames of reference. In different frames of reference the space coordinates, time, mass, energy, etc., to be associated to specified objects and events will be different . . .

At first sight the point of view described above may seem to be very different to that of "common sense" (as well as of the older Newtonian physics). For are we not in the habit of regarding the world as constituted of more or less permanent objects, satisfying certain permanent laws? That is to say, in everyday life we never talk about "invariant relationships," but rather we refer to tables, chairs, trees, buildings, people, etc., each of which is more or less unconsciously conceived as being a certain kind of object or entity, which, added to others, makes up the world as we know it. We do not regard these objects or entities as *relative invariants* which along with their properties, and the laws that they satisfy, have been abstracted from the total flux of change and movement. There appears then to be a striking difference between the way we conceive the world as observed in immediate experience (as well as in the domain of classical nonrelativistic physics) and the way it is conceived in relativity theory.

In this essay we shall show that the difference between the notions of common experience and those of relativity theory arise mainly because of certain habitual *ideas* concerning this experience, and that there is now a great deal of new, but fairly well confirmed, scientific evidence suggesting that our actual mode of *perception* of the world (seeing it, hearing it, touching it, etc.) is much closer in character and general structure to what is suggested by relativistic physics than it is to what is suggested by prerelativistic physics. In the light of this evidence it would seem that nonrelativistic notions appear more natural to us than relativistic notions, mainly because of our limited and inadequate understanding of the *domain of common experience*, rather than because of any inherent inevitability of our habitual mode of apprehending this domain.

2   THE DEVELOPMENT OF OUR COMMON NOTIONS IN INFANTS AND YOUNG CHILDREN

We shall begin with the fascinating studies of the development of intelligence in infants and young children carried out by Piaget.[2] On the basis of long and careful observations of children of all ages from birth up to 10 or more years, he was actually able to see the development of our customary ideas of space, time, the permanent objects, the permanent substance with the conserved total quantity, etc., and thus to trace the process in which such notions are built up until they seem natural and inevitable.

The very young infant does not behave as if he had the adult's concept of a world separate from himself, containing various more or less permanent objects in it. Rather, Piaget gives good evidence suggesting that the infant begins by experiencing an almost undifferentiated totality. That is to say he has not yet learned to distinguish between what arises inside of him and outside of him, nor to distinguish between the various aspects of either the "outer" or the "inner" worlds. Instead there is experienced only one world, in a state of continual flux of sensations, perceptions, feelings, etc., with nothing recognizable as permanent in it. However, the infant is endowed with certain inborn reflexes, connected with food, movements, etc. These reflexes can develop so as to selectively accommodate different aspects of the environment; and in this way the environment begins effectively to be differentiated to the extent of taking on certain "recognizable" features. But at this stage recognition is largely *functional* (e.g., some objects are "for eating," some "for drinking," some "for pulling" etc.), and there seems to be little or no development of the adult's ability to recognize an object by the shape, form, structure, or other perceived characteristics.

At first these reflexes and functions are carried out largely in the satisfaction of primary needs, indicated by sensations, such as hunger, etc. In the next stage, however, there develops the so-called "circular reflex," which is crucial to the development of intelligence. In such a reflex there is an outgoing impulse (e.g., leading to the movement of the hand) followed, not mainly by the satisfaction of need, but rather by some incoming sensory impulse (e.g., in the eye, ear, etc.). This may be said to be a beginning of real perception. For the most elementary way of coming into contact with something that is not just the immediate satisfaction of a bodily need is by incorporating it into a process in which a certain impulse toward action is accompanied by a certain sensation.

This principle of the circular reflex is carried along in all further developments. Thus, at a certain stage, the infant begins to take pleasure in operating such reflexes, in order, as Piaget puts it, "to produce interesting spectacles." He finds, for example, that pulling a certain cord will produce an interesting sensation of movement in front of him (e.g., if the cord is attached to a colored object). It must not be supposed that he understands the causal connection between the cord and the movement, or even that he foresees the sensation of movement in his imagination and then tries to realize it by some operation. Rather, he *discovers* that by doing such an operation he gets a pleasant sensation that is *recognizable*. In other words, recognition that a past event has been repeated comes first; the ability to call up this event in the memory comes only much later. Thus, at this stage, he only knows that a certain operation will lead to some recognizable experience that is pleasurable.

The ability to recognize something as similar to what was experienced before is certainly a necessary prerequisite for beginning to see something relatively permanent in the flux of process that is very probably the major element in the infant's early experiences. Another important prerequisite for this is the coordination of many different kinds of reflexes that are associated to a given object. Thus, at first the infant seems to have little or no realization that the object he sees is the same as the object he hears. Rather, there seem to be fairly separate reflexes, such as listening, looking with the eyes, etc. Later, however, these reflexes begin to be coordinated, so that he is finally able to understand that he sees what he hears, grasps what he sees, etc. This is an important step in the growth of intelligence, for in it is already implicit the notion that will finally develop – of a single object that is responsible for all of our different kinds of experience with it.

The infant is, however, as yet far from the notion of a permanent object, or of permanent causal relationships between such objects. Rather, his behavior at this stage suggests that when presented with

something familiar he now abstracts certain vaguely recognizable totalities of sensation and response, involving the coordination of hand, eye,
ear, etc. Thus there is a kind of a germ of the notion of the invariant
here; for in the total flux of experience he can now recognize certain
invariant combinations of features of the pattern. *These combinations
are themselves experienced as totalities,* so that the object is not recognized outside of its customary context.

Later the infant begins to follow a moving object with his eyes,
being able to recognize the invariance of its form, etc., despite its
movement. He is thus beginning to build up the reflexes needed for
perceiving the continuity of existence of certain objects, apart from their
customary contexts. However, he still has no notion of anything permanent. Rather, he behaves as if he believed that an object comes into
existence where he first sees it and passes out of existence where he last
sees it. Thus, if an object passes in front of him and disappears later
from his field of view, he looks for it, not in the direction where he has
last seen it, but rather toward the place where he *first* saw it, as if this
were regarded as the natural source of such objects. Thus, if an object
goes behind an obstacle, he does not seem to have any notion of looking
for it there. The realization that this can be done comes only later, after
the child has begun to work with what Piaget calls "groups of
operations."

The most elementary of these is the "group of two." That is, there
are operations such as turning something round and round, hiding it
behind an obstacle and bringing it back to view, shaking something
back and forth, etc., which have in common that there is an operation,
the result of which can be "undone" by a second operation, so that the
two operations following each other lead back to the original state of
affairs. It is only after he understands this possibility that the infant
begins to look for an object behind the obstacle where it vanished from
view. But his behavior suggests that he still does not have the idea of a
permanent object, existing even when he doesn't see it. Rather, he probably feels that he can "undo" the vanishing of an object, by means of the
"operation" of putting his hand behind the obstacle and bringing forth
the object in question.

In this connection we must recall that the infant still sees no clear
and permanent demarcation between himself and the world, or between
the various objects in it. However, he is building up the reflexes and
operations needed to conceive this demarcation later. Thus, he is beginning to develop the notion of causality, and the distinction of cause and
effect. At first he seems to regard causality as if it were a kind of

"sympathetic" magic. As long as the child views all aspects of his experience as a single totality, with no clear distinction of "within" and "without," there is nothing in his experience to deny the expectation of such sympathetic magical causality. Later, however, he begins to see the need for intermediate connections in causal relationships, and still later he is able to recognize other people, animals, and even objects as the causes of things that are happening in his field of experiencing.

Meanwhile, the notions of space and time are being built up. Thus as the child handles objects and moves his body he learns to coordinate his changing visual experiences with the tactile perceptions and bodily movements. At this stage, his notion of groups of movements is being extended from the "group of two" to more general groups. Thus, he is learning that he can go from one place A to another B by many different paths, and that all these paths lead him to the same place (or alternatively that if he goes from A to B by any one path, he can "undo" this and return to A by a large number of alternative paths). This may seem to be self-evident to us, but for an infant living in a flux of process it is probably a gigantic *discovery* to find out that in all of this movement there are certain things that he can always *return* to in a wide variety of ways. The notion of the *reversible* group of movements or operations thus provides a foundation on which he will later erect that of *permanent places* to which one can return, and *permanent objects*, which can always be brought back to something familiar and recognizable by means of suitable operations (e.g., rotations, displacements, etc.).

Meanwhile the child is gradually learning to call up images of the past, in some approximation to the sequence in which it occurred, and not merely to *recognize* something as familiar only after he sees it. Thus begins true *memory*, and with it the basis for the notion of the distinction of past time and present time (and later future time, when the child begins to form mental images of what he expects).

A really crucial step occurs when the child is able to form an image of an absent object, as existing even when he is not actually perceiving it. Just before he can do this he seems to deal with this problem as if he regarded the absent object as something that he (or other people) can *produce* or *create* with the aid of certain operations. But now he begins to form a mental image of the world, containing both perceived and unperceived things, each in its place. These objects, along with their places, are now conceived as permanently existing, and in a set of relationships corresponding perfectly to the groups of movements and operations already known to him (e.g., the picture of a space in which each point is connected to every other by many paths faithfully represents the invariant feature of his experience with groups of operations, in which he was able to go from one point to another by many routes).

At this stage it seems that the child begins to see clearly the distinction between himself and the rest of the world. Until now he could not make such a distinction, because there was only one field of experiencing what was actually present to his total set of perceptions. However, with his ability to create a mental image of the world, i.e., to *imagine* it, he now conceives a set of places which are permanent, these places being occupied by various permanent objects. But one of these objects is *himself*. In his new mental "map" of the world he can maintain a permanent distinction between himself and other objects. Everything on this map falls into two categories – what is "inside his skin" and what is not. He learns to associate various feelings, pleasures, pains, desires, etc., with what is "inside his skin," and thus he forms the concept of a "self," distinct from the rest of the world, and yet having its place in this world. He similarly attributes "selves" to the insides of other people's skins, as well as to animals. Each "self" is conceived as both initiating causal actions in the world and suffering the effects of causal actions originating outside of it. Eventually he learns to attribute to inanimate objects a lower and more mechanical kind of "selfhood" without feelings, aims, and desires, but still having a certain ability to initiate causal actions, and to suffer the effects of causes originating outside of it. In this way the general picture of a world in space (and time), constituted of separate and permanent entities which can act on each other causally, is formed.

The notions of an objective world and of a subject corresponding to one of the objects in the world are, as we have seen, thus formed together, in the same step. And this is evidently necessary, since the mental image of the world that serves as a kind of conceptual "map" requires the singling out of one of the objects on this "map" to represent the place of the observer, in order that his special perspective on the world at each moment can be taken into account. That is to say, just as the relativistic "map," in the form of the Minkowski diagram[3] must contain something in it to represent the place, time, orientation, velocity, etc., of the observer, so the mental map that is created by each person must have a corresponding representation of that person's relationship to the environment.

It must not be supposed, of course, that the child knows that he is making a mental image or map of the world. Rather, as Piaget brings out very well, young children often find it difficult to distinguish between what is imagined or remembered in thought and what is actually perceived through their senses (e.g., they may think that other people are able to see the objects that they are thinking about). Thus the child will take this mental map as equivalent to reality. And this habit is intensified with each new experience, because once the map is formed it *enters into and shapes all immediate perceptions*, thus interpenetrating the whole of experience and becoming inseparable from it. Indeed, it is well known that how we see something depends on what we know about it. (E.g., an extreme case is that of an ambiguous picture, subject to two interpretations, one obvious and the other less so. Once a person is told about the second interpretation, in many cases, he can no longer see the picture in the original way.) Thus, over a period of years we learn to see the world through a certain structure of ideas, with which we react immediately to each new experience before we even have time to think. In this way we come to believe that certain ways of conceiving and perceiving the world cannot be otherwise, although in fact they were discovered and built up by us when we were children, and have since then become habits that may well be appropriate only in certain domains of experience . . .

It will be relevant for our purposes here to discuss briefly the development of the child's concept of the constancy of the number of objects, and of the total quantity of matter in them, because these concepts have evidently played a fundamental role in physics. As Piaget demonstrates, a child who has recently begun to talk does not at first have the notion that a set of objects has a fixed number, independent of how they are moved and rearranged. Rather, he forms at each instant a general perceptual estimate of whether a given collection seems to be more, or less, or equal to another, and does not hesitate to say that two initially equal collections are unequal, after they have been subject to some rearrangements in space (even though the numbers of objects have actually remained constant).

The results described above will be seen to be not surprising, if one keeps in mind the fact that the child does not yet have the idea of the conservation of the number of objects as they move and change their relationships to each other and to the observer. Indeed, this notion is developed only in a series of stages. First, the child learns to establish a one-to-one *correspondence* between objects that are in simple relationship, such as parallel rows. When he loses sight of this correspondence (e.g., when the objects are rearranged and are no longer in rows) he cannot yet think of them as having the same number. Later, as he learns to put them into correspondence again, he forms an idea similar to that of the "reversible group," i.e., that certain sets of objects can be brought back by suitable operations into their original state of one-to-one correspondence. From here he forms a new concept or "mental map" of the objects as having at all times a fixed number, which faithfully portrays the structure of his operations with such sets as capable of being put back into correspondence. Then, gradually, he forgets the operations that establish correspondence and thinks of the number of objects as a fixed property belonging to a given total set, even when these move and rearrange.

The procedure of thinking of numbers as an inherent and permanent property of a set becomes so habitual that the problem "What is number?" is considered as being too obvious to require much discussion. Yet when modern mathematicians came to study this question, what they had to do was in effect to uncover the operational basis on which each child originally develops his concept of number. We see then that the deepest problems are often found in the study of what seems obvious, because the "obvious" is frequently merely a notion that summarizes the invariant features of a certain domain of experience which has become habitual and the basis of which has dropped out of consciousness. So to understand the obvious it is necessary very often to go to a broader point of view, in which one brings to light the basic operations, movements, and changes, within which certain characteristics have been found to be invariant.

A very similar problem arises with regard to conservation of the quantity of matter or substance. Thus, when a given quantity of liquid is distributed into many containers of various shapes, the young child does not hesitate to say that the total quantity of water has increased or decreased, according to the impressions that the new distribution produces in his immediate perceptions. Later, when he sees the possibility of bringing the water back into the original container, where it has the same volume as it had originally, he is led to the idea of a constant quantity of liquid. The necessity for this step in the development of the child's conceptions is evident. For *a priori* there is no reason to suppose that the quantity of a given substance is conserved. This idea comes forth only as a result of the need to understand certain kinds of experience. Then later, one forgets that such an idea had to be developed. It becomes habitual, and eventually it seems inevitable to suppose that the world is made of certain basic substances that are absolutely permanent in their total quantities. Then, when we do not find this absolute permanence in the level of common experience, we postulate it in the atomic level or somewhere else.

As in the case of numbers, some very deep problems arise here in the effort to understand what seems obvious. Nothing seems more obvious than the notion of a permanent quantity of substance. Yet, to understand this idea more deeply, we must go on to a broader context, in which such a notion need not apply. We can then see that such conceptions arise when the child discovers a kind of *relative invariance* under certain operations, e.g., of pouring the liquid back into the original container. So we find that in the *understanding* of immediate perception, one must do essentially what is done in the theory of relativity, i.e., to give up the concept of something that is absolutely permanent and constant, to see the constancy of certain relationships or properties in a broad domain of operations involved in observation, measurement, etc., in which the conditions, context, and perspective are altered.

To sum up the work of Piaget, then, we recall that the infant begins with some kind of totality of sensation, perception, feeling, etc., in a state of flux, in which there is little or no recognizable structure with permanent characteristics. The development of intelligence then arises in a series of operations, movements, etc., by which the child *learns* about the world. In particular, what he learns is always based on his ability to see invariant relationships in these operations and movements, e.g., an invariant kind of correspondence between what he sees and what he hears, etc., an invariant relationship between cause and effect, an invariant form to an object as he follows it with his eye, an invariant possibility of "undoing" certain changes by means of suitable operations, etc. The perception of each kind of invariance is then followed by the development of a corresponding mental image (and later a structure of organized ideas and language) which functions as a kind of "map" representing the invariance relationships correctly, in the sense that it implies invariant features similar to those disclosed in the operations (e.g., the mental image of a space with permanent positions connected by an infinity of possible paths corresponds to the operational experience of being able to reach the same place by many different routes). Very soon immediate perception takes on the structure of these "maps," and, after this, one is no longer aware that the map only *represents* what has been found to be invariant. Rather, the map begins to interpenetrate what is perceived in such a way that it seems to be an inevitable and necessary feature of the whole of experience, so obvious that it is very difficult to question its basic features.

The work of Piaget indicates that in order to understand the process of perception it is necessary to go beyond the habitual standpoint, in which one more or less confuses the general structural features of our mental "maps" with features of the world that cannot be otherwise, under any conceivable circumstances. Rather, one is led to consider the broader totality of our perceptive process as a kind of flux, in which certain *relatively invariant* features have emerged, to be represented by such "maps," in the sense that these faithfully portray the structure of such features. But a similar step is involved in going from a nonrelativistic point of view in physics to a relativistic point of view. For in doing this we cease to regard our concepts of space, time, mass, etc., as representing absolutely permanent and necessary features of the world, and, instead, we regard them as expressing the invariant relationships that actually exist in certain domains of investigation.

3 THE ROLE OF THE INVARIANT IN PERCEPTION

The work of Piaget, discussed in the previous section, shows that the development of intelligence seems to be based on the ability to realize what is invariant in a given domain of operations, changes, movements, etc., and to grasp these relationships by means of suitable mental images, ideas, verbal expressions, mathematical symbolism, etc., implying a structure similar to that which is actually encountered. We shall now cite some evidence coming from the direct study of the process of perception, which strongly confirms the implications of this point of view, and considerably extends their domain of applicability.

There is a common notion of perception as a sort of *passive* process, in which we simply allow sense impressions to come into us, there to be assembled into whole structures, recorded in memory, etc. Actually, however, the new studies make it clear that perception is, on the contrary, an active process, in which a person must do a great many things in the course of which actions he helps to supply a certain *general* structure to what he perceives. To be sure, this structure is objectively correct, in the sense that it is similar to the structure of the kind of things that are encountered in common experience. Yet the fact that a great deal of what we see is ordered and organized in a form determined by the functioning of our own bodies and nervous systems has very far reaching implications for the study of new domains of experience, whether in the field of immediate perception itself or in science (which generally depends on instrumentally aided perception, in order to reach new domains).

One can see the active role of the observer most clearly by first considering tactile perception. Thus, if one tries to find the shape of an unseen object simply by feeling it, one must *handle* the object, turn it round, touch it in various ways, etc. (This problem has been studied in detail by Gibson and his co-workers.[4])

In such operations one seldom notices the *individual sensations* on the fingers, wrist joints, etc. Rather, one directly perceives the general structure of the object, which emerges, somehow, out of a very complex change in all the sensations. This perception of the structure depends on two nervous currents of energy – not only the inward current of sensations to which we have referred above, but also an outward current determining movements of the hand. For knowledge of this structure is implicit in the *relationship* between the outward and the inward currents (e.g., in the response to certain movements of turning, pressing, etc.).

It is evident, then, that tactile perception is evidently inherently the result of a set of active operations, performed by the percipient. Nevertheless, the outgoing impulses leading to the movement of the hand and the influx of sensations are either not noticed or else they are only on the fringe of awareness. What is perceived most strongly is actually the structure of the object itself. It seems clear that out of a remarkably complex and variable flux of movement with their related sensual responses, the brain is able to abstract a *relatively invariant* structure of the object that is handled. This invariant structure is evidently not in the individual operations and sensations but can be abstracted only out of their totality over some period of time.

At first one might think that in *vision* the situation is basically different, and that one just passively "takes in" the picture of the world. But more careful studies show that vision involves a similar active role of the percipient, and that the structure of what one sees is abstracted out of similar invariant relations between certain movements and the changing sensations which are the eye's response to these movements.

One of the most elementary movements that is necessary for vision has been demonstrated by Ditchburn,[5] who has discovered that the eyeball is continually undergoing small and very rapid vibrations, which shift the image by a distance equal roughly to that between adjacent cells on the retina of the eye. In addition, it has a slower regular drift, followed by a "flick" which brings the image more or less back to its original center. Experiments in which a person looked at the whole field of vision through mirrors arranged to cancel the effects of this movement led at first to a distorted vision and soon to a complete breakdown of vision, in the sense that the viewer could see nothing at all, even though a clear image of the world was being focused on his retina.

Ditchburn has explained this phenomenon by appealing to the fact that when a constant stimulus is maintained on nerve cells for some time, they *accommodate*; i.e., the strength of their response tends to decrease, eventually falling below the threshold of what is perceptible. Under conditions in which the pattern of intensity of light on the whole retina is kept fixed by mirrors that compensate for the movements of the eyeball, it is then to be expected that such a process of accommodation will take place. In this way one can explain the distortion and eventual fading away of what is in the field of view, as observed in the experiments of Ditchburn. In normal vision, however, accommodation will be only partial, because the vibration and other movements of the eyeball will always be producing corresponding changes in the pattern of light on the retina. The response of the nerves connected to a given retinal cell will therefore depend less on the light intensity at the point in question than on the way in which this light intensity *changes* with position. This means that the excitation of the optic nerve does not correspond to the pattern of light on the retina, but rather to a modified pattern in which contrasts are heightened, and in which a strong impression is produced at the boundaries of objects, where the light intensity varies sharply with position. In this way one obtains an emphasis on the outlines and forms of objects which helps to lead to their being perceived as separate and distinct, a perception that would not be nearly so clear and noticeable if the eye were sensitive to the light intensity itself, rather than to its changes . . .

The essential point that we wish to emphasize in the work concerning the eye is that nothing is perceived without movements or variations in the image on the retina of the eye, and that the characteristics of these variations play a large part in determining the structure that is actually seen. It is important that such variations shall not only be a result of changes that take place naturally in the environment, but that (as in the case of tactile perception) they also can be produced actively by movements in the sense organs of the observer himself. These variations are not themselves perceived to any appreciable extent. What is perceived is something relatively invariant, e.g., the outline and form of an object, the straightness of lines, the sizes and shapes of things, etc. Yet the invariant could not be perceived unless the image were actively varied.

Experiments by Held and his co-workers and by Gibson[6] make it clear that movements of the body also play an essential role in optical perception, particularly the coordination between such movements and the resulting changes that are seen in the optical image of the world. For example, when people are furnished with distorting spectacles (which cause straight lines to appear to be curved) and allowed to enter a room patterned in a way that is not previously known to them, they eventually learn to "correct" the effects of this distortion by the spectacles, and *cease to see the curvature that must actually be present in the image of a straight line on the retinas of their eyes.* Later, when they take off the spectacles, they see straight lines as curved, at least for a while. (A more extreme case of such an experiment is to allow a person to see the world through spectacles that invert the image. After some time he sees it right side up, but when he takes off the spectacles he sees it upside down again, for a while.)

The interesting point of these experiments is that the "relearning" of what corresponds to a straight line depends very strongly on the ability to move the body actively. Thus people who are free to walk around are able to adjust their vision to their spectacles fairly rapidly, whereas people who passively undergo equivalent movements in chairs either never learn to do so or else are very much less effective in such learning. So it is clear that what is essential is not only that there shall be appropriate *variations* of the image on the eye, resulting from movement, but also that some of these variations shall be produced *actively* by the percipient. In other words, as in the case of tactile perception, what one actually sees is determined somehow by the abstraction of what is invariant from a set of variations in what is seen, this variation having been produced, at least in part, as an essential aspect of the process of observation itself . . .

The perceived picture is therefore not just an image or reflection of our momentary sense impressions, but rather it is the outcome of a complex process leading to an ever-changing (three-dimensional) *construction* which is present to our awareness in a kind of "inner show." This construction is based on the abstraction of what is invariant in the relationship between a set of movements produced actively by the percipient himself and the resulting changes in the totality of his sensual "inputs." Such a construction functions, in effect, as a kind of "hypothesis" compatible with the observed invariant features of the person's over-all experience with the environment in question . . .

Not only is the process of construction dependent on the abstraction of invariant relationships between movement and sense perceptions, as described above; it also depends on all that is *known* by the percipient. For example, if a person looks at a letter at a distance too great for clear distinct vision, he will see something very vague and indistinct in form. But if he is told what the letter is, its image will suddenly appear with comparatively great clarity. Or alternatively, he can drop a small coin on a highly patterned carpet, where he will generally find that it is lost to his sight. Then, if he catches a glint of reflected light, the coin that he *knows* that he has lost will suddenly stand out in his perception. Its image must have been on the retina of the eye all the time, but it did not enter the "inner show" of perception until the reflected glint contradicted the perception of a carpet with nothing on it, and also suggested the lost coin that he knows about . . .

Thus far we have been considering only the case in which a percipient moves in a relatively static environment. If movements are taking place in his environment as well, then there is the additional problem of knowing which of the observed changes are due to the movements of the observer and which are due to movements of what is in the environment. This problem is dealt with, in effect, by the capacity to abstract a higher order type of invariant, i.e., *a relatively invariant state of movement.*

Generally speaking, as a person moves in his environment his brain begins (largely unconsciously) to note those features which do not change significantly as a result of these movements. These are treated as a distant and relatively fixed background, against which other movements can be perceived. The closer objects do, of course, change their apparent sizes, shapes, etc., appreciably in a systematic way as a person walks, moves his head, etc. It seems that the brain has developed the ability to be sensitive to such apparent movements and changes in the nearby environment, especially when they are coordinated with movements produced by the percipient himself. This permits the elimination of the self-produced movements in the field of what is perceived, so that the construction of the "inner show" corresponds to a generally static world, in which the percipient himself is seen to be moving. Therefore, as a person walks around a room, he does not feel the room to be moving, whirling around, and changing its shape, etc. Rather, he perceives the room as fixed and himself as moving, in such a way as to explain all the variations in what he has perceived. But if, for example, he has suffered damage to the delicate balancing mechanism in the inner ear, he can no longer coordinate his mechanical perceptions with his optical perceptions. He may then suffer vertigo, and feel that the world is moving around him. The difference between these two modes of perception is very striking to anyone who has ever experienced it.

On the basis of the elimination of the movement of the percipient, the brain is then able to go to the next level of abstraction, in which it senses the movement of some part of the field of vision against a background that is perceived as fixed. The simplest case arises when a given object merely suffers a dislocation in space and perhaps also a rotation. In this case one is able to perceive the object as actually having a constant size and shape, despite the fact that its image on the retina is changing all the time. This perception is inextricably bound up with the ability to see such an object as possessing a *certain state of motion*, rather than as a series of "still" pictures of the object in question, each in a slightly different position. It is almost as if the brain were able to establish a co-moving reference frame, in which a moving object could be seen to have a constant shape. In this way, the brain seems to include in its construction process the ability to abstract a certain state of movement, which under the assumption of an object of a given shape is compatible with the changes that have been perceived in the appearance of the object over some period of time.

Of course, there will then be further kinds of changes which cannot be explained in this way (e.g., an object may actually grow in size, change its form, etc.). These will have to be perceived in terms of more subtle internal changes in the object in question.

The problem of how movement is perceived is far from being fully solved. Yet it is already clear that such perception cannot be based merely on "sense impressions" at a given moment. Rather, the "inner show" that we perceive embodies certain structural features, based not only on abstractions from immediate sensations, but also on a series of abstractions over a more or less extended set of earlier perceptions. Indeed, without such a series of abstractions we could not be able to see a world having some well-defined order, organization, structure, etc. Even a static environment is effectively presented in the "inner show" as a tentative and hypothetical structure, which when assumed to be invariant, will be compatible with the changing experiences that the percipient has had with this environment, in movements that he himself has produced. And an environment which is itself changing is presented in the "inner show" as a structure expressed in terms of invariant states of movement of parts of the environment which account for earlier changing experiences that are not explained by the movements of the percipient.

There may also arise an ambiguity in the attribution of movements to the observer or to various parts of the environment. Thus, if a person is sitting in a train that is not moving, and watches another moving train through the window, he may find that he perceives himself as moving, and that he even gets some of the physical (kinesthetic) sensations of movement. But when he fails to feel the expected shaking and vibration of the train, he begins to look more carefully, and can soon see in the environment certain further clues, suggesting that the other train is moving and that he is at rest. Suddenly his mode of perception of the world changes. This is a striking demonstration of how our perceptions of the world are a construction in the "inner show," based on the search for a hypothesis that is compatible with all that we have experienced in connection with a certain situation. So we do not perceive *just* what is before our eyes. We perceive it organized and structured through abstractions of what kind of invariant state of affairs (which may include invariant *states of movement*) will explain immediate experience and a wide range of earlier experiences that led up to it.

Results of the kind described above led Gibson[7] to suggest a new concept of what constitutes perception. He emphasizes the need to drop the idea that perception consists of passively gathering sense impressions, which are organized and structured through principles supplied only by the observer. Indeed, the isolated sense impression is seen to be an extremely high level abstraction, which does not play any significant part in the actual process of perception. Instead, we are sensitive directly to the structure of our environment itself. In the last analysis the observer therefore does not supply the structure of his perceptions, so much as he *abstracts* it. Or as Gibson himself puts this point, the structure of our environment is the *stimulus* that gives rise to what we perceive (i.e., to the construction in the "inner show" that is presented in our awareness). With regard to optical perception, for example, Gibson points out that through each region of space there passes an infinity of rays of light, going in all directions. These rays of light *implicitly* contain all the information about the structure of the world that we can obtain from vision.[8] But an eye fixed in a certain position cannot abstract this information. It must move in many ways, and at least some part of these movements must be produced by the observer himself, because (as was first brought out by Held and his co-workers) structural information is abstracted mainly from invariant relationships between the outgoing nervous excitations that give rise to these movements and the corresponding ingoing nervous excitations that result from them.

Gibson raises a related set of questions regarding the role of time in perception. A typical question is, for example, "When does a particular stimulus come to an end?" The older way of looking at this problem is to refer to what is called the "specious present." That is, it is found that there is an interval of time, of the order of a tenth of a second, which is "speciously" experienced as a single moment, in the sense that people do not seem to be able clearly to discriminate changes that take place in times less than this. From this notion it would follow that all our perceptions can in principle be uniquely ordered in time, within an accuracy of a tenth of a second or so. Nevertheless, Gibson raises questions which suggest that it is a source of confusion to try to understand the essential features of the process of perception by referring it in this way to such a time order.

To see why Gibson questions the simple time order of perceptions described above, let us recall that we do not perceive momentary sensations, to any appreciable extent. Rather, we perceive an over-all structure that is abstracted from these, a structure evidently built up over some period of time. We have already seen in connection with optical perception, for example, that clues obtained over some time may come together at a given moment and give rise to a new structure of what is perceived. It evidently makes no sense to say that this new structure is based only on the very last clue to be received. Rather, it is based on the whole set of clues. This means that a given stimulus to our perceptions is not restricted to the smallest time interval that can be discriminated. Rather, it may be said that some stimuli take place over much longer intervals.

In music the property of stimuli is much more clearly seen. As one is listening to a tune, the notes heard earlier continue to reverberate in the mind, while each new note comes in. One may suddenly understand (i.e., perceive the over-all structure) of a piece of music at a certain moment in this process. But evidently the very last note to be received is not the sole basis of such an understanding. Rather, it is the whole structure of tones reverberating in the mind. These tones have manifold relationships, which are not restricted to their time order. To grasp these relationships is essential to the understanding of the music. The effort to regard the essential content of the music in terms of its time order could then lead to too narrow a way of looking at the problem, which would tend to produce confusion.

In a similar way one can consider the problem of how one perceives rhythm. At any moment there is only one beat to be heard. But one beat is not a rhythm. Evidently it is the reverberation of a whole set of beats in the mind, all in a certain relationship that constitutes the perception of rhythm. The stimulus that constitutes a rhythm cannot then refer only to a single moment of time. So it seems important to realize that the essential features of perception will not always be understood by stringing out what is perceived in a time order.

Indeed, in many cases it is not possible to assign a unique moment of time to a given feature of what is perceived. While listening to a piece of music one may be appreciating a rhythm that is based on many seconds, a theme that may require a minute or more to be apprehended, and we may be looking at a stop-watch, seeing the movements of the hand that perhaps indicates some fraction of a second. When one says "now," what does one mean by this? Does it refer to the perception of a certain position of the indicator on the watch, the perception of a certain part of the rhythm, the perception of some part of the theme, or perhaps to something else?

It would seem then that the effort to order the totality of one's perceptions in terms of a single, unique time order must lead to confusion and absurdity. Certain perceptions can thus be ordered (e.g., those that are similar to seeing the indicator on the watch dial). But to understand the process of perception in a broader context, we must see that the structures that are perceived are not as rigidly related to such a time order as our customary notions might lead us to think. There is a loose time order, in the sense, for example, that today's perceptions are not strongly related to yesterday's events (although these do in fact still "reverberate" in us and help to shape present perceptions). Yet the hard and fast notion that each perception is uniquely ordered as earlier, later, or simultaneous with another (within the period of the "specious present") seems to lead to a kind of confusion, indicating that it probably has little relevance to the actual facts of perception.

It may perhaps be instructive to consider a simple example of a physical problem in which the attempt to regard the time order of events as basic to the understanding of a process leads to a type of confusion similar to that in which it results when applied to perception. Suppose, for the sake of our discussion, that there were beings on Mars, and that they had become interested in studying the radio signals coming from the Earth. When they came to observe television signals they would not be able to make a great deal of sense of them, if they supposed that the essential principle of these signals were some kind of formula or set of relationships determining their *time order*. The signals can in fact be understood properly only when it is realized that they originate in a series of *whole pictures*, which are then translated systematically into a time series of pulses. The principles governing the actual order of pulses are therefore to be grasped in terms of a spatial structure very different from that of the time order that is received in the radio signals. Or, to put it differently, *the order of the signals is not essentially related to the order of time*. In a similar way, the structure of our perceptual process may also not be essentially related to some hypothetical series of instants, but may be based on entirely different kinds of principles involving (like the television signal) the integration of what is received over suitable intervals of time, extending far beyond the period of the "specious present."

If a given perception integrates what comes in over such extended periods of time, does this mean that *memory* is the main factor that determines the general structure of what we perceive? (Memory being the ability, for example, to recall approximately the sensations, events, objects, etc., that were experienced in the past.) Gibson does not accept the notion that the structure in our perceptions comes *mainly* from memory, although of course memories do evidently have *some* influence in shaping such perceptions. He suggests that the main process is what he calls "attunement" to what one perceives. Thus, as one sees something new and unfamiliar he first vaguely perceives only a few *general* structural features. Then as he moves in relation to what he is looking at and perhaps probes as well, he starts to abstract more of the details of the structure, and his perceptions sharpen. Perhaps one could compare this process to a kind of skill, which is also not based simply on memory of all the steps by which the skill was acquired.

Both in the case of perception and in that of building a skill, a person must actively meet his environment in such a way that he coordinates his outgoing nervous impulses with those that are coming in. As a result the structure of his environment is, as it were, gradually incorporated into his outgoing impulses, so that he learns how to meet his environment with the right kind of response. With regard to learning a skill it is evident how this happens. But in a sense the perception of each kind of thing is also a skill, because it requires a person actively to meet the environment with the movements that are appropriate for the disclosure of the structure of that environment. (This fact would also be evident if it were not for our habitual notion that perception is a purely passive affair.)

If we learn the structure of things by "attunement" it seems clear that the very general features of our ability to apprehend the structure of the world will, in many cases, go back to what was learned in early childhood. It is here that the studies of the process of perception can link up with the work of Piaget, discussed in the previous section. For there we saw how the infant begins with a limited set of inborn reflexes. When these are developed into the "circular reflex" he has the most basic feature of perception, i.e., the ability to be sensitive to a relationship between outgoing and incoming nervous impulses, a relationship that is characteristic of what is to be perceived. From here on he is able to "attune" himself step by step with his environment, by abstracting from such relationships what is invariant in its general structure. In doing this he builds up his notions of space, time, causality, the division of the world into permanent objects (one of which is himself), the notion of permanent substance, permanent numbers of objects, etc. All of these notions are interwoven into the fabric of perception, in the sense that they help shape the structure of what appears in the "inner show" that is present in our awareness. So while we are able to "attune" ourselves to new kinds of structures when we meet something new, there seem to be certain general structural features, of the kind described above, which were first learned in childhood, and which are present in all that we perceive.

The over-all or general structure of our total perceptual process can be regarded not only from the standpoint of its development from infancy but can also be investigated directly in the adult. Such studies have been made by Hebb and his group,[9] by isolating individuals in environments in which there was little or nothing to be perceived. The extreme cases of such isolation involved putting people in tanks of water at a comfortable temperature, with nothing to be seen or heard, and with hands covered in such a way that nothing could be felt. Those individuals who were hardy enough to volunteer for such treatment found that after a while the structure of the perceptual field began to change. Hallucinations and other self-induced perceptions, as well as distortions of awareness of time, became more and more frequent. Finally, when these people emerged from isolation, it was found that they had undergone a considerable degree of general disorientation, not only in their emotions but also in their ability to perceive. For example, they often found themselves unable to see the shapes of objects clearly, or even to see their forms as fixed. They saw changing colors which were not there, etc. (In time, normal perception was, of course, regained.)

The results of these experiments were rather difficult to understand in detail, but their over-all implication was that the general structural "attunements," built into the brain since early childhood, tend to disintegrate when there is no appropriately structured environment for them to work on. If we compare these attunements to some kinds of skills, needed in meeting our typical environment, then perhaps it is not entirely unexpected that they should decay when they are not used. But what is still surprising is the extremely great speed with which such "skills" built up over a lifetime can deteriorate. To explain this it has been proposed that when there is no external environment for the brain to work on it starts to operate on the internal environment, i.e., on the impulses produced spontaneously on the nervous system itself. But these impulses do not seem actually to have a well-defined structure that is comprehensible to us. So in the active effort to "attune" to a structure that is either nonexistent or else incomprehensible to the people who actually did the experiment, the older adjustments, built up over a lifetime, are mixed up and broken down.

The above hypothesis has to some extent been confirmed by experiments in which people looked for a long time at a television screen containing a changing random (unstructured) pattern of spots. A disorientation of perception resulted which was similar to that obtained in the experiments in which subjects were isolated. Thus it could be argued that in the effort to adjust to a nonexistent or incomprehensible structure in its general environment, the brain began to break down the older structural "attunement" that was appropriate to the normal environment in which people generally live.

The implications of these experiments are so far-reaching as to be rather disturbing. Nevertheless, it can be seen that, on the whole, they tend to carry further what is already suggested in the work of Piaget and in the results that have been summarized in this section. For in all of this we have seen that in perception there is present an outgoing nervous impulse producing a movement, in response to which there is a coordinated incoming set of sensations. The ability to abstract an invariant relationship in these nervous impulses seems to be what is at the basis of intelligent perception. For the structure that is present in the "inner show" is determined by the need to account for what is invariant in the relationship of the outgoing movements and the incoming sensations. In this way the percipient is not only always learning about his environment but is also *changing himself.* That is, some reflection of the general structure of his environment is being built into his nervous system. As long as his general environment is not too different in structure from what has already thus been built into his nervous system he can make adjustments by "attuning" to the new features of the environment. But in an environment without such a perceptible structure, it seems that there is a tendency for this attunement to be lost, in the search for a structure which either does not exist or which has features that are beyond the ability of the percipient to grasp if it does exist.

These results lead us back to the old question first formulated by Kant, as to whether our general mode of apprehending the world as ordered and structured in space and time and through causal relationships, etc., is objectively inherent in the nature of the world, or whether it is *imposed* by our own minds. Kant proposed that these general principles constituted a kind of *a priori* knowledge, built into the mind, which was a necessary precondition for any recognizable experience at all, but which may not be a characteristic of "things in themselves." It would seem that Kant's proposal was right in some respects but basically wrong in that he had considered the problem in too narrow a framework. It is certainly true that at any given moment we meet new experience with a particular structural "attunement" in the brain that is a necessary condition for perception of recognizable aspects of the world. This "attunement" is responsible for our ability to see a more or less fixed set of things at each moment, organized in space, causally related, changing in a simply ordered time sequence, etc. When this "attunement" is broken down by long isolation from perception or by perception of an environment without visible structure, then the experiments cited above do indeed show that the process of recognizable experiencing of an environment is seriously interfered with.

On the other hand, a broader view of this problem shows that an adult's attunement to the general structure of the world has been built up in a development, starting with infancy. In the beginning of this development the child must *discover* the structure of his environment in a long process in which he *experiments* with it, operates on it, etc. His procedure in doing this is perhaps not basically different from that used in scientific research. He is *interested* in his environment, probing it, testing it, observing it, etc., and, as it were, always developing new perceptual "hypotheses" in the "inner show" that explain his experiences better. In doing this he is "attuning" himself to his environment, developing the right responses to perceive its structure adequately. As he gets older this whole process tends to fall into the domain of habit. But whenever he meets something strange and unexpected, he is able to abstract new structural features, by a continuation of the kind of interested experimentation and observation that is characteristic of early childhood.

Of course, a person finds it hard to change *very general* structural features, such as the organization of all experience in terms of space, time, causality, etc. Yet the experiments cited above suggest that there seems to be no inherent need to continue any particular structure, and that the brain probably is capable of abstracting a very wide variety of kinds of structural features that may be actually present in the part of the environment that is available to his senses, provided that there is appropriate interest, leading to the proper kind of experimentation, probing, etc. At any given moment the structure that we already know depends on past experiences, habits, etc.; this in turn is dictated in part by the general environment that people have actually lived in, and in part by the interests that determine to which structural features people will have paid a great deal of attention. So we do in fact approach new experiences, as Kant suggested, with some kind of already given general structural principles. Yet the experiments cited here suggest that Kant was wrong in regarding any particular set of principles of this kind as *inevitably* following *a priori* from the very nature of the human mind. Rather, along the lines suggested by Gibson, it would seem that a person might become "attuned" to any structural features of his environment to which his nervous system could respond, and in which he was sufficiently interested.

In terms of the notions described above we can see that while our perceptions do have a subjective side, dependent on the particular background and conditioning of each person, as well as on the general background and conditioning of the whole of humanity, they also have a kind of objective content, which can go beyond this particular and limited background. For the *general* structure of our perceptions (resulting from this background) can be regarded as a kind of *hypothesis*, with the aid of which we approach subsequent experiences in which things have changed not only of their own accord, but also because of our own movements, actions, and probings, which alter our own relationships to our environments. To the extent that the new experiences fit into the continuation of the old structure without contradictions, these hypotheses are effectively confirmed. But if we are alert, we will sense contradictions when they arise (as we have already seen, in numerous examples discussed earlier). When this happens the brain is sensitive to the discovery of new relationships, leading spontaneously to further hypotheses, which are embodied in the appearance of new structures in the "inner show." Anyone can see this happening as he approaches a distant object that is unknown to him, or as he approaches something unknown in the obscure light, for example, of the moon. He will see various forms, shapes, objects, etc., which appear and then disappear, because they are not compatible with further experiences resulting from his movements, probings, etc. So there is a continual process of "trial and error" in which what is shown to be false is continually being set aside, while new structures are continually being put forth for "criticism." Eventually there develops in this way a perception which stands up to further movements, probings, etc., in the sense that its predictive implications are actually borne out in such experiences. (Of course, even this is always tentative, in the sense that it can be contradicted later.)

The objective content of our perceptions is then implicit in the process of falsification and confirmation described above. Indeed, the very fact that our vision of the world can be falsified as a result of further movement, observation, probing, etc., implies that there is more in the world than what we have perceived and known. That is to say, we do not actually create the world. In fact, we only create an "inner show" of the world in response to our movements and sensations. It is, however, the possibility of confirmation of the "inner show" which demonstrates that there is more in it than merely a summary of past experiences. For this "inner show" is based on the abstraction of the *general structure* of these past experiences, the structure having predictive inferences for later experiences. For example, as we approach the front of an object such as a house, we (largely unconsciously) predict a great many structural features of the parts of the object that are not yet visible. Thus, on seeing the front and one side of the house, along with parts of its roof, we infer that it has other sides, that these have certain parallel lines, certain angles, etc. These inferences may come *partially* from memory, having gone round similar houses previously. But in large part they come, not from the simple recall of earlier experiences themselves, but from the general structural principles that have been abstracted from a very wide range of such experiences (e.g., the three-dimensionality of space, the existence of straight lines, parallel lines, and right angles, all of which together imply a certain general field of possibilities for the unseen sides of an object, independent of the particular memories of similar objects that we may possess).

A little reflection shows that there is an enormous number of cases in which the above-described kinds of predictive inferences based on the general structure of our perceptions have turned out to be correct. That is to say, the "world" that we see in immediate perception has, at a given moment, a *general* structure, which has withstood a long series of tests, in the observations that have led up to the moment in question. And as a rule it happens that the natural projection of this structure in accordance with the known state of movement of the observer and of what is in the field of perception will continue to be more or less in accord with later observations in a great many respects. This means that the general structure of our perceptions has a certain similarity to the general structure of what is actually in our environment. Yet, the similarity is not perfect, as is evidenced by the appearance of contradictions, unexpected events, etc., which necessitate continual changes in what is "constructed" in the field of perception, and are not merely the result of the natural projection from what was perceived earlier. In this way we are continually being confronted with what is not even implicitly contained in our earlier perceptions, thus we are being reminded that there is a reality beyond what we have already perceived, aspects of which are always in the process of being revealed in our further perceptions.

4 THE SIMILARITY BETWEEN THE PROCESS OF PERCEPTION AND THE PROCESS BY WHICH SCIENCE INVESTIGATES THE WORLD

In the previous sections of this essay we have discussed studies of the development of the process of perception in an individual human being from infancy, as well as direct studies of how this process takes place in adults. What comes out of these studies can be summed up in the statement that in the process of perception we learn about the world mainly by being sensitive to what is invariant in the relationships between our own movements, activities, probings, etc., and the resulting changes in what comes in through our sense organs. These invariant relationships are then presented immediately in our awareness as a kind of "construction" in an "inner show," embodying, in effect, a hypothesis that accounts for the invariant features that have been found in such experiences up to the moment in question. This hypothesis is, however, tentative in the sense that it will be replaced by another one, if in our subsequent movements, probings, etc., we encounter contradictions with the implications of our "constructions."

However, we have seen that research in physics has shown basic features very similar to those of perception described above, and that with the further development of physics, into its more modern forms (in particular, with the theory of relativity), this similarity has tended to become stronger. Thus, those aspects of Newtonian mechanics which eventually proved to be correct consisted of the discovery of the invariance of certain relationships (Newton's laws of motion), in a wide variety of systems, movements, changes of frames of reference, etc. On the other hand, those features of the theory which were considered to represent absolutes (i.e., absolute space, absolute time, the notion of permanent substances with fixed masses, etc.) were eventually shown to be unnecessary, and indeed important sources of confusion and error, in the effort to extend scientific knowledge of the laws of movement into broader domains.

Einstein's major steps were based on setting aside such ideas of an absolute, and on extending into broader domains the notion of the laws of physics as invariant relationships (e.g., so as to include velocities comparable to that of light). In doing this he was led also to drop the notion of fixed quantities of substances, having constant masses. Instead, mass was seen to be only a relatively invariant property, expressing a relationship between energy of a body and its inertial resistance to acceleration, along with its gravitational properties. Further developments in modern physics, including quantum theory and the studies of the transformations of the so-called "elementary" particles suggest that the notion of permanent entities constituted of substances with unchanging qualitative and quantitative properties may have to be dropped altogether, and that physics will be left with nothing but the study of what is relatively invariant in as wide as possible a variety of movements, transformations of coordinates, changes of perspective, etc.

Moreover, it seems that the notion that science is collecting absolute truths about nature, or even approaching such truths in a convergent fashion, is not in good accord with the facts concerning the actual development of scientific theories thus far, and has indeed also been a major source of confusion in scientific research. Rather, as Professor Popper has emphasized, science actually progresses through the putting forth of falsifiable hypotheses, which are confirmed up to a certain point and thereafter, as a rule, eventually falsified. New hypotheses are then put forth, which are criticized and tested by a process of "trial and error" very similar to that to which our immediate perceptions are continually being subjected.

The interesting point that has emerged from a simultaneous consideration of what has developed in modern science and of what has been disclosed in modern studies of the process of perception is that the new ideas required to understand both of them are rather similar. In this section we shall give some arguments in favor of the suggestion that this similarity is not accidental but rather has a deep reason behind it. The reason that we are proposing is that scientific investigation is basically a mode of extending our *perception* of the world, and not mainly a mode of obtaining *knowledge* about it. That is to say, while science does involve a search for knowledge, the essential role of this knowledge is that it is an adjunct to an extended perceptual process. And if science is basically such a mode of perception, then, as we shall try to show, it is quite reasonable that certain essential features of scientific research shall be rather similar to corresponding features of immediate perception.[10]

Since science has generally been regarded thus far as *basically* a search for knowledge, it will be necessary to begin by going more deeply into the question of the relationship between knowledge and immediate perception. Now, as we have seen, what appears in immediate perception already embodies a kind of abstraction of the general structure of what has been found to be invariant in an earlier active process of probing the environment that has led up to the perception in question. We propose that knowledge is a *higher-level abstraction*, based on what is found to be invariant in a wide range of experiences involving immediate perception.

We can perhaps explain this notion most directly by first referring to Piaget's account of the development of the child's concept of space (discussed in Section 2). At first the child discovers a group of operations, such that he can go from one place to another by a certain route, and return invariantly to the same place by a wide range of different routes. Later the child is able to *imagine* (i.e., to produce a mental image) of a space, containing even objects that are no longer in his field of immediate perception, and also an imagined object that corresponds to *himself*. The structure of this mental image faithfully corresponds to what has been found by the child to be invariant in his earlier experiences with groups of movements. This mental image therefore *abstracts* a kind of "higher-order invariant," i.e., something that has been invariant in a wide range of immediate perceptions. When we use the words "to abstract" we do not wish to suggest that there is merely a process of induction, or of taking out some kind of summation of what has been experienced earlier. Rather, each abstraction constitutes, as it were, a kind of "hypothesis," put forth to explain what has been found to be invariant in such earlier experiences. Only the abstractions which stand up to further tests and probings will be retained. Eventually, however, these become habitual, and we cease to be aware of their basically hypothetical and tentative character, regarding them instead as inherent and necessary features of all that exists, in every possible domain and field of experiencing and investigation.

Piaget then goes on to describe how with the development of language and logical thinking the child goes on to make still higher level abstractions, in which there are formed structures of words, ideas, concepts, etc., which express the invariant features of the world that he abstractly considers in his perceptions. Evidently there is in principle no limit to this process of abstraction. Thus science and mathematics may be said to form still higher level abstractions (formulated in words, diagrams, and mathematical symbols), expressing the invariant features of what has been found in experiments and observations (which latter are carried out in terms of the ordinary abstractions of everyday language and common sense). Thus all knowledge is a structure of abstractions, the ultimate test of the validity of which is, however, in the process of coming into contact with the world that takes place in immediate perception.

It can be seen that a crucial state in this over-all process of abstraction is the setting aside of certain parts of what appears in the "inner show" as not *directly* representing immediate perception. These are what we *imagine*, conceive, symbolize, think about, etc. These parts are then seen to be related to immediate perception as abstractions, representing the general structural features of this perception, much as a map represents the terrain of which it is a map.[11] However, as has been pointed out in Section 2, a young child does not readily distinguish between what has been imagined and what is seen in response to immediate perception. In this way, there arises the habit of confusing our abstract conceptual "maps" with reality itself, and of not noticing that they are only maps. When the child grows older he is able to avoid this confusion in superficial problems, but when it comes to fundamental concepts, such as space, time, causality, etc., it is much more difficult to do so. As a result, the adult continues the habit of looking, as it were, at his comparatively abstract conceptual maps, and seeing them as if they were inherent in the nature of things, rather than understanding that they are higher-level abstractions, having only a kind of structural similarity to what has been found to be invariant in lower levels. It is this confusion, based on habits of very long standing, which makes a clear discussion of such fundamental problems so difficult.

We can perhaps best illustrate these notions with the aid of a simple example. Suppose that we are looking at a circular disk. Now its immediate appearance to our eyes will be that of an ellipse, corresponding to its projection on the retina of the eyes (as would, for example, be portrayed by an artist, who was trying to draw it in perspective). Nevertheless, we *know* that it is really a circle. What is the basis of this knowledge?

What actually happens is, as we have indicated earlier, that the eye, the head, the body, etc., are always moving. In these movements the appearance of the disk is always changing, undergoing in fact a series of *projective transformations* that are related in a definite way to the movements in question. By various means (some of which are discussed in Section 3) the brain is able to abstract what is invariant in all this movement, change of perspective, etc. This abstraction, expressed in terms of the notion that a circular object accounts for all the changing views of it, is the basis of the "construction" of it that we perceive in the "inner show." The "hypothesis" that this object is really a circle is then further probed and tested in subsequent ways of coming in contact with it perceptually, and it is retained as long as it stands up to such probing and testing.

But the realization that the perceived object is a circle depends also on knowledge going beyond the level of immediate perception. Thus from early childhood a person has learned to *imagine* looking straight at the object in a perpendicular direction, and seeing its circular shape (as well as feeling it to be circular when his hands grasp it). He may also have learned further to imagine himself represented as a point on a diagram, and to follow the course of the light rays from the circle to his point of perspective, thus being able to see how the circular shape is transformed into an elliptical appearance. If he has been further educated, he can go to a still higher level of abstraction, by mathematically calculating the correct shape of the disk, from a knowledge of its appearance in several views and from a knowledge of the relationship of the observer to the disk in all of these views (distance, etc.). In carrying out this calculation he will do consciously on a higher level of abstraction what his brain does spontaneously on a lower level, i.e., to find a single structure that accounts for what is invariant in our changing relationships with the object under discussion.

We see then that there is no sharp break between the abstractions of immediate perception and those which constitute our knowledge, even if we carry this knowledge to the highest levels reached by science and mathematics. From the very first, our immediate perceptions express a "construction" in an "inner show," based on a preconscious abstraction of what is invariant in, or active process of coming into contact with, our environment. Each higher level of abstraction repeats a similar process of discovery of what is invariant in lower levels, which is then represented in the form of a picture, an image, a symbolic structure of words and formulas, etc. These higher-level abstractions then contribute to shaping the general structure of those at lower levels, even coming down to that of immediate perception. So between all the levels of abstraction there is a continual two-way interaction.

Consider, for example, the experience of looking out at the night sky. Ancient man abstracted from the stars the patterns of animals, men, and gods, and thereafter was unable to look at the sky without seeing such entities in it. Modern man knows that what is really behind this view is an immeasurable universe of stars, galaxies, galaxies of galaxies, etc., and that each person, having a particular place in this universe, obtains a certain perspective on it, which is what is seen in the night sky. Such a man does not see animals, gods, etc., in the sky, but he sees an immense universe there. But even the view of modern science is probably true only in a certain domain. So future man may form a very different notion of the invariant totality that is behind our view of the night sky, in which present notions will perhaps be seen as a simplification, approximation, and limiting case, but actually very far from being completely true. Can we not say then that at every stage man was extending his perception of the night sky, going from one level of abstraction to another, and in each stage thus being led to hypotheses on what is invariant, which are able to stand up better to further tests, probings, etc.? But if this is the case, then the most abstract and general scientific investigations are natural extensions of the very same process by which the young child learns to come into perceptual contact with his environment.

As we have pointed out on several occasions (e.g., in the discussion of Piaget's work in Section 2 and of the perception of movement in Section 3) one of the basic problems that has to be solved in every act of perception is that of taking into account the special point of view and perspective of the observer. The solution of this problem depends essentially on the use of a number of levels of abstraction, all properly related to each other. Thus a person not only perceives the immediate elliptical appearance of the disk in front of him. He can also perceive the changes in appearance of the disk, which result from certain movements which he himself actively undertakes. From these changes his brain is able to abstract information about his relationship to the disk (e.g., how far away it is). The essential point here is that through many levels of abstraction, all going on simultaneously in the mind, it is possible to perceive not only a projection of the object of interest but also the relationship of the observer to the object in question. From this it is always possible in principle to obtain an invariant notion as to what is actually going on. This is represented in a higher level of abstraction, for example, by *imagining* space containing the disk and the observer himself, in which both are represented in their proper relationships. When a person says that the object is *really* circular, he is then evidently not referring to an immediate sensation of the shape of the object but to this extended process of abstraction, the essential results of which are represented in this imagined space, containing both the object and himself.

A very similar problem arises in science. Here, the hands, body, and sense organs of the observer are generally, in effect, extended by means of suitable instruments, which are in *certain ways* more sensitive, more powerful, more accurate, as well as capable of new modes of making contact with the world. But in the essential point that the observer is *actively* probing and testing his environment, the situation is very similar to what it is in immediate perception, unaided by such instruments.

In such tests there is always some observable response to this probing and testing; and it is the relationship of variations in this response to known variations in the state of the instruments that constitutes the relevant information in what is observed (just as happens directly with the sense organs themselves).

As in the case of immediate perception, however, such an observation has very little significance until one knows the relationship of the instrument to the field that is under observation. It is possible to know this relationship with the aid of a series of abstractions. Thus in any experiment one not only knows the observed result; one knows the structure of the instrument, its mode of functioning, etc., all of which has been found out with the aid of earlier observations and actions of many kinds. In other words, in each process of observation there is always implicit an observation of the observing instrument itself, carried out in terms of different levels of conceptual abstractions. But to *understand* the observation one always needs certain modes of thinking about the problem, in which the instrument and what is observed are represented together, so that one can see "a total picture" in which an invariant field of what is being studied stands in a certain relationship to the instrument, this relationship determining, as it were, how what is in the field "projects" into some observable response of the instrument.

In the theory of relativity one uses the Minkowski diagram,[12] in which one can in principle represent all the events that happen in the whole of space-time. However, each example of such a diagram must contain a line corresponding to the world line of the observer whose results are under discussion. This is usually represented by the axis of the diagram. Then, if we wish to discuss the results of another observer, we must include in the diagram a representation of *his* world line. In a similar way we must choose a point to represent the place and time which determine the perspective of a given observation. By taking all of this into account we are able, from the response of the observing instruments (which is *relative* to their speed, time and place of functioning, etc.), to calculate the invariant properties of what is observed, in such a way that the different results of different observers are explained by their differing *relationships* to the process under investigation. It can be seen then that relativity theory approaches the universe in a way very similar to that in which a person approaches his environment in immediate perception. In both these fields all that is observed is based on the abstraction of what is invariant as seen in various movements, from various points of view, perspectives, frames of reference, etc. And in both the invariant is finally understood with the aid of various hypotheses, expressed in terms of higher levels of abstraction, which serve as a kind of "map," having an order, pattern, and structure similar to that of what is being observed.

The tendency for the use of such maps to become habitual is also common to scientific investigation and to immediate perception. When this happens a person's thinking is limited to what can fit into such maps, because he thinks that they contain all that can possibly happen, in every condition and domain of experience. For example, the common-sense notion of simultaneity of all that is co-present in our immediate perceptions is abstracted into the Newtonian concept of absolute time, with the result that it seems incomprehensible that two twins who are accelerated in different ways and then meet may experience different amounts of time.[13] But in Section 3, we saw that the notion of a single unique time order does not seem to apply without confusion in the field of our immediate perceptions either. The main reason that this has been so little noticed is probably our habit of taking seriously only what fits into our habitual perception of all that happens, both inwardly and outwardly, as being in such a unique and universal time order.

It may be remarked in passing that in the quantum theory the point of view described above is carried even further. The reason is basically the indivisibility of the quantum of action, which implies that when we observe something very precisely at the atomic level, it is found that there must be an irreducible disturbance of the observed system by the quanta needed for such an observation (the fact behind the derivation of Heisenberg's famous uncertainty principle). On the large-scale level the effects of these quanta can be neglected. Therefore, although the observer must engage in active movements and probings in order to perceive anything whatsoever, he can in principle (at least in large-scale optical perception) refrain from significantly disturbing what he is looking at. At the quantum level of accuracy, however, the situation is different. Here, the light quanta may be compared to a blind man's fingers, which can give information about an object only if they move and disturb the latter. The blind man is nevertheless able to abstract certain invariant properties of the object (e.g., size and shape), but in doing this, his brain spontaneously *takes into account* the movement which his perceptual operations impart to the object. Similarly, the physicist is still able to abstract certain invariant properties of atoms, electrons, protons, etc. (e.g., charge, mass, spin, etc.); but in so doing he must consciously take into account the *operations* involved in his observation process in a similar way. (To discuss this point in detail is, of course, beyond the scope of the present work; but these questions will be treated in subsequent publications.)

5 THE ROLE OF PERCEPTION IN SCIENTIFIC RESEARCH

In the previous discussion we have seen the close similarity between our modes of immediate perception of the world and our modes of approach to it in modern scientific investigations. We shall now go on to consider directly the centrally perceptual character of scientific research, which we suggested at the beginning of Section 4.

While man's scientific instruments do constitute, as we have seen, an effective extension of his body and his sense organs, there are no comparable *external* structures that substitute for the *inward* side of the perceptive process (in which the invariant features of what has been experienced are presented in the "inner show"). Thus, it is up to the *scientist himself* to be aware of contradictions between his hypotheses and what he observes, to be sensitive to new relationships in what he observes, and to put forth conjectures or hypotheses, which explain the known facts, embodying these new relationships, and have additional implications with regard to what is as yet unknown, so that they can be tested in further experiments and observations. So there is always finally a stage where an *essentially perceptual process* is needed in scientific research – a process taking place within the scientist himself.

The importance of the perceptual stage tends to be underemphasized, however, because scientists pay attention mostly to the next stage, in which hypotheses that have withstood a number of tests are incorporated into the body of currently accepted scientific knowledge. In effect they are thus led to suppose the essential activity of the scientist is the *accumulation of verified knowledge*, toward which goal all other activities of the scientist are ultimately directed.

If such knowledge could constitute a set of absolute truths, then it would make at least some kind of sense to regard its accumulation as the main purpose of science. As we have seen, however, it is the fate of all theories eventually to be falsified, so that they are relative truths, adequate in certain domains, including what has already been observed, along with some as yet further unknown region that can be delimited, to some extent at least, in future experiments and observations. But if this is the case then the accumulation of knowledge cannot be regarded as the *essential* purpose of scientific research, simply because the validity of all knowledge is relative to something that is not in the knowledge itself. So one will not be able to see what scientific research is really about without taking into account what it is to which even established and well-tested scientific knowledge must continually be further related, if we are to be able to discuss its (necessarily incompletely known) domain of validity.

There is also a similar relative validity of the knowledge that we gain in immediate perception. But in this field the reason for this is fairly evident. Indeed, the world is so vast and has so much that is unknown within it that we are not tempted to suppose that what we learn from immediate perception is a set of absolute truths, the implications of which could be expected to be valid in unlimited domains of future experience. Rather, we realize that immediate perception is actually a means of remaining in a kind of contact with a certain segment of the world, in such a way that we can be aware of the general structure of that segment, from moment to moment, if we carry out the process of perception properly. In this contact we are satisfied if we are able to keep up with what we see and perhaps, in some respects, get a little ahead of it (e.g., in driving an automobile, we can, to a certain extent, anticipate the movements of other automobiles, people, the turns in the road, etc.). Thus, in the process of immediate perception, one obtains a kind of knowledge, the *implications* of which are valid in the moment of contact and for some unpredictable period beyond this moment. The major significance of past knowledge of this kind is then in its *implications* for present and future perceptions, rather than in the accumulation of a store of truths, considered to be absolute.

Thus our knowledge of what happened yesterday is *in itself* of little significance because yesterday is gone and will never return. This knowledge will be significant, however, to the extent that its implications and the inferences that can be drawn from it may be valid today or at some later date.

Of course, scientific theories evidently have much broader domains of validity of their predictive inferences than do the "hypotheses" that arise in immediate perception (these broader domains being purchased, however, at the expense of the need to operate only at very high levels of abstraction). Because the domain of validity is so broad, it often takes a long time to demonstrate its limits. Nevertheless, what happens in scientific research is, in regard to the problem under discussion, not *fundamentally* different from what happens in immediate perception. For in science too the totality of the universe is too much to be grasped definitively in *any* form of knowledge, not only because it is so vast and immeasurable, but even more because in its many levels, domains, and aspects it contains an inexhaustible variety of structures, which escape any given conceptual "net" that we may use in trying to express their order and pattern. Therefore, as in the field of immediate perception, our knowledge is adequate for an original domain of contact with the world, extending in an unpredictable way into some further domains. Since the goal of obtaining absolutely valid knowledge has no relevance in such a situation, we are led to suggest that scientific research is basically to be regarded as a mode of extending man's perceptual contact with the world, and that the main significance of scientific *knowledge* is (as happens in immediate perception) that it is an adjunct to this process.

The basically perceptual character of scientific research shows up most strongly when the time comes to *understand* new facts, as distinct from merely accumulating further knowledge. Everyone has experienced such a process on various occasions in his life. Suppose something unfamiliar is being explained (e.g., a theorem in geometry). At first a person is able to take in only various bits of knowledge, the relationship of which is not yet clear. But at a certain stage, in a very rapid process often described as "click" or as a "flash," he *understands* what is being explained. When this happens he says "I see," indicating the basically *perceptual* character of such a process. (Of course, he does not see with optical vision but rather, as it were, with the "mind's eye.") But what is it that he sees? What he perceives is a new total structure in terms of which the older items of knowledge all fall into their proper places, naturally related, while many new and unsuspected relationships suddenly come into view. Later, to preserve this understanding, to communicate it to other people, to apply it, or to test its validity, he may translate it into words, formulas, diagrams, etc. But initially it seems to be a single act, in which older structures are set aside and a new structure comes into being in the mind . . .

There seems to be no limit to the possibility of the human mind for developing new structures in the way described above. And it is this possibility that seems to be behind our ability to put forth new theories and concepts, which lead to knowledge that goes beyond the facts that are accessible at the time when the theories are first developed. It should be recalled that this possibility exists as much in immediate perception as in scientific research, since very often what is constructed in the "inner show" leads, as we have seen earlier, to many correct predictive inferences for future perceptions. It is evident that such an ability cannot be due merely to some sort of mechanism that randomly puts forth "hypotheses" until one of them is confirmed. Rather, for reasons that are as yet not known, the human mind in its general process of perception, whether on the immediate level or on the highest level involved in understanding, can create structures that have a remarkably good chance of being correct in domains going beyond that on which the evidence for them is founded. On the basis of this possibility, the process of "trial and error" can efficiently weed out those structures that are inappropriate. At the same time it can help provide material, the criticism of which leads to a fresh act of understanding or perception, in which yet newer structures are put forth which are generally likely to have a broader domain of validity and better correspondence to the facts than the earlier ones had.

To sum up, the essential point is that through perception we are always in a process of coming into contact with the world, in such a way that we can be aware of the general structure of the segment with which we have been in contact. Science may then be regarded as a means of establishing new kinds of contacts with the world, in new domains, in new levels, with the aid of different instruments, etc. But these contacts would mean very little without the act of understanding, which corresponds on a very high level to that process by which what has been invariant is presented in terms of structure in the "inner show" of immediate perception. It need then no longer be puzzling that science does not lead to knowledge of an absolute truth. For the knowledge supplied by science is (like all other knowledge) basically an expression of the structure that has been revealed in our process of coming from moment to moment into contact with a world the totality of which is beyond our ability to grasp in terms of any given sets of percepts, ideas, concepts, notions, etc. Nevertheless, we can obtain a fairly good grasp of that with which we have thus far been in contact, which is also valid in some domain, either large or small, beyond what is based on this contact. By remaining alert to contradictions and sensitive to new relationships, thus permitting the growth of a fresh understanding, we can keep up with our contact with the world, and in some ways we can anticipate what is coming later.

In science this process takes place at a very high level of abstraction, on a scale of time involving years. In immediate perception it occurs on a lower level of abstraction, and it is very rapid. In science the process depends strongly on collective work, involving contributions of many people, and in immediate perception it is largely individual. But fundamentally both can be regarded as limiting cases of one over-all process, of a generalized kind of perception, in which no absolute knowledge is to be encountered.

Notes

1 The following image may be illustrative. On a dark and stormy night, a ship's captain struggles to navigate a circuitous harbor and reach port. In fact, he succeeds and arrives safely. But the question remains, did his route approximate the channel well enough to avoid being grounded (a good functional *fit*), or did his course exactly *match* the channel? (Ed.)

2 Piaget, J., *The Origins of Intelligence in the Child*, Routledge and Kegan Paul, London (1953); Piaget, J. and Inhelder, B., *The Child's Conception of Space*, Routledge and Kegan Paul, London (1956).

3 See Bohm, D., *The Special Theory of Relativity*, Routledge, London (1996), Ch. 31.

4 See Gibson, J. G., *Psychological Review*, **69**, 477 (1962).

5 Ditchburn, R. W., *Research*, **9**, 466 (1951), *Optica Actica*, **1**, 171 and **2**, 128 (1955).

6 For a discussion of these experiments see Held, R. and Freeman, S. J., *Science*, **142**, 455 (1963); *Psychology, A Study of Science*, S. Koch (ed.), McGraw-Hill, New York, 1959, p. 456; Held, R. and Rekosh, J., *Science*, **141**, 722 (1963).

7 Gibson, J. G., *American Psychologist*, **15**, 694 (1960).

8 The same principle applies to the radio telescope, which is in contact (as it were) with the structure of the whole universe, through a similar set of radio waves.

9 See survey by Held, R. and Freeman, S. J., *Science*, **142**, 455 (1963); also the Symposium on Sensory Deprivation, *The Journal of Nervous and Mental Diseases*, No. 1, January 1961.

10 The similarity between scientific research and perception has already been noted by several authors. See Hanson, N. H., *Pattern of Discovery*, Cambridge University Press, New York (1958), and Kuhn, T., *The Structure of Scientific Revolutions*, University of Chicago Press, Chicago (1963).

11 See Bohm, D., *The Special Theory of Relativity*, Chapter 31, where a similar role has been suggested for the Minkowski diagram in physics.

12 See *ibid.*

13 See *ibid.*, Ch. 28.

3 THE ENFOLDING–UNFOLDING UNIVERSE AND CONSCIOUSNESS (1980)

In this chapter Bohm gives an extended account of what is meant by an "implicate order," as well as analogies indicating how implicate orders manifest in the explicate, sensual world. The processes of matter, organic life, and consciousness are all seen as flowing from the reciprocal ordering principles of enfoldment and unfoldment. It is through these ordering principles that the holomovement – the "ground of all that is" – expresses itself in particular forms and experiences.

The basic notion of enfoldment is quite straightforward – as when baking a cake eggs are "folded into" the batter. What was once overt and explicit – the eggs – becomes implicit and enfolded in the batter. But in the case of the cake, the medium of the batter results in an essentially random process of enfoldment, lacking any sustainable or discernible order. Bohm suggests that a more penetrating image of enfoldment is a mixing device that enfolds drops of ink into a viscous solution like glycerine. The combination of viscous medium and ink illustrate a process of ordered enfoldment. Through describing multiple variations of this ordered enfoldment and unfoldment, Bohm draws analogies with the movements and discontinuities of electrons as understood in quantum physics.

Rather than suggesting a continuous entity that moves "through" time and space, the image of ordered enfoldment–unfoldment allows for a view of the electron as a perpetually emerging explicate structure, temporarily unfolding from an ordered implicate background, and then rapidly enfolding back into this background, in an ongoing cycle. By extension, the whole of

Extract from Chapter 7 of D. Bohm, *Wholeness and Implicate Order*, Routledge, London (1980).

experience can be understood as a flow of appearances resulting from such a cycle of enfoldment and unfoldment.

Bohm also uses enfoldment as a means of explaining non-local effects in the quantum domain. He presents the analogy of a fish in an aquarium, projected onto two TV screens via two separate cameras, from two different angles. As a result of this configuration, each movement of the actual fish produces two seemingly separate images on the screens. But these two images have a suspicious, instantaneous relationship to one another – much like the non-local relation between "entangled" particles at the quantum level. In this analogy, the crucial relationship is that of the three-dimensional "actual" fish to the two-dimensional TV images of the fish, these latter being seen as unfolded projections from the more fundamental three-dimensional reality. In similar fashion, claims Bohm, our three-dimensional world – including entangled particles in a laboratory – manifests as a projection from a yet more fundamental multi-dimensional reality.

The concept of a multi-dimensional reality – from which explicate "things" unfold and into which they enfold – is closely linked to Bohm's considerations of the nature of space. In this view, space is neither inert nor empty. The deep nature of space – of which our "regular" space is but a projection – is understood as a plenum, a highly varied structure-process which includes potentially infinite dimensions. Thus space itself is understood as a multi-dimensional ordering medium – a higher-order correlate to the mechanical analogy of the glycerine mixing device.

In outlining the relationship of consciousness to enfoldment and unfoldment, Bohm first considers consciousness as a "substantial" process. In this regard, our sense experiences, nervous system, and brain are understood as continuous with the whole of the material world. Though we normally assume a distinct separation between subject and object in our perceptual field, this distinction cannot be maintained when we consider the substantial media of light and sound which transmit much of our sensory input; there is really no ultimate break to be found between subject and object. Bohm suggests that the totality of these material phenomena may be enfolded and unfolded throughout the brain in a process not unlike that of a hologram, creating memory structures that bear a likeness to "original" perceptions.

Beyond this substantial aspect of consciousness, Bohm outlines his views of the "essence of conscious experience" – factors such as awareness, attention, and understanding. At the root of any such capacities, says Bohm, is a pre-conscious "undivided state of flowing movement" – the actual and immediate activity of the holomovement. The nature of this movement can be discerned in a number of common experiences, such as listening to music. A sequence of harmonious notes, says Bohm, does not sufficiently account for the experience of coherence we may feel when listening to music. If the sequence of notes was stretched out so that long lapses occurred between the notes, the sense of musical integrity would collapse. It is the co-present reverberations of multiple notes, in varying degrees of interpenetration and unfoldment, that give music a sense of meaning and wholeness.

Similarly, our overall explicate experience derives from an implicate structure of flowing movement, active transformations, and subtle interpenetrating meanings, not unlike that revealed by Piaget and discussed in Chapter 2. Bohm suggests that the "essence" of consciousness is just this flowing movement, arising from the depths of the holomovement itself. The substantial aspects of consciousness – memory, reification, and subject–object polarities – are understood as explicated forms of this deeper unitary movement, much as the two images of the fish unfold from a deeper, single actuality. Thus, the material structures of the objective world, the substantial aspects of consciousness, and the "essence" of consciousness can now be seen as a continuous flow, eliminating any absolute distinction between mind and matter.

In the appendix to this chapter, Bohm advances his argument for conceptualizing a new order in modern physics. To this end, the glycerine device and the hologram are explored in greater detail than in the main body of the chapter. While the glycerine device serves as an elementary, intuitive analogy for the structure of an implicate order, Bohm suggests that a hologram – though still three-dimensional – is yet closer to the nature of a multidimensional implicate order.

1   INTRODUCTION

Throughout this book the central underlying theme is the unbroken wholeness of the totality of existence as an undivided flowing movement without borders. The implicate order is particularly suitable for the understanding of such unbroken wholeness in flowing movement, for in the implicate order the totality of existence is enfolded within each region of space (and time). So, whatever part, element, or aspect we may abstract in thought, this still enfolds the whole and is therefore intrinsically related to the totality from which it has been abstracted. Thus, wholeness permeates all that is being discussed, from the very outset.

In this chapter we shall give a non-technical presentation of the main features of the implicate order, first as it arises in physics, and then as it may be extended to the field of consciousness, to indicate certain general lines along which it is possible to comprehend both cosmos and consciousness as a single unbroken totality of movement.[1]

2 RÉSUMÉ, CONTRASTING MECHANISTIC ORDER IN PHYSICS WITH IMPLICATE ORDER

It will be helpful to begin by giving a résumé of some of the main points that have been made earlier, contrasting the generally accepted mechanistic order in physics and the implicate order. Let us first consider the mechanistic order. The principal feature of this order is that the world is regarded as constituted of entities which are *outside of each other*, in the sense that they exist independently in different regions of space (and time) and interact through forces that do not bring about any changes in their essential natures. The machine gives a typical illustration of such a system of order. Each part is formed (e.g., by stamping or casting) independently of the others, and interacts with the other parts only through some kind of external contact. By contrast, in a living organism, for example, each part grows in the context of the whole, so that it does not exist independently, nor can it be said that it merely "interacts" with the others, without itself being essentially affected in this relationship.

Physics has become almost totally committed to the notion that the order of the universe is basically mechanistic. The most common form of this notion is that the world is assumed to be constituted of a set of separately existent, indivisible and unchangeable "elementary particles," which are the fundamental "building blocks" of the entire universe. Originally, these were thought to be atoms, but atoms were eventually divided into electrons, protons and neutrons. These latter were thought to be the absolutely unchangeable and indivisible constituents of all matter, but then, these were in turn found to be subject to transformation into hundreds of different kinds of unstable particles, and now even smaller particles called "quarks" and "partons" have been postulated to explain these transformations. Though these have not yet been isolated there appears to be an unshakable faith among physicists that either such particles, or some other kind yet to be discovered, will eventually make possible a complete and coherent explanation of everything.

The theory of relativity was the first significant indication in physics of the need to question the mechanistic order. It implied that no coherent concept of an independently existent particle is possible, neither one in which the particle would be an extended body, nor one in which it would be a dimensionless point.[2] Thus, a basic assumption underlying the generally accepted form of mechanism in physics has been shown to be untenable.

To meet this fundamental challenge, Einstein proposed that the particle concept no longer be taken as primary, and that instead reality be regarded from the very beginning as constituted of fields, obeying laws that are consistent with the requirements of the theory of relativity. A key new idea of this "unified field theory" of Einstein is that the field equations be *non-linear*. These equations could have solutions in the form of localized pulses, consisting of a region of intense field that could move through space stably as a whole, and that could thus provide a model of the "particle." Such pulses do not end abruptly but spread out to arbitrarily large distances with decreasing intensity. Thus the field structures associated with two pulses will merge and flow together in one unbroken whole. Moreover, when two pulses come close together, the original particle-like forms will be so radically altered that there is no longer even a resemblance to a structure consisting of two particles. So, in terms of this notion, the idea of a separately and independently existent particle is seen to be, at best, an abstraction furnishing a valid approximation only in a certain limited domain. Ultimately, the entire universe (with all its "particles," including those constituting human beings, their laboratories, observing instruments, etc.) has to be understood as a single undivided whole, in which analysis into separately and independently existent parts has no fundamental status.

However, Einstein was not able to obtain a generally coherent and satisfactory formulation of his unified field theory. Moreover (and perhaps more important in the context of our discussion of the mechanistic approach to physics) the field concept, which is his basic starting point, still retains the essential features of a mechanistic order, for the fundamental entities, the fields, are conceived as existing outside of each other, at separate points of space and time, and are assumed to be connected with each other only through external relationships which indeed are also taken to be local, in the sense that only those field elements that are separated by "infinitesimal" distances can affect each other.[3]

Though the unified field theory was not successful in this attempt to provide an ultimate mechanistic basis for physics in terms of the field concept, it nevertheless did show in a concrete way how consistency with the theory of relativity may be achieved by deriving the particle concept as an abstraction from an unbroken and undivided totality of existence. Thus, it helped to strengthen the challenge posed by relativity theory to the prevailing mechanistic order.

The quantum theory presents, however, a much more serious challenge to this mechanistic order, going far beyond that provided by the theory of relativity. The key features of the quantum theory that challenge mechanism are:

1   Movement is in general *discontinuous*, in the sense that action is constituted of *indivisible quanta* (implying also that an electron, for example, can go from one state to another, without passing through any states in between).

2   Entities, such as electrons, can show different properties (e.g., particle-like, wavelike, or something in between), depending on the environmental context within which they exist and are subject to observation.

3   Two entities, such as electrons, which initially combine to form a molecule and then separate, show a peculiar nonlocal relationship, which can best be described as a non-causal connection of elements that are far apart[4] (as demonstrated in the experiment of Einstein, Podolsky, and Rosen[5]).

It should be added of course that the laws of quantum mechanics are statistical and do not determine individual future events uniquely and precisely. This is, of course, different from classical laws, which do in principle determine these events. Such indeterminism is, however, not a serious challenge to a mechanistic order, i.e., one in which the fundamental elements are independently existent, lying outside each other, and connected only by external relationships. The fact that (as in a pinball machine) such elements are related by the rules of chance (expressed mathematically in terms of the theory of probability) does not change the basic externality of the elements[6] and so does not essentially affect the question of whether the fundamental order is mechanistic or not.

The three key features of the quantum theory given do, however, clearly show the inadequacy of mechanistic notions. Thus, if all actions are in the form of discrete quanta, the interactions between different entities (e.g., electrons) constitute a single structure of indivisible links, so that the entire universe has to be thought of as an unbroken whole. In this whole, each element that we can abstract in thought shows basic properties (wave or particle, etc.) that depend on its overall environment, in a way that is much more reminiscent of how the organs constituting living beings are related, than it is of how parts of a machine interact. Further, the non-local, non-causal nature of the relationships of elements distant from each other evidently violates the requirements of separateness and independence of fundamental constituents that is basic to any mechanistic approach.

It is instructive at this point to contrast the key features of relativistic and quantum theories. Relativity theory requires continuity, strict causality (or determinism) and locality. On the other hand, quantum theory requires non-continuity, non-causality and non-locality. So the basic concepts of relativity and quantum theory directly contradict each other. It is therefore hardly surprising that these two theories have never been unified in a consistent way. Rather, it seems most likely that such a unification is not actually possible. What is very probably needed instead is a qualitatively new theory, from which both relativity and quantum theory are to be derived as abstractions, approximations and limiting cases.

The basic notions of this new theory evidently cannot be found by beginning with those features in which relativity and quantum theory stand in direct contradiction. The best place to begin is with what they have basically in common. This is undivided wholeness. Though each comes to such wholeness in a different way, it is clear that it is this to which they are both fundamentally pointing.

To begin with undivided wholeness means, however, that we must drop the mechanistic order. But this order has been, for many centuries, basic to all thinking on physics. The mechanistic order is most naturally and directly expressed through the Cartesian grid. Though physics has changed radically in many ways, the Cartesian grid (with minor modifications, such as the use of curvilinear coordinates) has remained the one key feature that has not changed. Evidently, it is not easy to change this, because our notions of order are pervasive, for not only do they involve our thinking but also our senses, our feelings, our intuitions, our physical movement, our relationships with other people and with society as a whole and, indeed, every phase of our lives. It is thus difficult to "step back" from our old notions of order sufficiently to be able seriously to consider new notions of order.

To help make it easier to see what is meant by our proposal of new notions of order that are appropriate to undivided wholeness, it is therefore useful to start with examples that may directly involve sense perception, as well as with models and analogies that illustrate such notions in an imaginative and intuitive way. We have noted elsewhere[7] that the photographic lens is an instrument that has given us a very direct kind of sense perception of the meaning of the mechanistic order, for by bringing about an approximate correspondence between points on the object and points on the photographic image, it very strongly calls attention to the separate elements into which the object can be analysed. By making possible the point-to-point imaging and recording of things that are too small to be seen with the naked eye, too big, too fast, too slow, etc., it leads us to believe that eventually everything can be *perceived* in this way. From this grows the idea that there is nothing that cannot also be *conceived* as constituted of such localized elements. Thus, the mechanistic approach was greatly encouraged by the development of the photographic lens.

We then went on to consider a new instrument, called the *hologram*. As explained in the appendix to this chapter, this makes a photographic record of the interference pattern of light waves that have come off an object. The key new feature of this record is that each part contains information about the *whole object* (so that there is no point-to-point correspondence of object and recorded image). That is to say, the form and structure of the entire object may be said to be *enfolded* within each region of the photographic record. When one shines light on any region, this form and structure are then *unfolded* to give a recognizable image of the whole object once again.

We proposed that a new notion of order is involved here, which we call the *implicate order* (from a Latin root meaning "to enfold" or "to fold inward"). In terms of the implicate order one may say that everything is enfolded into everything. This contrasts with the *explicate order* now dominant in physics in which things are *unfolded* in the sense that each thing lies only in its own particular region of space (and time) and outside the regions belonging to other things.

The value of the hologram in this context is that it may help to bring this new notion of order to our attention in a sensibly perceptible way; but of course, the hologram is only an instrument whose function is to make a static record (or "snapshot") of this order. The actual order itself which has thus been recorded is in the complex movement of electromagnetic fields, in the form of light waves. Such movement of light waves is present everywhere and in principle enfolds the entire universe of space (and time) in each region (as can be demonstrated in any such region by placing one's eye or a telescope there, which will "unfold" this content).

This enfoldment and unfoldment takes place not only in the movement of the electromagnetic field but also in that of other fields, such as the electronic, protonic, sound waves, etc. There is already a whole host of such fields that are known, and any number of additional ones, as yet unknown, that may be discovered later. Moreover, the movement is only approximated by the classical concept of fields (which is generally used for the explanation of how the hologram works). More accurately, these fields obey quantum-mechanical laws, implying the properties of discontinuity and non-locality, which we have already mentioned (and which we shall discuss again later in this chapter). As we shall see later, even the quantum laws may only be abstractions from still more general laws, of which only some outlines are now vaguely to be seen. So the totality of movement of enfoldment and unfoldment may go immensely beyond what has revealed itself to our observations thus far.

We have called this totality by the name *holomovement*. Our basic proposal is that *what is* is the holomovement, and that everything is to be explained in terms of forms derived from this holomovement. Though the full set of laws governing its totality is unknown (and, indeed, probably unknowable) nevertheless these laws are assumed to be such that from them may be abstracted relatively autonomous or independent sub-totalities of movement (e.g., fields, particles, etc.) having a certain recurrence and stability of their basic patterns of order and measure. Such sub-totalities may then be investigated, each in its own right, without our having first to know the full laws of the holomovement. This implies, of course, that we are not to regard what we find in such investigations as having an absolute and final validity, but rather we have always to be ready to discover the limits of independence of any relatively autonomous structure of law, and from this to go on to look for new laws that may refer to yet larger relatively autonomous domains of this kind.

Up till now we have contrasted implicate and explicate orders, treating them as separate and distinct, but the explicate order can be regarded as a particular or distinguished case of a more general set of implicate orders from which latter it can be derived. What distinguishes the explicate order is that what is thus derived is a set of recurrent and relatively stable elements that are *outside* of each other. This set of elements (e.g., fields and particles) then provides the explanation of that domain of experience in which the mechanistic order yields an adequate treatment. In the prevailing mechanistic approach, however, these elements, assumed to be separately and independently existent, are taken as constituting the basic reality. The task of science is then to start from such parts and to derive all wholes through abstraction, explaining them as the results of interactions of the parts. On the contrary, when one works in terms of the implicate order, one begins with the undivided wholeness of the universe, and the task of science is to derive the parts through abstraction from the whole, explaining them as approximately separable, stable and recurrent, but externally related elements making up relatively autonomous sub-totalities, which are to be described in terms of an explicate order.

3 THE IMPLICATE ORDER AND THE GENERAL STRUCTURE OF MATTER

We shall now go on to give a more detailed account of how the general structure of matter may be understood in terms of the implicate order. To do this we shall begin by considering a device, which serves as an analogy, illustrating certain essential features of the implicate order.[8] (It must be emphasized, however, that it is only an analogy and that, as will be brought out in more detail later, its correspondence with the implicate order is limited.)

This device consists of two concentric glass cylinders, with a highly viscous fluid such as glycerine between them, which is arranged in such a way that the outer cylinder can be turned very slowly, so that there is negligible diffusion of the viscous fluid. A droplet of insoluble ink is placed in the fluid, and the outer cylinder is then turned, with the result that the droplet is drawn out into a fine thread-like form that eventually becomes invisible. When the cylinder is turned in the opposite direction the thread-form draws back and suddenly becomes visible as a droplet, essentially the same as the one that was there originally (see Figure 3.1). It is worthwhile to reflect carefully on what is actually happening in the process described above. First, let us consider an element of fluid. The parts at larger radii will move faster than those at smaller radii. Such an element will therefore be deformed, and this explains why it is eventually drawn out into a long thread. Now, the ink droplet consists of an aggregate of carbon particles that are initially suspended in such an element of fluid. As the element is drawn out the ink particles will be carried with it. The set of particles will thus spread out over such a large volume that their density falls below the minimum threshold that is visible. When the movement is reversed, then (as is known from the physical laws governing viscous media) each part of the fluid retraces its path, so that eventually the thread-like fluid element draws back to its original form. As it does so, it carries the ink particles with it, so that eventually they, too, draw together and become dense enough to pass the threshold of perceptibility, so emerging once again as visible droplets.

When the ink particles have been drawn out into a long thread, one can say that they have been *enfolded* into the glycerine, as it might be said that an egg can be folded into a cake. Of course, the difference is that the droplet can be unfolded by reversing the motion of the fluid, while there is no way to unfold the egg (this is because the material here undergoes irreversible diffusive mixing).

The analogy of such enfoldment and unfoldment to the implicate

---

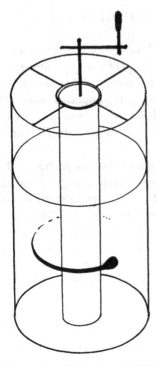

Figure 3.1 Illustration by Kevin Shluker

order introduced in connection with the hologram is quite good.[9] To develop this analogy further, let us consider two ink droplets close to each other, and to make visualization easier we will suppose that the ink particles in one droplet are red, while those in the other are blue. If the outer cylinder is then turned, each of the two separate elements of fluid in which the ink particles are suspended will be drawn out into a thread-like form, and the two thread-like forms will, while remaining separate and distinct, weave through each other in a complex pattern too fine to be perceptible to the eye (rather like the interference pattern that is recorded on the hologram, which has, however, quite a different origin). The ink particles in each droplet will of course be carried along by the fluid motions, but with each particle remaining in its own thread of fluid. Eventually, however, in any region that was large enough to be visible to the eye, red particles from the one droplet and blue particles from the other will be seen to intermingle, apparently at random. When the fluid motions are reversed, however, each thread-like element of fluid will draw back into itself until eventually the two gather into clearly separated regions once again. If one were able to watch what is happening more closely (e.g., with a microscope) one would see red and blue particles that were close to each other beginning to separate, while particles of a given colour that were far from each other would begin to come together. It is almost as if distant particles of a given colour had "known" that they had a common destiny, separate from that of particles of the other colour, to which they were close.

Of course, there is in this case actually no such "destiny." Indeed, all that has happened mechanically, through the complex movements of the fluid elements in which the ink particles are suspended. But we have to recall here that this device is only an analogy, intended to illustrate a new notion of order. To allow this new notion to stand out clearly, it is necessary to begin by focusing our attention on the ink particles alone, and to set aside the consideration of the fluid in which they are suspended, at least for the moment. When the sets of ink particles from each droplet have been drawn out into an invisible thread, so that particles of both colours intermingle, one can nevertheless say that *as an ensemble* each set is, in a certain way, distinct from the other. This distinction is not in general evident to the senses, but it has a certain relationship to the total situation out of which the ensembles have come. This situation includes the glass cylinders, the viscous fluid and its movements, and the original distribution of ink particles. It may then be said that each ink particle belongs to a certain distinct ensemble and that it is bound up with the others in this ensemble by the force of an overall necessity, inherent in this total situation, which can bring the whole set to a common end (i.e., to reconstitute the form of a droplet).

In the case of this device, the overall necessity operates mechanically as the movement of fluid, according to certain well-known laws of hydrodynamics. As indicated earlier, however, we will eventually drop this mechanical analogy and go on to consider the holomovement. In the holomovement, there is still an overall necessity (which we have elsewhere called "holonomy"[10]) but its laws are no longer mechanical. Rather, as pointed out in Section 2 of this chapter, its laws will be in a first approximation those of the quantum theory, while more accurately they will go beyond even these, in ways that are at present only vaguely discernible. Nevertheless, certain similar principles of distinction will prevail in the holomovement as in the analogy of the device made up of glass cylinders. That is to say, ensembles of elements which intermingle or interpenetrate in space can nevertheless be distinguished, but only in the context of certain total situations in which the members of each ensemble are related through the force of an overall necessity, inherent in these situations, that can bring them together in a specifiable way.

Now that we have established a new kind of distinction of ensembles that are enfolded together in space, we can go on to put these distinctions into an *order*. The simplest notion of order is that of a sequence or succession. We shall start with such a simple idea and develop it later to much more complex and subtle notions of order.

The essence of a simple, sequential order is in the series of relationships among distinct elements:

$$A : B :: B : C :: C : D. \ldots$$

For example, if A represents one segment of a line, B the succeeding one, etc., the sequentiality of segments of the line follows from the above set of relationships.

Let us now return to our ink-in-fluid analogy, and suppose that we have inserted into the fluid a large number of droplets, set close to each other and arranged in a line (this time we do not suppose different colours). These we label as $A, B, C, D \ldots$ We then turn the outer cylinder many times, so that each of the droplets gives rise to an ensemble of ink particles, enfolded in so large a region of space that particles from all the droplets intermingle. We label the successive ensembles $A', B', C', D' \ldots$

It is clear that, in some sense, an entire linear order has been enfolded into the fluid. This order may be expressed through the relationships

$$A' : B' :: B' : C' :: C' : D' \ldots$$

This order is not present to the senses. Yet its reality may be demonstrated by reversing the motion of the fluid, so that the ensembles, $A', B', C', D' \ldots$, will unfold to give rise to the original linearly arranged series of droplets, $A, B, C, D \ldots$

In the above, we have taken a pre-existent explicate order, consisting of ensembles of ink particles arranged along a line, and transformed it into an order of enfolded ensembles, which is in some key way similar. We shall next consider a more subtle kind of order, not derivable from such a transformation.

Suppose now that we insert an ink droplet, A, and turn the outer cylinder n times. We then insert a second ink droplet, B, at the same place, and again turn the cylinder n times. We keep up this procedure with further droplets, $C, D, E \ldots$ The resulting ensembles of ink particles, $a, b, c, d, e, \ldots$, will now differ in a new way, for, when the motion of the fluid is reversed, the ensembles will successively come together to form droplets in an order opposite to the one in which they were put in. For example, at a certain stage the particles of ensemble d will come together (after which they will be drawn out into a thread again). This will happen to those of c, then to b, etc. It is clear from this that ensemble d is related to c as c is to b, and so on. So these ensembles form a certain sequential order. However, this is in no sense a transformation of a linear order in space (as was that of the sequence A', B', C', D' ... that we considered earlier), for in general only one of these ensembles will unfold at a time; when any one is unfolded, the rest are still enfolded. In short, we have an order which cannot all be made explicate at once and which is nevertheless real, as may be revealed when successive droplets become visible as the cylinder is turned.

We call this an *intrinsically implicate order*, to distinguish it from an order that may be enfolded but which can unfold all at once into a single explicate order. So we have here an example of how, as stated in Section 2, an explicate order is a particular case of a more general set of implicate orders.

Let us now go on to combine both of the above-described types of order.

We first insert a droplet, A, in a certain position and turn the cylinder n times. We then insert a droplet, B, in a slightly different position and turn the cylinder n more times (so that A has been enfolded by $2n$ turns). We then insert C further along the line AB and turn n more times, so that A has been enfolded by $3n$ turns, B $2n$ turns, and C by n turns. We proceed in this way to enfold a large number of droplets. We then move the cylinder fairly rapidly in the reverse direction. If the rate of emergence of droplets is faster than the minimum time of resolution of the human eye, what we will see is apparently a particle moving continuously and crossing the space.

Such enfoldment and unfoldment in the implicate order may evidently provide a new model of, for example, an electron, which is quite different from that provided by the current mechanistic notion of a particle that exists at each moment only in a small region of space and that changes its position continuously with time. What is essential to this new model is that the electron is instead to be understood through a total set of enfolded ensembles, which are generally not localized in space. At any given moment one of these may be unfolded and therefore localized, but in the next moment, this one enfolds to be replaced by the one that follows. The notion of continuity of existence is approximated by that of very rapid recurrence of similar forms, changing in a simple and regular way (rather as a rapidly spinning bicycle wheel gives the impression of a solid disc, rather than of a sequence of rotating spokes). Of course, more fundamentally, the particle is only an abstraction that is manifest to our senses. *What is* is always a totality of ensembles, all present together, in an orderly series of stages of enfoldment and unfoldment, which intermingle and inter-penetrate each other in principle throughout the whole of space.

It is further evident that we could have enfolded any number of such "electrons," whose forms would have intermingled and interpenetrated in the implicate order. Nevertheless, as these forms unfolded and became manifest to our senses, they would have come out as a set of "particles" clearly separated from each other. The arrangement of ensembles could have been such that these particle-like manifestations came out "moving" independently in straight lines, or equally well, along curved paths that were mutually related and dependent, as if there had been a force of interaction between them. Since classical physics traditionally aims to explain everything in terms of interacting systems of particles, it is clear that in principle one could equally well treat the entire domain that is correctly covered by such classical concepts in terms of our model of ordered sequences of enfolding and unfolding ensembles.

What we are proposing here is that in the quantum domain this model is a great deal better than is the classical notion of an interacting set of particles. Thus, although successive localized manifestations of an electron, for example, may be very close to each other, so that they approximate a continuous track, this need not always be so. In principle, discontinuities may be allowed in the manifest tracks – and these may, of course, provide the basis of an explanation of how, as stated in Section 2, an electron can go from one state to another without passing through states in between. This is possible, of course, because the "particle" is only an abstraction of a much greater totality of structure. This abstraction is what is manifest to our senses (or instruments) but evidently there is no reason why it has to have continuous movement (or indeed continuous existence).

Next, if the total context of the process is changed, entirely new modes of manifestation may arise. Thus, returning to the ink-in-fluid analogy, if the cylinders are changed, or if obstacles are placed in the fluid, the form and order of manifestation will be different. Such a dependence – the dependence of what manifests to observation on the total situation – has a close parallel to a feature which we have also mentioned in Section 2, i.e., that according to the quantum theory electrons may show properties resembling either those of particles or those of waves (or of something in between) in accordance with the total situation involved in which they exist and in which they may be observed experimentally.

What has been said thus far indicates that the implicate order gives generally a much more coherent account of the quantum properties of matter than does the traditional mechanistic order. What we are proposing here is that the implicate order therefore be taken as fundamental. To understand this proposal fully, however, it is necessary to contrast it carefully with what is implied in a mechanistic approach based on the explicate order; for, even in terms of this latter approach, it may of course be admitted that in a certain sense at least, enfoldment and unfoldment can take place in various specific situations (e.g., such as that which happens with the ink droplet). However, this sort of situation is not regarded as having a fundamental kind of significance. All that is primary, independently existent, and universal is thought to be express-ible in an explicate order, in terms of elements that are externally related (and these are usually thought to be particles, or fields, or some com-bination of the two). Whenever enfoldment and unfoldment are found actually to take place, it is therefore assumed that these can ultimately be explained in terms of an underlying explicate order through a deeper mechanical analysis (as, indeed, does happen with the ink-droplet device).

Our proposal to start with the implicate order as basic, then, means that what is primary, independently existent, and universal has to be expressed in terms of the implicate order. So we are suggesting that it is the implicate order that is autonomously active while, as indicated earl-ier, the explicate order flows out of a law of the implicate order, so that it is secondary, derivative, and appropriate only in certain limited con-texts. Or, to put it another way, the relationships constituting the fun-damental law are between the enfolded structures that interweave and interpenetrate each other, throughout the whole of space, rather than between the abstracted and separated forms that are manifest to the senses (and to our instruments).

What, then, is the meaning of the appearance of the apparently independent and self-existent "manifest world" in the explicate order? The answer to this question is indicated by the root of the word "mani-fest," which comes from the Latin "manus," meaning "hand." Essen-tially, what is manifest is what can be held with the hand – something solid, tangible, and visibly stable. The implicate order has its ground in the holomovement which is vast, rich, and in a state of unending flux of enfoldment and unfoldment, with laws most of which are only vaguely known, and which may even be ultimately unknowable in their totality. Thus it cannot be grasped as something solid, tangible and stable to the senses (or to our instruments). Nevertheless, as has been indicated earlier, the overall law (holonomy) may be assumed to be such that in a certain sub-order, within the whole set of implicate order, there is a totality of forms that have an approximate kind of recurrence, stability and separability. Evidently, these forms are capable of appearing as the relatively solid, tangible, and stable elements that make up our "manifest world." The special distinguished sub-order indicated above, which is the basis of the possibility of this manifest world, is then, in effect, what is meant by the explicate order.

We can, for convenience, always picture the explicate order, or imagine it, or represent it to ourselves, as the order present to the senses. The fact that this order is actually more or less the one appearing to our senses must, however, be explained. This can be done only when we bring consciousness into our "universe of discourse" and show that matter in general and consciousness in particular may, at least in a certain sense, have this explicate (manifest) order in common. This question will be explored further when we discuss consciousness in Sections 7 and 8.

4 QUANTUM THEORY AS AN INDICATION OF A MULTIDIMENSIONAL IMPLICATE ORDER

Thus far we have been presenting the implicate order as a process of enfoldment and unfoldment taking place in the ordinary three-dimensional space. However, as pointed out in Section 2 the quantum theory has a fundamentally new kind of non-local relationship, which may be described as a non-causal connection of elements that are distant from each other, which is brought out in the experiment of Einstein, Podolsky, and Rosen.[11] For our purposes, it is not necessary to go into the technical details concerning this non-local relationship. All that is important here is that one finds, through a study of the implications of the quantum theory, that the analysis of a total system into a set of independently existent but interacting particles breaks down in a radically new way. One discovers, instead, both from consideration of the meaning of the mathematical equations and from the results of the actual experiments, that the various particles have to be taken literally as projections of a higher-dimensional reality which cannot be accounted for in terms of any force of interaction between them.[12]

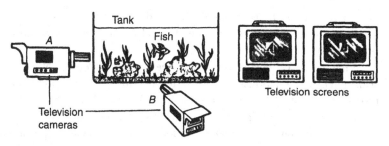

Tank

Fish

A

Television screens

Television
cameras

B

Figure 3.2

We can obtain a helpful intuitive sense of what is meant by the notion of projection here, through the consideration of the following device. Let us begin with a rectangular tank full of water, with transparent walls (see Figure 3.2).

Suppose further that there are two television cameras, A and B, directed at what is going on in the water (e.g., fish swimming around) as seen through the two walls at right angles to each other. Now let the corresponding television images be made visible on screens A and B in another room. What we will see there is a certain *relationship* between the images appearing on the two screens. For example, on screen A we may see an image of a fish, and on screen B we will see another such image. At any given moment each image will generally look different from the other. Nevertheless the differences will be related, in the sense that when one image is seen to execute certain movements, the other will be seen to execute corresponding movements. Moreover, content that is mainly on one screen will pass into the other, and vice versa (e.g., when a fish initially facing camera A turns through a right angle, the image that was on A is now to be found on B). Thus at all times the image content on the other screen will correlate with and reflect that of the other.

Of course, we know that the two images do not refer to independently existent though interacting actualities (in which, for example, one image could be said to "cause" related changes in the other). Rather, they refer to a single actuality, which is the common ground of both (and this explains the correlation of images without the assumption that they causally affect each other). This actuality is of higher dimensionality than are the separate images on the screens; or, to put it differently, the images on the screens are two-dimensional *projections* (or facets) of a three-dimensional reality. In some sense this three-dimensional reality holds these two-dimensional projections within it. Yet, since these projections exist only as abstractions, the three-dimensional reality is neither of these, but rather it is something else, something of a nature beyond both.

What we are proposing here is that the quantum property of a non-local, non-causal relationship of distant elements may be understood through an extension of the notion described above. That is to say, we may regard each of the "particles" constituting a system as a projection of a "higher-dimensional" reality, rather than as a separate particle, existing together with all the others in a common three-dimensional space. For example, in the experiment of Einstein, Podolsky, and Rosen, which we have mentioned earlier, each of two atoms that initially combine to form a single molecule are to be regarded as three-dimensional projections of a six-dimensional reality. This may be demonstrated experimentally by causing the molecule to disintegrate and then observing the two atoms after they have separated and are quite distant from each other, so that they do not interact and therefore have no causal connections. What is actually found is that the behaviour of the two atoms is correlated in a way that is rather similar to that of the two television images of the fish, as described earlier. Thus (as is, indeed, further shown by a more careful consideration of the mathematical form of the quantum laws involved here), each electron acts as if it were a projection of a higher-dimensional reality.

Under certain conditions,[13] the two three-dimensional projections corresponding to the two atoms may have a relative independence of behaviour. When these conditions are satisfied it will be a good approximation to treat both atoms as relatively independent but interacting particles, both in the same three-dimensional space. More generally, however, the two atoms will show the typical non-local correlation of behaviour which implies that, more deeply, they are only three-dimensional projections of the kind described above.

A system constituted of N "particles" is then a $3N$-dimensional reality, of which each "particle" is a three-dimensional projection. Under the ordinary conditions of our experience, these projections will be close enough to independence so that it will be a good approximation to treat them in the way that we usually do, as a set of separately existing particles all in the same three-dimensional space. Under other conditions this approximation will not be adequate. For example, at low temperatures an aggregate of electrons shows a new property of superconductivity, in which electrical resistance vanishes, so that electric current can flow indefinitely. This is explained by showing that the electrons enter a different kind of state, in which they are no longer relatively independent. Rather, each electron acts as a projection of a single higher-dimensional reality and all these projections share a non-local, non-causal correlation, which is such that they go round obstacles "co-operatively" without being scattered or diffused, and therefore without resistance. (One could compare this behaviour to a ballet dance, while the usual behaviour of electrons could be compared to that of an agitated crowd of people, moving in a helter-skelter way.)

What follows from all this is that basically the implicate order has to be considered as a process of enfoldment and unfoldment in a higher-dimensional space. Only under certain conditions can this be simplified as a process of enfoldment and unfoldment in three dimensions. Thus far, we have indeed used this sort of simplification, not only with the ink-in-fluid analogy but also with the hologram. Such a treatment, though, is only an approximation, even for the hologram. Indeed, as has already been pointed out earlier in this chapter, the electromagnetic field, which is the ground of the holographic image, obeys the laws of the quantum theory, and when these are properly applied to the field it is found that this, too, is actually a multidimensional reality which can only under certain conditions be simplified as a three-dimensional reality.

Quite generally, then, the implicate order has to be extended into a multidimensional reality. In principle this reality is one unbroken whole, including the entire universe with all its "fields" and "particles." Thus we have to say that the holomovement enfolds and unfolds in a multidimensional order, the dimensionality of which is effectively infinite. However, as we have already seen, relatively independent sub-totalities can generally be abstracted, which may be approximated as autonomous. Thus the principle of relative autonomy of sub-totalities which we introduced earlier as basic to the holomovement is now seen to extend to the multidimensional order of reality.

5 COSMOLOGY AND THE IMPLICATE ORDER

From our consideration of how the general structure of matter can be understood in terms of the implicate order, we now come to certain new notions of cosmology that are implicit in what is being done here.

To bring these out, we first note that when the quantum theory is applied to fields (in the manner discussed in the previous section) it is found that the possible states of energy of this field are discrete (or quantized). Such a state of the field is, in some respects, a wavelike excitation spreading out over a broad region of space. Nevertheless, it also has somehow a discrete quantum of energy (and momentum) proportional to its frequency, so that in other respects it is like a particle[14] (e.g., a photon). However, if one considers the electromagnetic field in empty space, for example, one finds from the quantum theory that each such "wave-particle" mode of excitation of the field has what is called a "zero-point" energy, below which it cannot go, even when its energy falls to the minimum that is possible. If one were to add up the energies of all the "wave-particle" modes of excitation in any region of space, the result would be infinite, because an infinite number of wavelengths is present. However, there is good reason to suppose that one need not keep on adding the energies corresponding to shorter and shorter wavelengths. There may be a certain shortest possible wavelength, so that the total number of modes of excitation, and therefore the energy, would be finite.

Indeed, if one applies the rules of quantum theory to the currently accepted general theory of relativity, one finds that the gravitational field is also constituted of such "wave-particle" modes, each having a minimum "zero-point" energy. As a result the gravitational field, and therefore the definition of what is to be meant by distance, cease to be completely defined. As we keep on adding excitations corresponding to shorter and shorter wavelengths to the gravitational field, we come to a certain length at which the measurement of space and time becomes totally undefinable. Beyond this, the whole notion of space and time as we know it would fade out, into something that is at present unspecifiable. So it would be reasonable to suppose, at least provisionally, that this is the shortest wavelength that should be considered as contributing to the "zero-point" energy of space.

When this length is estimated it turns out to be about 10^{-33} cm. This is much shorter than anything thus far probed in physical experiments (which have got down to about 10^{-17} cm or so). If one computes the amount of energy that would be in one cubic centimetre of space, with this shortest possible wavelength, it turns out to be very far beyond the total energy of all the matter in the known universe.[15]

What is implied by this proposal is that what we call empty space contains an immense background of energy, and that matter as we know it is a small, "quantized" wavelike excitation on top of this background, rather like a tiny ripple on a vast sea. In current physical theories, one avoids the explicit consideration of this background by calculating only the difference between the energy of empty space and that of space with matter in it. This difference is all that counts in the determination of the general properties of matter as they are presently accessible to observation. However, further developments in physics may make it possible to probe the above-described background in a more direct way. Moreover, even at present, this vast sea of energy may play a key part in the understanding of the cosmos as a whole.

In this connection it may be said that space, which has so much energy, is full rather than empty. The two opposing notions of space as empty and space as full have indeed continually alternated with each other in the development of philosophical and physical ideas. Thus, in Ancient Greece, the School of Parmenides and Zeno held that space is a plenum. This view was opposed by Democritus, who was perhaps the first seriously to propose a world view that conceived of space as emptiness (i.e., the void) in which material particles (e.g., atoms) are free to move. Modern science has generally favoured this latter atomistic view, and yet, during the nineteenth century, the former view was also seriously entertained, through the hypothesis of an *ether* that fills all space. Matter, thought of as consisting of special recurrent stable and separable forms in the ether (such as ripples or vortices), would be transmitted through this plenum as if the latter were empty.

A similar notion is used in modern physics. According to the quantum theory, a crystal at absolute zero allows electrons to pass through it without scattering. They go through as if the space were empty. If the temperature is raised, inhomogeneities appear, and these scatter electrons. If one were to use such electrons to observe the crystal (i.e. by focusing them with an electron lens to make an image) what would be visible would be just the inhomogeneities. It would then appear that the inhomogeneities exist independently and that the main body of the crystal was sheer nothingness.

It is being suggested here, then, that what we perceive through the senses as empty space is actually the plenum, which is the ground for the existence of everything, including ourselves. The things that appear to our senses are derivative forms and their true meaning can be seen only when we consider the plenum, in which they are generated and sustained, and into which they must ultimately vanish.

This plenum is, however, no longer to be conceived through the idea of a simple material medium, such as an ether, which would be regarded as existing and moving only in a three-dimensional space. Rather, one is to begin with the holomovement, in which there is the immense "sea" of energy described earlier. This sea is to be understood in terms of a multidimensional implicate order, along the lines sketched in Section 4, while the entire universe of matter as we generally observe it is to be treated as a comparatively small pattern of excitation. This excitation pattern is relatively autonomous and gives rise to approximately recurrent, stable and separable projections into a three-dimensional explicate order of manifestation, which is more or less equivalent to that of space as we commonly experience it.

With all this in mind let us consider the current generally accepted notion that the universe, as we know it, originated in what is almost a single point in space and time from a "big bang" that happened some ten thousand million years ago. In our approach this "big bang" is to be regarded as actually just a "little ripple." An interesting image is obtained by considering that in the middle of the actual ocean (i.e., on the surface of the Earth) myriads of small waves occasionally come together fortuitously with such phase relationships that they end up in a certain small region of space, suddenly to produce a very high wave which just appears as if from nowhere and out of nothing. Perhaps something like this could happen in the immense ocean of cosmic energy, creating a sudden wave pulse, from which our "universe" would be born. This pulse would explode outward and break up into smaller ripples that spread yet further outward to constitute our "expanding universe." The latter would have its "space" enfolded within it as a special distinguished explicate and manifest order.[16]

In terms of this proposal it follows that the current attempt to understand our "universe" as if it were self-existent and independent of the sea of cosmic energy can work at best in some limited way (depending on how far the notion of a relatively independent sub-totality applies to it). For example, the "black holes" may lead us into an area in which the cosmic background of energy is important. Also, of course, there may be many other such expanding universes.

Moreover, it must be remembered that even this vast sea of cosmic energy takes into account only what happens on a scale larger than the critical length of 10^{-33} cm, to which we have referred earlier. But this length is only a certain kind of limit on the applicability of ordinary notions of space and time. To suppose that there is nothing beyond this limit at all would indeed be quite arbitrary. Rather, it is very probable that beyond it lies a further domain, or set of domains, of the nature of which we have as yet little or no idea.

What we have seen thus far is a progression from explicate order to simple three-dimensional implicate order, then to a multidimensional implicate order, then to an extension of this to the immense "sea" in what is sensed as empty space. The next stage may well lead to yet further enrichment and extension of the notion of implicate order, beyond the critical limit of 10^{-33} cm mentioned above; or it may lead to some basically new notions which could not be comprehended even within the possible further developments of the implicate order. Nevertheless, whatever may be possible in this regard, it is clear that we may assume that the principle of relative autonomy of sub-totalities continues to be valid. Any sub-totality, including those which we have thus far considered, may up to a point be studied in its own right. Thus, without assuming that we have already arrived even at an outline of absolute and final truth, we may at least for a time put aside the need to consider what may be beyond the immense energies of empty space, and go on to bring out the further implications of the sub-totality of order that has revealed itself thus far.

6 THE IMPLICATE ORDER, LIFE AND THE FORCE OF OVERALL NECESSITY

In this section we shall bring out the meaning of the implicate order by first showing how it makes possible the comprehension of both inanimate matter and life on the basis of a single ground, common to both, and then we shall go on to propose a certain more general form for the laws of the implicate order.

Let us begin by considering the growth of a living plant. This growth starts from a seed, but the seed contributes little or nothing to the actual material substance of the plant or to the energy needed to make it grow. This latter comes almost entirely from the soil, the water, the air and the sunlight. According to modern theories the seed contains *information*, in the form of DNA, and this information somehow "directs" the environment to form a corresponding plant.

In terms of the implicate order, we may say that even inanimate matter maintains itself in a continual process similar to the growth of plants. Thus, recalling the ink-in-fluid model of the electron, we see that such a "particle" is to be understood as a recurrent stable order of unfoldment in which a certain form undergoing regular changes manifests again and again, but so rapidly that it appears to be in continuous existence. We may compare this to a forest, constituted of trees that are continually dying and being replaced by new ones. If it is considered on a long time-scale, this forest may be regarded likewise as a continuously existent but slowly-changing entity. So when understood through the implicate order, inanimate matter and living beings are seen to be, in certain key respects, basically similar in their modes of existence.

When inanimate matter is left to itself the above-described process of enfoldment and unfoldment just reproduces a similar form of inanimate matter, but when this is further "informed" by the seed, it begins to produce a living plant instead. Ultimately, this latter gives rise to a new seed, which allows the process to continue after the death of this plant.

As the plant is formed, maintained and dissolved by the exchange of matter and energy with its environment, at which point can we say that there is a sharp distinction between what is alive and what is not? Clearly, a molecule of carbon dioxide that crosses a cell boundary into a leaf does not suddenly "come alive" nor does a molecule of oxygen suddenly "die" when it is released to the atmosphere. Rather, life itself has to be regarded as belonging in some sense to a totality, including plant and environment.

It may indeed be said that life is enfolded in the totality and that, even when it is not manifest, it is somehow "implicit" in what we generally call a situation in which there is no life. We can illustrate this by considering the ensemble of all the atoms that are now in the environment but that are eventually going to constitute a plant that will grow from a certain seed. This ensemble is evidently, in certain key ways, similar to that considered in Section 3, of ink particles forming a droplet. In both cases the elements of the ensemble are bound together to contribute to a common end (in one case an ink droplet and in the other case a living plant).

The above does not mean, however, that life can be reduced completely to nothing more than that which comes out of the activity of a basis governed by the laws of inanimate matter alone (though we do not deny that *certain* features of life may be understood in this way). Rather, we are proposing that as the notion of the holomovement was enriched by going from three-dimensional to multidimensional implicate order and then to the vast "sea" of energy in "empty" space, so we may now enrich this notion further by saying that in its totality the holomovement includes the principle of life as well. Inanimate matter is then to be regarded as a relatively autonomous sub-totality in which, at least as far as we now know, life does not significantly manifest. That is to say, inanimate matter is a secondary, derivative, and particular abstraction from the holomovement (as would also be the notion of a "life force" entirely independent of matter). Indeed, the holomovement which is "life implicit" is the ground both of "life explicit" and of "inanimate" matter, and this ground is what is primary, self-existent and universal. Thus we do not fragment life and inanimate matter, nor do we try to reduce the former completely to nothing but an outcome of the latter.

Let us now put the above approach in a more general way. What is basic to the law of the holomovement is, as we have seen, the possibility of abstraction of a set of relatively autonomous sub-totalities. We can now add that the laws of each such abstracted sub-totality quite generally operate under certain conditions and limitations defined only in a corresponding total situation (or set of similar situations). This operation will in general have these three key features:

1   A set of implicate orders.
2   A special distinguished case of the above set, which constitutes an explicate order of manifestation.
3   A general relationship (or law) expressing a force of necessity which binds together a certain set of the elements of the implicate order in such a way that they contribute to a common explicate end (different from that to which another set of inter-penetrating and intermingling elements will contribute).

The origin of this force of necessity cannot be understood solely in terms of the explicate and implicate orders belonging to the type of situation in question. Rather, at this level, such necessity has simply to be accepted as inherent in the overall situation under discussion. An understanding of its origin would take us to a deeper, more comprehensive and more inward level of relative autonomy which, however, would also have its implicate and explicate orders and a correspondingly deeper and more inward force of necessity that would bring about their transformation into each other.[17]

In short, we are proposing that this form of the law of a relatively autonomous sub-totality, which is a consistent generalization of all the forms that we have studied thus far, is to be considered as universal; and that in our subsequent work we shall explore the implicates of such a notion, at least tentatively and provisionally.

7   CONSCIOUSNESS AND THE IMPLICATE ORDER

At this point it may be said that at least some outlines of our notions of cosmology and of the general nature of reality have been sketched (though, of course, to "fill in" this sketch with adequate detail would require a great deal of further work much of which still remains to be done). Let us now consider how consciousness may be understood in relation to these notions.

We begin by proposing that in some sense, consciousness (which we take to include thought, feeling, desire, will, etc.) is to be comprehended in terms of the implicate order, along with reality as a whole. That is to say, we are suggesting that the implicate order applies both to matter (living and non-living) and to consciousness, and that it can therefore make possible an understanding of the general relationship of these two, from which we may be able to come to some notion of a common ground of both (rather as was also suggested in the previous section in our discussion of the relationship of inanimate matter and life).

To obtain an understanding of the relationship of matter and consciousness has, however, thus far proved to be extremely difficult, and this difficulty has its root in the very great difference in their basic qualities as they present themselves in our experience. This difference has been expressed with particularly great clarity by Descartes, who described matter as "extended substance" and consciousness as "thinking substance." Evidently, by "extended substance" Descartes meant something made up of distinct forms existing in space, in an order of extension and separation basically similar to the one that we have been calling explicate. By using the term "thinking substance" in such sharp contrast to "extended substance" he was clearly implying that the various distinct forms appearing in thought do not have their existence in such an order of extension and separation (i.e., some kind of space), but rather in a different order, in which extension and separations have no fundamental significance. The implicate order has just this latter quality, so in a certain sense Descartes was perhaps anticipating that consciousness has to be understood in terms of an order that is closer to the implicate than it is to the explicate.

However, when we start, as Descartes did, with extension and separation in space as primary for matter, then we can see nothing in this notion that can serve as a basis for a relationship between matter and consciousness, whose orders are so different. Descartes clearly understood this difficulty and indeed proposed to resolve it by means of the idea that such a relationship is made possible by God, who being outside of and beyond matter and consciousness (both of which He has indeed created) is able to give the latter "clear and distinct notions" that are currently applicable to the former. Since then, the idea that God takes care of this requirement has generally been abandoned, but it has not commonly been noticed that thereby the possibility of comprehending the relationship between matter and consciousness has collapsed.

In this chapter, we have, however, shown in some detail that matter as a whole can be understood in terms of the notion that the implicate order is the immediate and primary actuality (while the explicate order can be derived as a particular, distinguished case of the implicate order). The question that arises here, then, is that of whether or not (as was in a certain sense anticipated by Descartes) the actual "substance" of consciousness can be understood in terms of the notion that the implicate order is also its primary and immediate actuality. If matter and consciousness could in this way be understood together, in terms of the same general notion of order, the way would be opened to comprehending their relationship on the basis of some common ground.[18] Thus we could come to the germ of a new notion of unbroken wholeness, in which consciousness is no longer to be fundamentally separated from matter.

Let us now consider what justification there is for the notion that matter and consciousness have the implicate order in common. First, we note that matter in general is, in the first instance, the object of our consciousness. However, as we have seen throughout this chapter, various energies such as light, sound, etc., are continually enfolding information in principle concerning the entire universe of matter into each region of space. Through this process, such information may of course enter our sense organs, to go on through the nervous system to the brain. More deeply, all the matter in our bodies, from the very first, enfolds the universe in some way. Is this enfolded structure, both of information and of matter (e.g., in the brain and nervous system), that which primarily enters consciousness?

Let us first consider the question of whether information is actually enfolded in the brain cells. Some light on this question is afforded by certain work on brain structure, notably that of Pribram.[19] Pribram has given evidence backing up his suggestion that memories are generally recorded all over the brain in such a way that information concerning a given object or quality is not stored in a particular cell or localized part of the brain, but rather that all the information is enfolded over the whole. This storage resembles a hologram in its function, but its actual structure is much more complex. We can then suggest that when the "holographic" record in the brain is suitably activated, the response is to create a pattern of nervous energy constituting a partial experience similar to that which produced the "hologram" in the first place. But it is also different in that it is less detailed, in that memories from many different times may merge together, and in that memories may be connected by association and by logical thought to give a certain further order to the whole pattern. In addition, if sensory data is also being attended to at the same time, the whole of this response from memory will, in general, fuse with the nervous excitation coming from the senses to give rise to an overall experience in which memory, logic, and sensory activity combine into a single unanalysable whole.

Of course, consciousness is more than what has been described above. It also involves awareness, attention, perception, acts of understanding, and perhaps yet more. We have suggested in the first chapter [of *Wholeness and Implicate Order*] that these must go beyond a mechanistic response (such as that which the holographic model of brain function would by itself imply). So in studying them we may be coming closer to the essence of actual conscious experience than is possible merely by discussing patterns of excitation of the sensory nerves and how they may be recorded in memory.

It is difficult to say much about faculties as subtle as these. However, by reflecting on and giving careful attention to what happens in certain experiences, one can obtain valuable clues. Consider, for example, what takes place when one is listening to music. At a given moment a certain note is being played but a number of the previous notes are still "reverberating" in consciousness. Close attention will show that it is the simultaneous presence and activity of all these reverberations that is responsible for the direct and immediately felt sense of movement, flow and continuity. To hear a set of notes so far apart in time that there is no such reverberation will destroy altogether the sense of a whole unbroken, living movement that gives meaning and force to what is heard.

It is clear from the above that one does not experience the actuality of this whole movement by "holding on" to the past, with the aid of a memory of the sequence of notes, and comparing this past with the present. Rather, as one can discover by further attention, the "reverberations" that make such an experience possible are not memories but are rather *active transformations* of what came earlier, in which are to be found not only a generally diffused sense of the original sounds, with an intensity that falls off, according to the time elapsed since they were picked up by the ear, but also various emotional responses, bodily sensations, incipient muscular movements, and the evocation of a wide range of yet further meanings, often of great subtlety. One can thus obtain a direct sense of how a sequence of notes is enfolding into many levels of consciousness, and of how at any given moment, the transformations flowing out of many such enfolded notes interpenetrate and intermingle to give rise to an immediate and primary feeling of movement.

This activity in consciousness evidently constitutes a striking parallel to the activity that we have proposed for the implicate order in general. Thus in Section 3, we have given a model of an electron in which, at any instant, there is a co-present set of differently transformed ensembles which interpenetrate and intermingle in their various degrees of enfoldment. In such enfoldment, there is a radical change, not only of form but also of structure, in the entire set of ensembles (which change we have elsewhere called a metamorphosis[20]); and yet, a certain totality of order in the ensembles remains invariant, in the sense that in all these changes a subtle but fundamental similarity of order is preserved.[21]

In the music, there is, as we have seen, a basically similar transformation (of notes) in which a certain order can also be seen to be preserved. The key difference in these two cases is that for our model of the electron an enfolded order is grasped in *thought*, as the presence together of many different but interrelated degrees of transformations of ensembles, while for the music, it is *sensed immediately* as the presence together of many different but interrelated degrees of transformations of tones and sounds. In the latter, there is a feeling of both tension and harmony between the various co-present transformations, and this feeling is indeed what is primary in the apprehension of the music in its undivided state of flowing movement.

In listening to music, *one is therefore directly perceiving an implicate order*. Evidently this order is *active* in the sense that it continually flows into emotional, physical, and other responses, that are inseparable from the transformations out of which it is essentially constituted.

A similar notion can be seen to be applicable for vision. To bring this out, consider the sense of motion that arises when one is watching the cinema screen. What is actually happening is that a series of images, each slightly different, is being flashed on the screen. If the images are separated by long intervals of time, one does not get a feeling of continuous motion, but rather, one sees a series of disconnected images perhaps accompanied by a sense of jerkiness. If, however, the images are close enough together (say a hundredth of a second) one has a direct and immediate experience, as if from a continuously moving and flowing reality, undivided and without a break.

This point can be brought out even more clearly by considering a well-known illusion of movement, produced with the aid of a stroboscopic device, illustrated in Figure 3.3.

Two discs, A and B, enclosed in a bulb, can be caused to give off light by means of electrical excitation. The light is made to flash on and off so rapidly that it appears to be continuous, but in each flash it is arranged that B will come on slightly later than A. What one actually feels is a sense of "flowing movement" between A and B, but that paradoxically

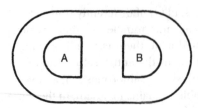

Figure 3.3

nothing is flowing out of B (contrary to what would be expected if there had been a real process of flow). This means that a sense of flowing movement is experienced when, on the retina of the eye, there are two images in neighbouring positions one of which comes on slightly later than the other. (Closely related to this is the fact that a blurred photograph of a speeding car, containing a sequence of overlaid images in slightly different positions, conveys to us a much more immediate and vivid sense of movement than does a sharp picture, taken with a high-speed camera.)

It seems evident that the sense of unbroken movement described above is basically similar to that arising from a sequence of musical notes. The main difference between music and visual images, in this regard, is that the latter may arrive so close together in time that they cannot be resolved in consciousness. Nevertheless, it is clear that visual images must also undergo active transformation as they "enfold" into the brain and nervous system (e.g., they give rise to emotional, physical and other more subtle responses of which one may be only dimly conscious as well as to "after images" that are in certain ways similar to the reverberations in musical notes). Even though the time difference of two such images may be small, the examples cited above make it clear that a sense of movement is experienced through the intermingling and interpenetration of the co-present transformations to which these images must give rise as they penetrate the brain and nervous system.

All of this suggests that quite generally (and not merely for the special case of listening to music), there is a basic similarity between the order of our immediate experience of movement and the implicate order as expressed in terms of our thought. We have in this way been brought to the possibility of a coherent mode of understanding the immediate experience of motion in terms of our thought (in effect thus resolving Zeno's paradox concerning motion).

To see how this comes about, consider how motion is usually thought of, in terms of a series of points along a line. Let us suppose that at a certain time t_1, a particle is at a position x_1, while at a later time t_2, it is at another position x_2. We then say that this particle is moving and that its velocity is:

$$v = \frac{x_2 - x_1}{t_2 - t_1}$$

Of course, this way of thinking does not in any way reflect or convey the immediate sense of motion that we may have at a given moment, for example, with a sequence of musical notes reverberating in consciousness (or in the visual perception of a speeding car). Rather, it is only an abstract symbolization of movement, having a relation to the actuality of motion, similar to that between a musical score and the actual experience of the music itself.

If, as is commonly done, we take the above abstract symbolization as a faithful representation of the actuality of movement, we become entangled in a series of confused and basically insoluble problems. These all have to do with the image in which we represent time, as if it were a series of points along a line that are somehow all present together, either to our conceptual gaze or perhaps to that of God. Our actual experience is, however, that when a given moment, say t_2, is present and actual, an earlier moment, such as t_1 is past. That is to say, it is *gone*, non-existent, never to return. So if we say that the velocity of a particular *now* (at t_2) is $(x_2 - x_1)/(t_2 - t_1)$ we are trying to relate *what is* (i.e., x_2 and t_2) to *what is not* (i.e., x_1, and t_1). We can of course do this *abstractly and symbolically* (as is, indeed, the common practice in science and mathematics), but the further fact, not comprehended in this abstract symbolism, is that the velocity *now* is active *now* (e.g., it determines how a particle will act from now on, in itself, and in relation to other particles). How are we to understand the *present* activity of a position (x_1) that is now non-existent and gone for ever?

It is commonly thought that this problem is resolved by the differential calculus. What is done here is to let the time interval, $\Delta t = t_2 - t_1$ become vanishingly small, along with $\Delta x = x_2 - x_1$. The velocity *now* is defined as the limit of the ratio $\Delta x/\Delta t$ as Δt approaches zero. It is then implied that the problem described above no longer arises, because x_2 and x_1 are in effect taken at the same time. They may thus be present together and related in an activity that depends on both.

A little reflection shows, however, that this procedure is still as abstract and symbolic as was the original one in which the time interval was taken as finite. Thus one has no immediate experience of a time interval of zero length, nor can one see in terms of reflective thought what this could mean.

Even as an abstract formalism, this approach is not fully consistent in a logical sense, nor does it have a universal range of applicability. Indeed, it applies only within the area of *continuous* movements and then only as a technical algorithm that happens to be correct for this sort of movement. As we have seen, however, according to the quantum theory, movement is *not* fundamentally continuous. So even as an algorithm its current field of application is limited to theories expressed in terms of classical concepts (i.e., in the explicate order) in which it provides a good approximation for the purpose of calculating the movements of material objects.

When we think of movement in terms of the implicate order,[22] however, these problems do not arise. In this order, movement is comprehended in terms of a series of interpenetrating and intermingling elements in different degrees of enfoldment *all present together*. The activity of this movement then presents no difficulty, because it is an outcome of this whole enfolded order, and is determined by relationships of co-present elements, rather than by the relationships of elements that exist to others that no longer exist.

We see, then, that through thinking in terms of the implicate order, we come to a notion of movement that is logically coherent and that properly represents our immediate experience of movement. Thus the sharp break between abstract logical thought and concrete immediate experience, that has pervaded our culture for so long, need no longer be maintained. Rather, the possibility is created for an unbroken flowing movement from immediate experience to logical thought and back, and thus for an ending to this kind of fragmentation.

Moreover we are now able to understand in a new and more consistent way our proposed notion concerning the general nature of reality, that *what is* is movement. Actually, what tends to make it difficult for us to work in terms of this notion is that we usually think of movement in the traditional way as an active relationship of what is to what is not. Our traditional notion concerning the general nature of reality would then amount to saying that *what is* is an active relationship of what is to what is not. To say this is, at the very least, confused. In terms of the implicate order, however, movement is a relationship of certain phases of *what is* to other phases of *what is*, that are in different stages of enfoldment. This notion implies that the essence of reality as a whole is the above relationship among the various phases in different stages of enfoldment (rather than, for example, a relationship between various particles and fields that are all explicate and manifest).

Of course, actual movement involves more than the mere immediate intuitive sense of unbroken flow, which is our mode of directly experiencing the implicate order. The presence of such a sense of flow generally implies further that, in the next moment, the state of affairs will actually change – i.e., it will be different. How are we to understand this fact of experience in terms of the implicate order?

A valuable clue is provided by reflecting on and giving careful attention to what happens when, in our thinking, we say that one set of ideas *implies* an entirely different set. Of course, the word "imply" has the same root as the word "implicate" and thus also involves the notion of enfoldment. Indeed, by saying that something is *implicit* we generally mean more than merely to say that this thing is an inference following from something else through the rules of logic. Rather, we usually mean that from many different ideas and notions (of some of which we are explicitly conscious) a new notion emerges that somehow brings all these together in a concrete and undivided whole.

We see, then, that each moment of consciousness has a certain *explicit* content, which is a foreground, and an *implicit* content, which is a corresponding background. We now propose that not only is immediate experience best understood in terms of the implicate order, but that thought also is basically to be comprehended in this order. Here we mean not just the *content* of thought for which we have already begun to use the implicate order. Rather, we also mean that the actual *structure, function* and *activity* of thought is in the implicate order. The distinction between implicit and explicit in thought is thus being taken here to be essentially equivalent to the distinction between implicate and explicate in matter in general.

To help clarify what this means, let us recall briefly the basic form of the law of a sub-totality (discussed in Sections 3 and 6), i.e., that the enfolded elements of a characteristic ensemble (e.g., of ink particles or of atoms) that are going to constitute the next stage of enfoldment are bound by a force of overall necessity, which brings them together, to contribute to a common end that emerges in the next phase of the process under discussion. Similarly, we propose that the ensemble of elements enfolded in the brain and nervous system that are going to constitute the next stage of development of a line of thought are likewise bound through a force of overall necessity, which brings them together to contribute to the common notion that emerges in the next moment of consciousness.

In this study, we have been using the idea that consciousness can be described in terms of a series of moments. Attention shows that a given moment cannot be fixed exactly in relation to time (e.g., by the clock) but rather, that it covers some vaguely defined and somewhat variable extended period of duration. As pointed out earlier, each moment is experienced directly in the implicate order. We have further seen that through the force of necessity in the overall situation, one moment gives rise to the next, in which content that was previously implicate is now explicate while the previous explicate content has become implicate (e.g., as happened in the analogy of the ink droplets).

The continuation of the above process gives an account of how *change* takes place from one moment to another. In principle, the change in any moment may be a fundamental and radical transformation. However, experience shows that in thought (as in matter in general) there is usually a great deal of recurrence and stability leading to the possibility of relatively independent sub-totalities.

In any such sub-totality, there is the possibility of the continuation of a certain line of thought that enfolds in a fairly regularly changing way. Evidently, the precise character of such a sequence of thoughts, as it enfolds from one moment to the next, will generally depend on the content of the implicate order in earlier moments. For example, a moment containing a sense of movement tends quite generally to be followed by a change in the next moment which is greater the stronger the sense of movement that was originally present (so that, as in the case of the stroboscopic device discussed earlier, when this does not happen we feel that something surprising or paradoxical is taking place).

As in our discussion of matter in general, it is now necessary to go into the question of how in consciousness the explicate order is what is manifest. As observation and attention show (keeping in mind that the word "manifest" means that which is recurrent, stable and separable) the manifest content of consciousness is based essentially on memory, which is what allows such content to be held in a fairly constant form. Of course, to make possible such constancy it is also necessary that this content be organized, not only through relatively fixed associations but also with the aid of the rules of logic, and of our basic categories of space, time, causality, universality, etc. In this way an overall system of concepts and mental images may be developed, which is a more or less faithful representation of the "manifest world."

The process of thought is not, however, merely a *representation* of the manifest world; rather, it makes an important *contribution* to how we experience this world, for, as we have already pointed out earlier, this experience is a fusion of sensory information with the "replay" of some of the content of memory (which latter contains thought built into its very form and order). In such experience, there will be a strong background of recurrent stable, and separable features, against which the transitory and changing aspects of the unbroken flow of experience will be seen as fleeting impressions that tend to be arranged and ordered mainly in terms of the vast totality of the relatively static and fragmented content of recordings from the past.

One can, in fact, adduce a considerable amount of scientific evidence showing how much of our conscious experience is a construction based on memory organized through thought, in the general way described above.[23] To go into this subject in detail would, however, carry us too far afield. It may nevertheless be useful here to mention that Piaget[24] has made it clear that a consciousness of what to us is the familiar order of space, time, causality, etc. (which is essentially what we have been calling the explicate order) operates only to a small extent in the earliest phases of life of the human individual. Rather, as he shows from careful observations, for the most part infants *learn* this content first in the area of sensori-motor experience, and later as they grow older they connect such experience with its expression in language and logic. On the other hand, there seems to be an immediate awareness of movement from the very earliest. Recalling that movement is sensed primarily in the implicate order, we see that Piaget's work supports the notion that the experiencing of the implicate order is fundamentally much more immediate and direct than is that of the explicate order, which, as we have pointed out above, requires a complex construction that has to be learned.

One reason why we do not generally notice the primacy of the implicate order is that we have become so habituated to the explicate order, and have emphasized it so much in our thought and language, that we tend strongly to feel that our primary experience is of that which is explicate and manifest. However, another reason, perhaps more important, is that the activation of memory recordings whose content is mainly that which is recurrent, stable, and separable, must evidently focus our attention very strongly on what is static and fragmented.

This then contributes to the formation of an experience in which these static and fragmented features are often so intense that the more transitory and subtle features of the unbroken flow (e.g., the "transformations" of musical notes) generally tend to pale into such seeming insignificance that one is, at best, only dimly conscious of them. Thus,

an illusion may arise in which the manifest static and fragmented content of consciousness is experienced as the very basis of reality and from
this illusion one may apparently obtain a proof of the correctness of that
mode of thought in which this content is taken to be fundamental.[25]

8   MATTER, CONSCIOUSNESS AND THEIR COMMON GROUND

At the beginning of the previous section we suggested that matter and
consciousness can both be understood in terms of the implicate order.
We shall now show how the notions of implicate order that we have
developed in connection with consciousness may be related to those
concerning matter, to make possible an understanding of how both may
have a common ground.

We begin by noting that current relativistic theories in physics
describe the whole of reality in terms of a process whose ultimate element is a point event, i.e., something happening in a relatively small
region of space and time. We propose instead that the basic element be a
moment which, like the moment of consciousness, cannot be precisely
related to measurements of space and time, but rather covers a somewhat vaguely defined region which is extended in space and has duration in time. The extent and duration of a moment may vary from
something very small to something very large, according to the context
under discussion (even a particular century may be a "moment" in the
history of mankind). As with consciousness, each moment has a certain
explicate order, and in addition it enfolds all the others, though in its
own way. So the relationship of each moment in the whole to all the
others is implied by its total content: the way in which it "holds" all the
others enfolded within it.

In certain ways this notion is similar to Leibniz's idea of monads,
each of which "mirrors" the whole in its own way, some in great detail
and others rather vaguely. The difference is that Leibniz's monads had a
permanent existence, whereas our basic elements are only moments
and are thus not permanent. Whitehead's idea of "actual occasions" is
closer to the one proposed here, the main difference being that we use
the implicate order to express the qualities and relationships of our
moments, whereas Whitehead does this in a rather different way.

We now recall that the laws of the implicate order are such that
there is a relatively independent, recurrent, stable sub-totality which
constitutes the explicate order, and which, of course, is basically the
order that we commonly contact in common experience (extended in

certain ways by our scientific instruments). This order has room in it for something like memory, in the sense that previous moments generally leave a trace (usually enfolded) that continues in later moments, though this trace may change and transform almost without limit. From this trace (e.g., in the rocks) it is in principle possible for us to unfold an image of past moments, similar in certain ways, to what actually happened; and by taking advantage of such traces, we design instruments such as photographic cameras, tape recorders, and computer memories, which are able to register actual moments in such a way that much more of the content of what has happened can be made directly and immediately accessible to us than is generally possible from natural traces alone.

One may indeed say that our memory is a special case of the process described above, for all that is recorded is held enfolded within the brain cells and these are part of matter in general. The recurrence and stability of our own memory as a relatively independent sub-totality is thus brought about as part of the very same process that sustains the recurrence and stability in the manifest order of matter in general.

It follows, then, that the explicate and manifest order of consciousness is not ultimately distinct from that of matter in general. Fundamentally these are essentially different aspects of the one overall order. This explains a basic fact that we have pointed out earlier – that the explicate order of matter in general is also in essence the sensuous explicate order that is presented in consciousness in ordinary experience.

Not only in this respect but, as we have seen, also in a wide range of other important respects, consciousness and matter in general are basically the same order (i.e., the implicate order as a whole). As we have indicated earlier this order is what makes a relationship between the two possible; but more specifically, what are we to say about the nature of this relationship?

We may begin by considering the individual human being as a relatively independent sub-totality, with a sufficient recurrence and stability of his total process (e.g., physical, chemical, neurological, mental, etc.) to enable him to subsist over a certain period of time. In this process we know it to be a fact that the physical state can affect the content of consciousness in many ways. (The simplest case is that we can become conscious of neural excitations as sensations.) Vice versa, we know that the content of consciousness can affect the physical state (e.g., from a conscious intention nerves may be excited, muscles may move, the heart-beat change, along with alterations of glandular activity, blood chemistry, etc.).

This connection of the mind and body has commonly been called psychosomatic (from the Greek "psyche," meaning "mind," and "soma," meaning "body." This word is generally used, however, in such a way as to imply that mind and body are separately existent but connected by some sort of interaction. Such a meaning is not compatible with the implicate order. In the implicate order we have to say that mind enfolds matter in general and therefore the body in particular. Similarly, the body enfolds not only the mind but also in some sense the entire material universe (in the manner explained earlier in this section, both through the senses and through the fact that the constituent atoms of the body are actually structures that are enfolded in principle throughout all space).

This kind of relationship has in fact already been encountered in Section 4, where we introduced the notion of a higher-dimensional reality which projects into lower-dimensional elements that have not only a non-local and non-causal relationship but also just the sort of mutual enfoldment that we have suggested for mind and body. So we are led to propose further that the more comprehensive, deeper, and more inward actuality is neither mind nor body but rather a yet higher-dimensional actuality, which is their common ground and which is of a nature beyond both. Each of these is then only a relatively independent subtotality and it is implied that this relative independence derives from the higher-dimensional ground in which mind and body are ultimately one (rather as we find that the relative independence of the manifest order derives from the ground of the implicate order).

In this higher-dimensional ground the implicate order prevails. Thus, within this ground, *what is* is movement which is represented in thought as the co-presence of many phases of the implicate order. As happens with the simpler forms of the implicate order considered earlier, the state of movement at one moment unfolds through a more inward force of necessity inherent in this overall state of affairs, to give rise to a new state of affairs in the next moment. The projections of the higher-dimensional ground, as mind and body, will in the later moment both be different from what they were in the earlier moment, though these differences will of course be related. So we do not say that mind and body causally affect each other, but rather that the movements of both are the outcome of related projections of a common higher-dimensional ground.

Of course, even this ground of mind and body is limited. At the very least we have evidently to include matter beyond the body if we are to give an adequate account of what actually happens and this must eventually include other people, going on to society and to mankind as a whole. In doing this, however, we will have to be careful not to slip back into regarding the various elements of any given total situation as having anything more than relative independence. In a deeper and generally more suitable way of thinking, each of these elements is a projection, in a sub-totality of yet higher "dimension." So it will be ultimately misleading and indeed wrong to suppose, for example, that each human being is an independent actuality who interacts with other human beings and with nature. Rather, all these are projections of a single totality. As a human being takes part in the process of this totality, he is fundamentally changed in the very activity in which his aim is to change that reality which is the content of his consciousness. To fail to take this into account must inevitably lead one to serious and sustained confusion in all that one does.

From the side of mind we can also see that it is necessary to go on to a more inclusive ground. Thus, as we have seen, the easily accessible explicit content of consciousness is included within a much greater implicit (or implicate) background. This in turn evidently has to be contained in a yet greater background which may include not only neuro-physiological processes at levels of which we are not generally conscious but also a yet greater background of unknown (and indeed ultimately unknowable) depths of inwardness that may be analogous to the "sea" of energy that fills the sensibly perceived "empty" space.[26]

Whatever may be the nature of these inward depths of consciousness, they are the very ground, both of the explicit content and of that content which is usually called implicit. Although this ground may not appear in ordinary consciousness, it may nevertheless be present in a certain way. Just as the vast "sea" of energy in space is present to our perception as a *sense* of emptiness or nothingness so the vast "unconscious" background of explicit consciousness with all its implications is present in a similar way. That is to say, it may be *sensed* as an emptiness, a nothingness, within which the usual content of consciousness is only a vanishingly small set of facets.

Let us now consider briefly what may be said about time in this total order of matter and consciousness.

First, it is well known that, as directly sensed and experienced in consciousness, time is highly variable and relative to conditions (e.g., a given period may be felt to be short or long by different people, or even by the same person, according to the interests of the different people concerned). On the other hand it seems in common experience that physical time is absolute and does not depend on conditions. However, one of the most important implications of the theory of relativity is that physical time is in fact relative, in the sense that it may vary according to the speed of the observer. (This variation is, however, significant only as we approach the speed of light and is quite negligible in the domain of ordinary experience.) What is crucial in the present context is that, according to the theory of relativity, a sharp distinction between space and time can not be maintained (except as an approximation, valid at velocities small compared with that of light). Thus, since the quantum theory implies that elements that are separated in space are generally non-causally and non-locally related projections of a higher-dimensional reality, it follows that moments separated in time are also such projections of this reality.

Evidently, this leads to a fundamentally new notion of the meaning of time. Both in common experience and in physics, time has generally been considered to be a primary, independent and universally applicable order, perhaps the most fundamental one known to us. Now, we have been led to propose that it is secondary and that, like space (see Section 5), it is to be derived from a higher-dimensional ground, as a particular order. Indeed, one can further say that many such particular interrelated time orders can be derived for different sets of sequences of moments, corresponding to material systems that travel at different speeds. However, these are all dependent on a multidimensional reality that cannot be comprehended fully in terms of any time order, or set of such orders.

Similarly, we are led to propose that this multidimensional reality may project into many orders of sequences of moments in consciousness. Not only do we have in mind here the relativity of psychological time discussed above, but also much more subtle implications. Thus, for example, people who know each other well may separate for a long time (as measured by the sequence of moments registered by a clock) and yet they are often able to "take up from where they left off" as if no time had passed. What we are proposing here is that sequences of moments that "skip" intervening spaces are just as allowable forms of time as those which seem continuous.[27]

The fundamental law, then, is that of the immense multidimensional ground; and the projections from this ground determine whatever time orders there may be. Of course, this law may be such that in certain limiting cases the order of moments corresponds approximately to what would be determined by a simple causal law. Or, in a different limiting case, the order would be a complex one of a high degree which would approximate what is usually called a random order.[28] These two alternatives cover what happens for the most part in the domain of ordinary experience as well as in that of classical physics. Nevertheless, in the quantum domain as well as in connection with consciousness and probably with the understanding of the deeper more inward essence of life, such approximations will prove to be inadequate. One must then go on to a consideration of time as a projection of multidimensional reality into a sequence of moments.

Such a projection can be described as creative, rather than mechanical, for by creativity one means just the inception of new content, which unfolds into a sequence of moments that is not completely derivable from what came earlier in this sequence or set of such sequences. What we are saying is, then, that movement is basically such a creative inception of new content as projected from the multidimensional ground. In contrast, what is mechanical is a relatively autonomous subtotality that can be abstracted from that which is basically a creative movement of unfoldment.

How, then, are we to consider the evolution of life as this is generally formulated in biology? First, it has to be pointed out that the very word "evolution" (whose literal meaning is "unrolling") is too mechanistic in its connotation to serve properly in this context. Rather, as we have already pointed out above, we should say that various successive living forms unfold creatively. Later members are not completely derivable from what came earlier, through a process in which effect arises out of cause (though in some approximation such a causal process may explain certain limited aspects of the sequence). The law of this unfoldment cannot be properly understood without considering the immense multidimensional reality of which it is a projection (except in the rough approximation in which the implications of the quantum theory and of what is beyond this theory may be neglected).

Our overall approach has thus brought together questions of the nature of the cosmos, of matter in general, of life, and of consciousness. All of these have been considered to be projections of a common ground. This we may call the ground of all that is, at least in so far as this may be sensed and known by us, in our present phase of unfoldment of consciousness. Although we have no detailed perception or knowledge of this ground it is still in a certain sense enfolded in our consciousness, in the ways in which we have outlined, as well as perhaps in other ways that are yet to be discovered.

Is this ground the absolute end of everything? In our proposed views concerning the general nature of "the totality of all that is" we regard even this ground as a mere stage, in the sense that there could in principle be an infinity of further development beyond it. At any particular moment in this development each such set of views that may arise will constitute at most a *proposal*. It is not to be taken as an *assumption* about what the final truth is supposed to be, and still less as a *conclusion* concerning the nature of such truth. Rather, this proposal becomes itself an *active factor* in the totality of existence which includes ourselves as well as the objects of our thoughts and experimental investigations. Any further proposals on this process will, like those already made, have to be *viable*. That is to say, one will require of them a general self-consistency as well as consistency in what flows from them in life as a whole. Through the force of an even deeper, more inward necessity in this totality, some new state of affairs may emerge in which both the world as we know it and our ideas about it may undergo an unending process of yet further change.

With this we have in essence carried the presentation of our cosmology and our general notions concerning the nature of the totality to a natural (though of course only a temporary) stopping point. From here on we can further survey it as a whole and perhaps fill in some of the details that have been left out in this necessarily sketchy treatment before going on to new developments of the kinds indicated above.

APPENDIX
IMPLICATE AND EXPLICATE ORDER IN PHYSICAL LAW

A1 INTRODUCTION

Previously we have called attention to the emergence of new orders throughout the history of physics.[29] A general feature of the development of this subject has been a tendency to regard certain basic notions of order as permanent and unchangeable. The task of physics was then taken to be to *accommodate* new observations by means of *adaptations* within these basic notions of order, so as to fit the new facts. This kind of adaptation began with the Ptolemaic epicycles, which continued from ancient times until the advent of the work of Copernicus, Kepler, Galileo, and Newton. As soon as the basic notions of order in classical physics had been fairly clearly expressed, it was supposed that further work in physics would consist of adaptation within this order to accommodate new facts. This continued until the appearance of relativity and the quantum theory. It can accurately be said that since then the main line of work in physics has been adaptation within the general orders underlying these theories, to accommodate the facts to which these in turn have led.

It may thus be inferred that accommodation within already existing frameworks of order has generally been considered to be the main activity to be emphasized in physics, while the perception of new orders has been thought of as something that happens only occasionally, perhaps in revolutionary periods, during which what is regarded as the normal process of accommodation has broken down.[30]

It is pertinent to this subject to consider Piaget's[31] description of

Extract from Chapter 6 of D. Bohm, *Wholeness and Implicate Order*, Routledge, London (1980).

all intelligent perception in terms of two complementary movements, *accommodation* and *assimilation*. From the roots "mod," meaning "measure," and "com," meaning "together," one sees that to accommodate means "to establish a common measure." Examples of accommodation are fitting, cutting to a pattern, adapting, imitating, conforming to rules, etc. On the other hand, "to assimilate" is "to digest" or to make into a comprehensive and inseparable whole (which includes oneself). Thus, to assimilate means "to understand."

It is clear that in intelligent perception, primary emphasis has in general to be given to assimilation, while accommodation tends to play a relatively secondary role in the sense that its main significance is as an aid to assimilation.

Of course, we are able in certain sorts of contexts just to accommodate something that we observe within known orders of thought, and in this very act it will be adequately assimilated. However, it is necessary in more general contexts to give serious attention to the possibility that the old orders of thought may cease to be relevant, so that they can no longer coherently be adapted to fit the new fact. One may then have to see the irrelevance of old differences, and the relevance of new differences, and thus one may open the way to the perception of new orders, new measures, and new structures.

Clearly, such perception can appropriately take place at almost any time, and does not have to be restricted to unusual and revolutionary periods in which one finds that the older orders can no longer be conveniently adapted to the facts. Rather, one may be continually ready to drop old notions of order at various contexts, which may be broad or narrow, and to perceive new notions that may be relevant in such contexts. Thus, understanding the fact by assimilating it into new orders can become what could perhaps be called the normal way of doing scientific research.

To work in this way is evidently to give primary emphasis to something similar to *artistic perception.* Such perception begins by observing the whole fact in its full individuality, and then by degree articulates the order that is proper to the assimilation of this fact. It does not begin with abstract preconceptions as to what the order has to be, which are then adapted to the order that is observed.

What, then, is the proper role of accommodation of facts within known theoretical orders, measures and structures? Here, it is important to note that facts are not to be considered as if they were independently existent objects that we might find or pick up in the laboratory. Rather, as the Latin root of the word "facere" indicates, the fact is "what has been made" (e.g., as in "manufacture"). Thus, in a certain sense, we "make" the fact. That is to say, beginning with immediate perception of an actual situation, we develop the fact by giving it further order, form and structure with the aid of our theoretical concepts. For example, by using the notions of order prevailing in ancient times, men were led to "make" the fact about planetary motions by describing and measuring in terms of epicycles. In classical physics, the fact was "made" in terms of the order of planetary orbits, measured through positions and times. In general relativity, the fact was "made" in terms of the order of Riemannian geometry, and of the measure implied by concepts such as "curvature of space." In the quantum theory, the fact was made in terms of the order of energy levels, quantum numbers, symmetry groups, etc., along with appropriate measures (e.g. scattering cross-sections, charges, and masses of particles, etc.).

It is clear, then, that changes of order and measures in the theory ultimately lead to new ways of doing experiments and to new kinds of instruments, which in turn lead to the "making" of correspondingly ordered and measured facts of new kinds. In this development, the experimental fact serves in the first instance as a test for theoretical notions. Thus, the general form of theoretical explanation is that of a generalized kind of ratio of reason. "As *A* is to *B* in our structure of thinking, so it is in fact." This ratio or reason constitutes a kind of "common measure" or "accommodation" between theory and fact.

As long as such a common measure prevails, then of course the theory used need not be changed. If the common measure is found not to be realized, then the first step is to see whether it can be re-established by means of adjustments within the theory without a change in its underlying order. If, after reasonable efforts, a proper accommodation of this kind is not achieved, then what is needed is a fresh perception of *the whole fact*. This now includes not only the results of experiments but also the *failure of certain lines of theory to fit the experimental results in a "common measure."* Then, as has been indicated earlier, one has to be very sensitively aware of all the relevant differences which underly the main orders in the old theory, to see whether there is room for a change of overall order. It is being emphasized here that this kind of perception should properly be inter-woven continually with the activities aimed at accommodation, and should not have to be delayed for so long that the whole situation becomes confused and chaotic, apparently requiring the revolutionary destruction of the old order to clear it up.

As relativity and quantum theory have shown that it has no meaning to divide the observing apparatus from what is observed, so the considerations discussed here indicate that it has no meaning to separate the observed fact (along with the instruments used to observe it) from the theoretical notions of order that help to give "shape" to this fact. As we go on to develop new notions of order going beyond those of relativity and quantum theory, it will thus not be appropriate to try immediately to apply these notions to current problems that have arisen in the consideration of the present set of experimental facts. Rather, what is called for in this context is very broadly to assimilate the whole of the fact in physics into the new theoretical notions of order. After this fact has generally been "digested," we can begin to glimpse new ways in which such notions of order can be tested and perhaps extended in various directions. We have to proceed slowly and patiently here or else we may become confused by "undigested" facts.

Fact and theory are thus seen to be different aspects of one whole in which analysis into separate but interacting parts is not relevant. That is to say, not only is undivided wholeness implied in the *content* of physics (notably relativity and quantum theory) but also in the *manner of working* in physics. This means that we do not try *always* to force the theory to fit the kinds of facts that may be appropriate in currently accepted general orders of description, but that we are also ready when necessary to consider changes in what is meant by fact, which may be required for assimilation of such fact into new theoretical notions of order.

A2 UNDIVIDED WHOLENESS – THE LENS AND THE HOLOGRAM

The undivided wholeness of modes of observation, instrumentation, and theoretical understanding indicated above implies the need to consider a *new order of fact*, i.e., the fact about the way in which modes of theoretical understanding and of observation and instrumentation are related to each other. Until now, we have more or less just taken such a relationship for granted, without giving serious attention to the manner in which it arises, very probably because of the belief that the study of the subject belongs to "the history of science" rather than to "science proper." However, it is now being suggested that the consideration of this relationship is essential for an adequate understanding of science itself, because the content of the observed fact cannot coherently be regarded as separate from modes of observation and instrumentation and modes of theoretical understanding.

An example of the very close relationship between instrumentation

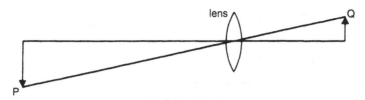

Figure 3.4

and theory can be seen by considering the *lens*, which was indeed one of the key features behind the development of modern scientific thought. The essential feature of a lens is, as indicated in Figure 3.4, that it forms an *image* in which a given point P in the object corresponds (in a high degree of approximation) to a point Q in the image. By thus bringing the correspondence of specified features of object and image into such sharp relief, the lens greatly strengthened man's awareness of the various parts of the object and of the relationship between these parts. In this way, it furthered the tendency to think in terms of analysis and synthesis. Moreover, it made possible an enormous extension of the classical order of analysis and synthesis to objects that were too far away, too big, too small, or too rapidly moving to be thus ordered by means of unaided vision. As a result, scientists were encouraged to extrapolate their ideas and to think that such an approach would be relevant and valid no matter how far they went, in all possible conditions, contexts, and degrees of approximation.

However, relativity and quantum theory imply undivided wholeness, in which analysis into distinct and well-defined parts is no longer relevant. Is there an instrument that can help give a certain immediate perceptual insight into what can be meant by undivided wholeness, as the lens did for what can be meant by analysis of a system into parts? It is suggested here that one can obtain such insight by considering the *hologram*. (The name is derived from the Greek words "holo," meaning "whole," and "gram" meaning "to write." Thus, the hologram is an instrument that, as it were, "writes the whole.")

As shown in Figure 3.5, coherent light from a laser is passed through a half-silvered mirror. Part of the beam goes on directly to a photographic plate, while another part is reflected so that it illuminates a certain whole structure. The light reflected from this whole structure also reaches the plate, where it interferes with that arriving there by a direct path. The resulting interference pattern which is recorded on the plate is not only very complex but also usually so fine that it is not even visible

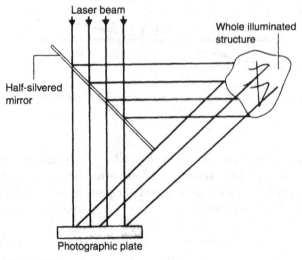

Laser beam

Whole illuminated structure

Half-silvered mirror

Photographic plate

Figure 3.5

to the naked eye. Yet, it is somehow relevant to the whole illuminated structure, though only in a highly implicit way.

This relevance of the interference pattern to the whole illuminated structure is revealed when the photographic plate is illuminated with laser light. As shown in Figure 3.6, a wavefront is then created which is very similar in form to that coming off the original illuminated structure. By placing the eye in this way, one in effect sees the whole of the original structure, in three dimensions, and from a range of possible points of view (as if one were looking at it through a window). If we then illuminate only a small region *R* of the plate, we *still see the whole structure*, but in somewhat less sharply defined detail and from a decreased range of possible points of view (as if we were looking through a smaller window).

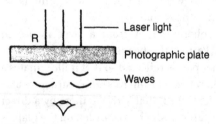

Laser light

R

Photographic plate

Waves

Figure 3.6

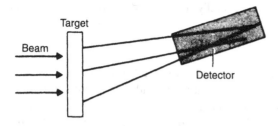

Figure 3.7

It is clear, then, that there is no one-to-one correspondence between parts of an "illuminated object" and parts of an "image of this object on the plate." Rather, the interference pattern in each region R of the plate is relevant to the whole structure, and each region of the structure is relevant to the whole of the interference pattern on the plate.

Because of the wave properties of light, even a lens cannot produce an exact one-to-one correspondence. A lens can therefore be regarded as a limiting case of a hologram.

We can, however, go further and say that in their overall ways of indicating the meaning of observations, typical experiments as currently done in physics (especially in the "quantum" context) are more like the general case of a hologram than like the special case of a lens. For example, consider a scattering experiment. As shown in Figure 3.7, what can be observed in the detector is generally relevant to the whole target, or at least to an area large enough to contain a great many atoms.

Moreover, although one might in principle try to make an image of a particular atom, the quantum theory implies that to do this would have little or no significance. Indeed, as the discussion of the Heisenberg microscope experiment shows,[32] the formation of an image is just what is *not* relevant in a "quantum" context; at most a discussion of image formation serves to indicate the limits of applicability of classical modes of description.

So we may say that in current research in physics, an instrument tends to be relevant to a whole structure, in a way rather similar to what happens with a hologram. To be sure, there are certain differences. For example, in current experiments with electron beams or with X-rays, these latter are seldom coherent over appreciable distances. If, however, it should ever prove to be possible to develop something like an electron laser or an X-ray laser, then experiments will directly reveal "atomic" and "nuclear" structures without the need for complex chains of inference of the sort now generally required, as the hologram does for ordinary large-scale structures.

A3 IMPLICATE AND EXPLICATE ORDER

What is being suggested here is that the consideration of the difference between lens and hologram can play a significant part in the perception of a new order that is relevant for physical law. As Galileo noted the distinction between a viscous medium and a vacuum and saw that physical law should refer primarily to the order of motion of an object in a vacuum, so we might now note the distinction between a lens and a hologram and consider the possibility that physical law should refer primarily to an order of undivided wholeness of the content of a description similar to that indicated by the hologram rather than to an order of analysis of such content into separate parts indicated by a lens.

However, when Aristotle's ideas on movement were dropped, Galileo and those who followed him had to consider the question of how the new order of motion was to be described in adequate details. The answer came in the form of Cartesian coordinates extended to the language of the calculus (differential equations, etc.). But this kind of description is of course appropriate only in a context in which analysis into distinct and autonomous parts is relevant, and will therefore in turn have to be dropped. What, then, will be the new kind of description appropriate to the present context?

As happened with Cartesian coordinates and the calculus, such a question cannot be answered immediately in terms of definite prescriptions as to what to do. Rather, one has to observe the new situation very broadly and tentatively and to "feel out" what may be the relevant new features. From this, there will arise a discernment of the new order, which will articulate and unfold in a natural way (and not as a result of efforts to make it fit well-defined and preconceived notions as to what this order should be able to achieve).

We can begin such an inquiry by noting that in some subtle sense, which does not appear in ordinary vision, the interference pattern in the whole plate can distinguish different orders and measures in the whole illuminated structure. For example, the illuminated structure may contain all sorts of shapes and sizes of geometric forms (indicated in Figure 3.8a), as well as topological relationships, such as inside and outside (indicated in Figure 3.8b), and intersection and separation (indicated in Figure 3.8c). All of these lead to different interference patterns and it is this difference that is somehow to be described in detail.

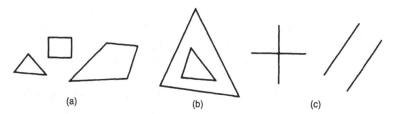

Figure 3.8

The differences indicated above are, however, not only in the plate. Indeed, the latter is of secondary significance, in the sense that its main function is to make a relatively permanent "written record" of the interference pattern of the light that is present in each region of space. More generally, however, in each such region, the movement of the light implicitly contains a vast range of distinctions of order and measure, appropriate to a whole illuminated structure. Indeed, in principle, this structure extends over the whole universe and over the whole past, with implications for the whole future. Consider, for example, how on looking at the night sky, we are able to discern structures covering immense stretches of space and time, which are in some sense contained in the movements of light in the tiny space encompassed by the eye (and also how instruments, such as optical and radio telescopes, can discern more and more of this totality, contained in each region of space).

There is the germ of a new notion of order here. This order is not to be understood solely in terms of a regular arrangement of *objects* (e.g., in rows) or as a regular arrangement of *events* (e.g. in a series). Rather, a *total order is* contained, in some *implicit* sense, in each region of space and time.

Now, the word "implicit" is based on the verb "to implicate." This means "to fold inward" (as multiplication means "folding many times"). So we may be led to explore the notion that in some sense each region contains a total structure "enfolded" within it.

It will be useful in such an exploration to consider some further examples of enfolded or *implicate* order. Thus, in a television broadcast, the visual image is translated into a time order, which is "carried" by the radio wave. Points that are near each other in the visual image are not necessarily "near" in the order of the radio signal. Thus, the radio wave carries the visual image in an implicate order. The function of the receiver is then to *explicate* this order, i.e., to "unfold" it in the form of a new visual image.

A more striking example of implicate order can be demonstrated in the laboratory, with a transparent container full of a very viscous fluid, such as treacle, and equipped with a mechanical rotator that can "stir" the fluid very slowly but very thoroughly. If an insoluble droplet of ink is placed in the fluid and the stirring device is set in motion, the ink drop is gradually transformed into a thread that extends over the whole fluid. The latter now appears to be distributed more or less at "random" so that it is seen as some shade of grey. But if the mechanical stirring device is now turned in the opposite direction, the transformation is reversed, and the droplet of dye suddenly appears, reconstituted.

When the dye was distributed in what appeared to be a random way, it nevertheless had *some kind* of order which is different, for example, from that arising from another droplet originally placed in a different position. But this order is *enfolded* or *implicated* in the "grey mass" that is visible in the fluid. Indeed, one could thus "enfold" a whole picture. Different pictures would look indistinguishable and yet have different implicate orders, which differences would be revealed when they were explicated, as the stirring device was turned in a reverse direction.

What happens here is evidently similar in certain crucial ways to what happens with the hologram. To be sure there are differences. Thus, in a fine enough analysis, one could see that the parts of the ink droplet remain in a one-to-one correspondence as they are stirred up and the fluid moves continuously. On the other hand, in the functioning of the hologram there is no such one-to-one correspondence. So in the hologram (as also in experiments in a "quantum" context), there is no way ultimately to reduce the implicate order to a finer and more complex type of explicate order.

All this calls attention to the relevance of a new distinction between implicate and explicate order. Generally speaking, the laws of physics have thus far referred mainly to the explicate order. Indeed, it may be said that the principle function of Cartesian coordinates is just to give a clear and precise description of explicate order. Now, we are proposing that in the formulation of the laws of physics, primary relevance is to be given to the implicate order, while the explicate order is to have a secondary kind of significance (e.g., as happened with Aristotle's notion of movement, after the development of classical physics). Thus, it may be expected that a description in terms of Cartesian coordinates can no longer be given a primary emphasis, and that a new kind of description will indeed have to be developed for discussing the laws of physics.

A4 THE HOLOMOVEMENT AND ITS ASPECTS

To indicate a new kind of description appropriate for giving primary relevance to implicate order, let us consider once again the key feature of the functioning of the hologram, i.e., in each region of space, the order of a whole illuminated structure is "enfolded" and "carried" in the movement of light. Something similar happens with a signal that modulates a radio wave (see Figure 3.9). In all cases, the content or meaning that is "enfolded" and "carried" is primarily an order and a measure, permitting the development of a structure. With the radio wave, this structure can be that of a verbal communication, a visual image, etc., but with the hologram far more subtle structures can be involved in this way (notably three-dimensional structures, visible from many points of view).

More generally, such order and measure can be "enfolded" and "carried" not only in electromagnetic waves but also in other ways (by electron beams, sound, and in other countless forms of movement). To generalize so as to emphasize undivided wholeness, we shall say that what "carries" an implicate order is the *holomovement*, which is an unbroken and undivided totality. In certain cases, we can abstract particular aspects of the holomovement (e.g., light, electrons, sound, etc.), but more generally, all forms of the holomovement merge and are inseparable. Thus, in its totality, the holomovement is not limited in any specifiable way at all. It is not required to conform to any particular order, or to be bounded by any particular measure. Thus, *the holomovement is undefinable and immeasurable.*

To give primary significance to the undefinable and immeasurable holomovement implies that it has no meaning to talk of a *fundamental* theory, on which *all* of physics could find a *permanent* basis, or to which *all* the phenomena of physics could ultimately be reduced. Rather, each theory will abstract a certain aspect that is *relevant* only in some limited context, which is indicated by some appropriate measure.

In discussing how attention is to be called to such aspects, it is useful to recall that the word "relevant" is a form obtained from the verb "to relevate" which has dropped out of common usage, and which

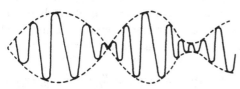

Figure 3.9

means "to lift up" (as in "elevate"). We can thus say in a particular context that may be under consideration, the general modes of description that belong to a given theory serve to *relevate* a certain content, i.e., to lift it into attention so that it stands out "in relief." If this content is pertinent in the context under discussion, it is said to be *relevant*, and otherwise, *irrelevant*.

To illustrate what it means to relevate certain aspects of the implicate order in the holomovement, it is useful to consider once again the example of the mechanical device for stirring a viscous fluid, as described in the previous section. Suppose that we first put in a droplet of dye and turn the stirring mechanism *n* times. We could then place another droplet of dye nearby and stir once again through *n* turns. We could repeat this process indefinitely, with a long series of droplets, arranged more or less along a line, as shown in Figure 3.10.

● ● ● ● ● ● ● ● ● ● ● ● ● ● ● ● ●

Figure 3.10

Suppose, then, that after thus "enfolding" a large number of droplets, we turn the stirring device in a reverse direction, but so rapidly that the individual droplets are not resolved in perception. Then we will see what appears to be a "solid" object (e.g. a particle) moving continuously through space. This form of a moving object appears in immediate perception primarily because the eye is not sensitive to concentrations of dye lower than a certain minimum, so that one does not directly see the "whole movement" of the dye. Rather, such perception *relevates a certain aspect.* That is to say, it makes this aspect stand out "in relief" while the rest of the fluid is seen only as a "grey background" within which the related "object" seems to be moving.

Of course, such an aspect has little interest *in itself*, i.e. apart from its *broader meaning.* Thus, in the present example, one possible meaning is that there *actually is* an autonomous object moving through the fluid. This would signify, of course, that the whole order of movement is to be regarded as similar to that in the immediately perceived aspect. In some contexts, such a meaning is pertinent and adequate (e.g., if we are dealing in the ordinary level of experience with a rock flying through the air). However, in the present context, a very different meaning is indicated, and this can be communicated only through a very different kind of description.

Such a description has to start by *conceptually* relevating certain broader orders of movement, going beyond any that are similar to those relevated in immediate perception. In doing this, one always begins with the holomovement, and then one abstracts special aspects which involve a totality broad enough for a proper description in the context under discussion. In the present example, this totality should include the whole movement of the fluid and the dye as determined by the mechanical stirring device, and the movement of the light, which enables us visually to perceive what is happening, along with the movement of the eye and nervous system, which determines the distinctions that can be perceived in the movement of light.

It may then be said that the content relevated in immediate perception (i.e., the "moving object") is a kind of *intersection* between two orders. One of these is the order of movement that brings about the possibility of a direct perceptual contact (in this case, that of the light and the response of the nervous system to this light), and the other is an order of movement that determines the detailed content that is perceived (in this case, the order of movement of the dye in the fluid). Such a description in terms of intersection of orders is evidently very generally applicable.[33]

It has already been seen that, in general, the movement of light is to be described in terms of "the enfolding and carrying" of implicate orders that are relevant to a whole structure, in which analysis into separate and autonomous parts is not applicable (though, of course, in certain limited contexts, a description in terms of explicate orders will be adequate). In the present example, however, it is also appropriate to describe the movement of *the dye* in similar terms. That is to say, in the movement, certain implicate orders (in the distribution of dye) become explicate, while explicate orders become implicate.

To specify this movement in more detail, it is useful here to introduce a new *measure*, i.e., an "implication parameter," denoted by T. In the fluid, this would be the number of turns needed to bring a given droplet of dye into explicate form. The total structure of dye present at any moment can then be regarded as a ordered series of substructures, each corresponding to a single droplet N with its implication parameter T_N.

Evidently, we have here a new notion of structure, for we no longer build structures solely as ordered and measured arrangements on which we join separate things, all of which are explicate together. Rather, we can now consider structures in which aspects of different degrees of implication (as measured by T) can be arranged in a certain order.

Such aspects can be quite complex. For example, we could implicate a "whole picture" by turning the stirring device n times. We could then implicate a slightly different picture, and so on indefinitely. If the stirring device were turned rapidly in the reverse direction, we could see a "three-dimensional scene" apparently consisting of a "whole system" of objects in continuous movement and interaction.

In this movement, the "picture" present at any given moment would consist only of aspects that can be explicated together (i.e., aspects corresponding to a certain value of the implication parameter T). As events happening at the same time are said to be *synchronous*, so aspects that can be explicated together can be called *synordinate*, while those that cannot be explicated together may then be called *asynordinate*. Evidently, the new notions of structure under discussion here involve *asynordinate* aspects, whereas previous notions involve only *synordinate* aspects.

It has to be emphasized here that the order of implication, as measured by the parameter T, has no necessary relationship to the order of time (as measured by *another* parameter, t). These two parameters are only related in a *contingent* manner (in this case by the rate of turning of the stirring device). It is the T parameter that is directly relevant to the description of the implicate structure, and not the t parameter.

When a structure is *asynordinate* (that is, constituted of aspects with different degrees of implication), then evidently the time order is not in general the primary one that is pertinent for the expression of law. Rather, as one can see by considering the previous examples, the *whole implicate order* is present at any moment, in such a way that the entire structure growing out of this implicate order can be described without giving any primary role to time. The law of the structure will then just be a law relating aspects with various degrees of implication. Such a law will, of course, not be deterministic *in time*. But, as has been indicated elsewhere,[34] determinism in time is not the only form of ratio or reason; and as long as we can find ratio or reason in the orders that are primarily relevant, this is all that is needed for law.

One can see in the "quantum context" a significant similarity to the orders of movement that have been described in terms of the simple examples discussed above. Thus, as shown in Figure 3.11, "elementary particles" are generally observed by means of tracks that they are supposed to make in detecting devices (photographic emulsions, bubble chambers, etc.). Such a track is evidently to be regarded as no more than an *aspect* appearing in immediate perception (as was done with the moving sequence of droplets of dye indicated in Figure 3.10). To describe it as the track of a "particle" is then to assume in addition that the primarily relevant order of movement is similar to that in the immediately perceived aspect.

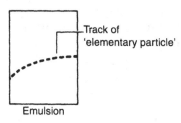

Track of
'elementary particle'

Emulsion

Figure 3.11

However, the whole discussion of the new order implicit in the quantum theory shows that such a description cannot coherently be maintained. For example, the need to describe movement discontinuously in terms of "quantum jumps" implies that the notion of a well-defined orbit of a particle that connects the visible marks constituting the track cannot have any meaning. In any case, the wave-particle properties of matter show that the overall movement depends on the total experimental arrangement in a way that is not consistent with the idea of autonomous motion of localized particles; and, of course, the discussion of the Heisenberg microscope experiment indicates the relevance of a new order of undivided wholeness in which it has no meaning to talk about an observed object as if it were separate from the entire experimental situation in which observation takes place. So the use of the descriptive term "particle" in this "quantum" context is very misleading.

Evidently, we have here to deal with something that is similar in certain important ways to the example of stirring a dye into a viscous fluid. In both cases, there appears in immediate perception an explicate order that cannot consistently be regarded as autonomous. In the example of the dye, the explicate order is determined as an intersection of the implicate order of "the whole movement" of the fluid and an implicate order of distinctions of density of dye that are relevated in sense perception. In the "quantum" context, there similarly will be an intersection of an implicate order of some "whole movement" corresponding to what we have called, for example, "the electron," and another implicate order of distinctions that are relevated (and recorded) by our instruments. Thus, the word "electron" should be regarded as no more than a name by which we call attention to a certain aspect of the holomovement, an aspect that can be discussed only by taking into account the entire experimental situation and that cannot be specified in terms of localized objects moving autonomously through space. And, of course, every kind of "particle" which in current physics is said to be a basic constituent of matter will have to be discussed in the same sort of terms (so that such "particles" are no longer considered as autonomous and separately existent). Thus, we come to a new general physical description in which "everything implicates everything" in an order of undivided wholeness.

A5 LAW IN THE HOLOMOVEMENT

We have seen that in the "quantum" context, the order in every immediately perceptible aspect of the world is to be regarded as coming out of a more comprehensive implicate order, in which all aspects ultimately merge in the undefinable and immeasurable holomovement. How, then, are we to understand the fact that descriptions involving the analysis of the world into autonomous components do actually work, at least in certain contexts (e.g., those in which classical physics is valid)?

To answer the question, we first note that the word "autonomy" is based on two Greek words: "auto," meaning "self," and "nomos" meaning "law." So, to be autonomous is to be *self-ruling.*

Evidently, nothing is "a law unto itself." At most, something may behave with a *relative and limited degree* of autonomy, under certain conditions and in certain degrees of approximation. Indeed, at the very least, each relatively autonomous thing (e.g., a particle) is limited by other such relatively autonomous things. Such a limitation is currently described in terms of *interaction.* However, we shall introduce here the word "heteronomy" to call attention to a law in which many relatively autonomous things are related in this way, i.e., externally and more or less mechanically.

Now, what is characteristic of heteronomy is the applicability of *analytic descriptions.* (The root of the word "analysis" is the Greek "lysis" meaning "to dissolve" or "to loosen." Since the prefix "ana" means "above," it may be said that "to analyse" is to "loosen from above," i.e., to obtain a broad view as if from a great height in terms of components that are regarded as autonomous and separately evident though in mutual interaction.)

As has been seen, however, in sufficiently broad contexts such analytic descriptions cease to be adequate. What is then called for is *holonomy,* i.e., the law of the whole. Holonomy does not totally deny the relevance of analysis in the sense discussed above. Indeed, "the law of the whole" will generally include the possibility of describing the "loosening" of aspects from each other, so that they will be relatively autonomous in limited contexts (as well as the possibility of describing the interactions of these aspects in a system of heteronomy). However, any form of relative autonomy (and heteronomy) is ultimately limited by holonomy, so that in a broad enough context such forms are seen to be merely aspects, relevated in the holomovement, rather than disjoint and separately existent things in interaction.

Scientific investigations have generally tended to begin by relevating apparently autonomous aspects of the totality. The study of the laws of these aspects has generally been emphasized at first, but as a rule this kind of study has led gradually to an awareness that such aspects are related to others originally thought to have no significant bearing on the subject of primary interest.

From time to time, a wide range of aspects has been comprehended within a "new whole." But of course the general tendency until now has been to fix on this "new whole" as a finally valid general order that is henceforth to be adapted (in the manner discussed in Section A1) to fit any further facts that may be observed or discovered.

It is implied here, however, that even such a "new whole" will itself be revealed as an aspect in yet another new whole. Thus, holonomy is not to be regarded as a fixed and final goal of scientific research, but rather as a movement in which "new wholes" are continually emerging. And of course this further implies that the total law of the undefinable and immeasurable holomovement could never be known or specified or put into words. Rather, such a law has necessarily to be regarded as *implicit*.

Notes

1   See *Re-Vision*, vol. 3, no. 4, 1978, for a treatment of this subject in a different way. (Published at 20 Longfellow Road, Cambridge, Mass. 02148, USA.)

2   See Bohm, D., *Wholeness and the Implicate Order*, Routledge, London (1980), Ch. 5.

3   See Bohm, D., *Causality and Chance in Modern Physics*, Routledge and Kegan Paul, London (1957), Ch. 2, for a further discussion of this point.

4   For a more detailed discussion of this point, see, for example, Bohm, D. and Hiley, B., *Foundations of Physics*, vol. 5, 1975, p. 93.

5   For a detailed discussion on this experiment see Bohm, D., *Quantum Theory*, Prentice-Hall, Englewood Cliffs, New Jersey (1951), Ch. 22.

6   See Bohm, *Causality and Chance in Modern Physics*, Ch. 2, for a discussion of this feature of "indeterminate mechanism."

7   See Appendix to this chapter.

8   See Appendix to this chapter for further discussion of this device.

9   See Appendix to this chapter.

10   See Appendix to this chapter.

11   See Bohm and Hiley, *Foundations of Physics*, vol. 5, 1975, p. 93, and Bohm, *Quantum Theory*, for a more detailed treatment of this feature of the quantum theory.

12   Mathematically one derives all the properties of the system from a $3N$-dimensional wave function (where N is the number of particles) which cannot be represented in three-dimensional space alone. Physically one actually finds the non-local, non-causal relationship of distant elements described above, which correspond very well with what is implied by the mathematical equations.

13   Notably those in which the "wave function" of the combined system can be factored approximately into two separate three-dimensional wave functions (as shown in Bohm and Hiley, *Foundations of Physics*).

14   This is just an example of the combination of wavelike and particle-like properties of matter described in Section 2.

15   This sort of calculation is suggested in Bohm, *Causality and Chance in Modern Physics*, p. 163.

16   In Section 8 we shall see that time, as well as space, may be enfolded in this way.

17   Compare with the idea of sub-system, system, and super-system, suggested in Bohm and Hiley, *Foundations of Physics*.

18   This notion has already been suggested in a preliminary way in Bohm, *Wholeness and the Implicate Order*, Ch. 3.

19   See Pribram, K., *Languages of the Brain*, G. Globus, *et al.* (eds), (1971); *Consciousness and the Brain*, Plenum, New York (1976).

20   Bohm, *Wholeness and the Implicate Order*, Ch. 6.

21   E.g., as shown in Section 3, a linearly ordered array of droplets may be enfolded together in such a way that this order is still subtly held in the whole set of ensembles of ink particles.

22   As shown in Bohm, *Wholeness and the Implicate Order*, Appendix to Ch. 6, on the implicate order the basic algorithm is an *algebra* rather than the calculus.

23   For a more detailed discussion, see Ch. 2 of the present volume, originally in Bohm, D., *The Special Theory of Relativity*, Benjamin, New York (1965), Appendix.

24   See *ibid.*

25   This illusion is essentially the one discussed in Bohm, *Wholeness and the Implicate Order*, Chs 1 and 2, in which the whole of existence is seen as constituted of basically static fragments.

26   In some ways this idea of an "unconscious" background is similar to that of Freud. However, in Freud's point of view the unconscious has a fairly definite and limited kind of content and is thus not comparable to the immensity of the background that we are proposing. Perhaps Freud's "oceanic feeling" would be somewhat closer to the latter than would be his notion of the unconscious.

27   This corresponds to the quantum theoretical requirement that electrons may go from one state in space to another without passing through intermediate states.

28   See Bohm, *Wholeness and the Implicate Order*, Ch. 5.

29   See *ibid.*

30   For a very clear presentation of this view, see Kuhn, T., *The Structure of Scientific Revolutions*, University of Chicago Press, Chicago (1955).

31   Piaget, J., *The Origin of Intelligence in the Child*, Routledge and Kegan Paul, London (1956).

32   See Bohm, *Wholeness and the Implicate Order*, Ch. 5.

33   See Bohm, D., Hiley, B., and Stuart, A., *Progressive Theoretical Physics*, vol. 3. 1970, p. 171, where this description of a perceived content considered as the intersection of two orders is treated in a different context.

34   See Bohm, *Wholeness and the Implicate Order*, Ch. 5.

4 THE SUPER-IMPLICATE ORDER (1986)

The following conversation with David Bohm was conducted by Renée Weber, Ph.D., Professor of Philosophy at Rutgers University, currently Affiliate Professor of Philosophy at University of Washington. Here Bohm refines his original model of the implicate order, addressing how the unfoldment of an implicate order results in manifest order and structure. Why is the multidimensional medium of space (the vacuum, or plenum) capable of unfolding the forms we experience in the sensual, three-dimensional world? Why does the plenum of space correspond more to the analogies of ink enfolded into glycerine, or to light enfolded in a hologram, than to the diffuse and random order of eggs enfolded into cake batter, as discussed in Chapter 3?

It is the activity of a super-implicate *order, says Bohm, that accounts for this ordered manifestation. Bohm derives the super-implicate order by applying his causal interpretation to quantum field theory (a concise account of this application is provided in Chapter 6). This super-implicate order infuses the implicate order of space with active information, which generates various levels of organization, structure, and meaning. To illustrate the relationship of active information to these levels of organization, Bohm introduces the principle of* soma-significance, *whereby meaning and form are transmitted throughout a hierarchical continuum of matter and consciousness.*

The final portion of the conversation is concerned with the nature of light. Bohm suggests an intimate relation, even an identity, between light and the "ocean" of energy that emanates from multi-dimensional space, or the vacuum. The movement and activity of light can be understood as a first

Extract from Chapter 2 of R. Weber, *Dialogues with Scientists and Sages: The Search for Unity*, Routledge and Kegan Paul, London (1986).

phase in explicate manifestation, foundationally necessary for the subsequent emergence of all material structures. Indeed, Bohm proposes that all matter is a form of "frozen" light, resulting from the oscillation of intersecting light rays.

By way of explaining the concept of a super-implicate order, Bohm gives a brief sketch of Louis de Broglie's model of the electron, first put forward in 1923. This model is particularly significant as it is the basis for what eventually becomes the causal-ontological interpretation, Bohm and Hiley's quantum formalism for the implicate order.

Weber   You've recently proposed a super-implicate order. How does it differ from the implicate order?

Bohm   If you apply this model, the enfoldment is now seen on two levels: first, an enfolded order of the vacuum with ripples on it that unfold; and second, a super-information field of the whole universe, a super-implicate order which organizes the first level into various structures and is capable of tremendous development of structure. The point about the super-implicate order is that if we take the holographic theory, though we have an implicate order, nothing organizes it. It is what's called "linear," and it just passes through itself and diffuses around; special devices can unfold it but it does not have an intrinsic capacity to unfold an order. The super-implicate order, which is the so-called higher field (the implicate order would be a wave function), would be a function of the wave function, a higher order, a super-wave function. The super-implicate order makes the implicate order non-linear and organizes it into relatively stable forms with complex structures.

Weber   Is there a super super-implicate order?

Bohm   There might be an implicate order even beyond that one. I'd like to propose that we are making a series of abstractions and any level of thought must cut off somewhere. Even if we put more in, there is still more left out. It's inherent in thought that it is not going to grasp the actual totality. But the holistic part of thought would be thought which does not make a break, thought which is unbroken.

Weber   It's a continuum of ordering principles.

Bohm   That's right. Even when we say that we have made a break, we realize that it really shades off into the unknown. That's essential for this quality of wholeness.

Weber   It has no endpoint and no origin?

Bohm   We produce an arbitrary distinction, merely for the sake of thought.

Weber   But the reason for postulating a super-implicate order is that you want an organizing and active principle.

Bohm   I don't need to postulate it. As soon as you make this model of de Broglie and extend it to the quantum mechanical field, you've got it.

Weber   In physics?

Bohm   That is exactly what is implied by quantum mechanical field theory. When seen in this way and through this model, this theory is exactly what I have described.

Weber   Would conventional physicists accept that?

Bohm   They *have* accepted it, but they say, "What is the use? It does not produce anything different from what we've already done. We only care about the empirical results." The other is philosophy and poetry as far as they are concerned.

Weber   They are only attuned to the explicate and not to the enfolded source of it.

Bohm   Yes, of course, that's the whole purpose of the operation. They say, "The essential thing of truth is to produce theoretical, mathematical ideas which actually predict the results of experiments, and in that act you are grasping truth."

Weber   How it got there is irrelevant?

Bohm   They say that's not interesting, and they dismiss it.

Weber   One of your reservations about the direction of contemporary physics is that it does not tie its findings enough to the philosophical meaning.

Bohm   The imaginative side, the intuitive side.

Weber   Making models?

Bohm   Not necessarily new models, but new forms of imagination. I regard the implicate order as a new form of imagination. They would say, "It's no use if it doesn't produce an empirical pay-off. We will look at that as soon as it is producing an empirical pay-off – we'll look at anything." That's one of the errors of science, which is just part of the error of our society, that the empirical pay-off is the main point of the operation, that it is truth. They feel that the empirical agreement is what is meant by truth, provided that you have a logical mathematical argument behind it.

Weber   You are saying that an imaginative model like the implicate order may, in fact, get us closer to the truth of things, even though at the moment we may not be sure what the empirical pay-off is.

Bohm   Yes. In fact, I think any new idea must involve the free play of the mind without thinking too much about the empirical pay-off.

Weber Einstein got to the idea of relativity because as a child he imagined what the universe looked like riding on a beam of light.

Bohm That's right. It took him ten years even to work out the theory of relativity.

Weber So you are endorsing that as one possible avenue to discovering physics. When I asked a colleague in physics for his reaction to the implicate order, he said: it is an interesting idea, but does Bohm have any shred of evidence for it?

Bohm I would say, what shred of evidence is there for the present interpretation as opposed to the one I propose? I have heard that argument before and I say it is fortuitous that this interpretation of quantum mechanics and this way of doing it have developed. For example, de Broglie proposed quite a different approach which was squelched by the leading physicists of the time. Had that been adopted and had people got used to that, then people would have asked the same question, "What is the point of this present approach?" It is nothing different from de Broglie.

Weber You are saying that the accepted model of quantum physics is accepted because of familiarity.

Bohm Yes, and that it was chosen for reasons that are fortuitous as far as science itself is concerned.

Weber The implicate order, as you said elsewhere, is highly compatible with the equations, more so perhaps.

Bohm Yes. Even de Broglie's idea extended in the way I did it would have been a more imaginative way of looking at the thing and it would have been easier to reach. Had people adopted it, they would have regarded this current way [of interpreting physics] as terribly obscure.

Weber Is it too technical to give a brief picture of de Broglie's theory?

Bohm It was the idea that basically an electron is a particle (I'll simplify it very much) and that it has a field around it, a new kind of quantum mechanical field which in some ways is similar to old kinds of fields, in some ways different. The key difference was that its activity did not depend on its intensity. That's like saying that it did not act by mechanical pressure on the particle, but it acted from the information content which carried information about the whole experimental arrangement. So the meaning of an experimental result and the form of the experimental conditions were no longer separable, they were a whole, as even Bohr said. This was immediately obvious in de Broglie's interpretation, whereas it's a deep, impenetrable mystery in Bohr's language.

Weber   Did Bohr not accept it?

Bohm   Bohr did. Bohr had the insight to see this, and this is the basis of his work on interpretation, that the form of the experimental conditions and the meaning of the results are a whole, not further analyzable. He has a very complex way of putting it which very few physicists understand. In fact, one of the points is that the only consistent interpretation of that kind is Bohr's, and the number who understand it is very small. Most physicists are just using it, taking it for granted that Bohr has done it right.

Weber   When you say "he," you have been referring to Bohr, not to de Broglie.

Bohm   Yes, de Broglie, even before Bohr, had proposed another interpretation. What happened was that scientists ignored his picture and just took the mathematical formula.

Weber   If the community of physicists had taken the de Broglie model, would that have moved them closer to an awareness of the unity of things?

Bohm   Yes, because nowadays no physicist understands this at all except by very complicated mathematical arguments which are so distant from his intuition that he regards it as significant only in connection with his work, but not connected with anything else. It's so complicated that very few physicists hear about it and each time new textbooks are written, more and more of it is lost so that by now textbooks don't refer to it all; they just present quantum mechanics as a set of formulae you've got to learn how to use. Because of the lack of imaginative understanding, this result was very likely to come about.

Weber   You are assigning a creative and constructive role to imagination, whereas earlier you cautioned against its abuse.

Bohm   Coleridge has proposed two kinds of imagination, primary and secondary. The primary imagination is the direct expression of the creative intent within, what we may call the display in the mind. The imagination is an unfoldment of some deeper operation of the mind which is displayed as if coming from the senses, and you can grasp it as if looking at it directly as a whole.

Weber   It reveals.

Bohm   Yes. Reveal and display have much the same meaning here. But the secondary imagination arises when you keep on repeating an image from the primary display and it becomes automatic.

Weber   It becomes self-referring. It no longer reveals but becomes a fantasy.

Bohm That's exactly what Coleridge called it. He called it "fancy," which is the same as fantasy.

Weber Imagination, then, in the creative scientific sense, is our attempt to verbalize deep insights about nature.

Bohm Or make a picture.

Weber So you are arguing for imaginative models that would be multi-leveled, mutually supportive, and, most importantly, show their interconnectedness. That is not being done in physics?

Bohm Well, they simply ignore it and say that it's out of date. It doesn't produce an empirical pay-off.

Weber This model would produce only understanding!

Bohm Yes. But they say, "What does it mean to understand unless you can predict something empirically?"

Weber So they have equated understanding with empirical prediction and control and you are diverging from that. You are saying to understand means to grasp it clearly and to see it connected to everything else. Is that right?

Bohm That's right. Comprehension is the word. To comprehend, to hold it all together.

Weber Is talk of a super-implicate order something new, or was it already implied in the implicate order?

Bohm In talking of a super-implicate order, I am not making further assumptions beyond what is implied in physics today. Once we extend this model of de Broglie to the quantum mechanical field rather than just to the particle, that picture immediately is the super-implicate order. So this is not just speculation, it is the picture which is implied by present quantum mechanics if you look at it imaginatively.

Weber Including the claim that there is an ordering principle?

Bohm That's right. Mathematically it is called non-linear equations.

Weber When you say we should look at it imaginatively, let's be clear: you are not proposing that we introduce vague fantasy.

Bohm No.

Weber You mean imaginative interpretations and models for the mathematical equations, creative imagination.

Bohm Imagination which directly displays the meaning of the mathematics, the mathematics which is being used by all the leading physicists now in whatever they are doing.

Weber You are drawing out the consequences of quantum mechanical mathematics.

Bohm Yes. Further, I am saying that by not doing this you fail to see the full meaning of that mathematics and are able to restrict it to making empirical predictions.

Weber This super-implicate order is not the end. It can go as far as thought can take it, a super super-implicate order, and so on?

Bohm That's right. That was one point I wanted to make. The second point is that if we remain with the holographic model, this essentially sticks to the implicate order and leaves out the super-implicate order. In other words, it's a tremendous simplification of quantum mechanics to make the holographic model; that is good enough in the classical sense where you use the holograph. But as a model for organizing the implicate order through the informational field – the quantum information potential – it leaves out what is very interesting, namely that this implicate order now actively organizes itself. This is crucial to understanding thought and the mind.

Weber So it's the self-organizing universe, and it makes clear that consciousness can't be divorced from matter because it resides within it in some way.

Bohm Yes, that's right. The relationship of the super-implicate order to the implicate order is similar to the relationship of consciousness to matter as we know it. There is a kind of analogy.

Weber The super-implicate order would be the conscious aspect and the implicate order would be the material aspect?

Bohm The neuro-physiological aspect, which is still enfolded relative to what we ordinarily see.

Weber So these pairs occur on many different levels.

Bohm Yes, in fact there is a principle I once thought of, I called it "soma-significance," instead of "psychosomatic." The word psychosomatic emphasizes two entities, mind and soma (or body), but I want to emphasize two sides of *one process*. Any process can be treated either as somatic or as significant. A very elementary case is the printed paper: it's somatic in that it's just printed ink, and it also has significance. I say all along the line any part of the body or the body processes is somatic, it's the nerves moving chemically and physically; and in addition it has a meaning which is active. The essential point about intelligence is the activity of significance, right? In computers, we have begun to imitate that to some extent. I am trying to say that all of nature is organized according to the activity of significance. This, however, can be conceived somatically in a more subtle form of matter which, in turn, is organized by a still more subtle form of significance. So in that way every level is both somatic and significant.

Weber   That is very much like Spinoza. Would you extend this all the way into the heart of matter to the atom and the sub-atomic particles?

Bohm   Yes, because what we call the atom is organized by the super or the quantum field of information, which gives it its significance.

Weber   Is the significance something that *we* impute to an otherwise neutral domain?

Bohm   No, the essential point is that if we merely imputed significance to it, it wouldn't be *active*. Do we impute significance to the activity of the computer? There is information-content stored in a program. That significance is active because it determines the activity of the computer and all sorts of activities that flow out of it.

Weber   One might question that analogy because the significance of the information-content in the computer is what we've put in and then we read it back out. But here we're dealing with nature as a whole.

Bohm   But quantum mechanics is indicating that that order of activity or that order of relationship is what is actually present, and that we are merely imitating nature.

Weber   We're imitating nature in our cognitive processes and creative acts?

Bohm   Yes, and extending it in some ways.

Weber   But are these processes truly similar?

Bohm   If we take the basic structure as similar, we can say that the super or information-potential is related to the implicate order of matter as the subtle aspects of consciousness are related to the material movements of hormones and electrical currents in the nerves.

Weber   It's almost the old hermetic principle, "as above, so below." We're the mirror image of larger processes.

Bohm   Yes, you could say that essentially it's the principle of what we call similarities and differences: the differences within one field are paralleled by similar differences in the other field. The quantum field contains information about the whole environment and about the whole past, which regulates the present activity of the electron in much the same way that information about the whole past and our whole environment regulates our own activity as human beings, through consciousness.

Weber   Is that like saying that nature thinks?

Bohm   Not exactly, but nature has active information as we have; at least at the level of *unconscious* thought it's similar.

Weber   And part of that active information is derived from its own past?

Bohm Yes, or from elsewhere.

Weber Is the super-implicate order a euphemism for God?

Bohm I don't know what the meaning of the question is since the super-implicate order is in turn part of a still greater implicate order. It's not a euphemism for God because it's limited.

Weber Then let's shift the question to the ultimate super super-implicate order.

Bohm But we can't grasp that in thought. We're not saying that any of this is another word for God. I would put it another way: people had insight in the past about a form of intelligence that had organized the universe and they personalized it and called it God. A similar insight can prevail today without personalizing it and without calling it a personal God.

Weber Still, it's a kind of super-intelligence and you've said elsewhere that that is benevolent and compassionate, not neutral.

Bohm Well, we can propose that.

Weber To you the notion of creativity entails building larger wholes.

Bohm We cannot in the end do anything but destroy if we have a fragmentary approach.

Weber How do you order these various levels?

Bohm To say that the higher level simply transcends the lower level altogether. It's immensely greater and has an entirely different set of relationships out of which the lower level is obtained as a very small part, in an abstraction.

Weber It has wholeness, more power, more energy, more insight?

Bohm Yes, and it contains the lower level in some sense.

Weber And not vice-versa?

Bohm The lower level will be the unfoldment of the higher level.

Weber In space and time.

Bohm Yes.

Weber So in a sense they contain each other but in another sense the higher one contains the whole and the lower one is more linear.

Bohm Yes. The higher one is called non-linear, mathematically, and the lower one is linear. That means of course that the linear organization of time and thought characteristic of the ordinary level will not necessarily be characteristic of the higher level. Therefore what is beyond time may have an order of its own, not the same as the simple linear order of time.

Weber In that case we have this all upside down: we foist our limited version of space and time on these higher levels and think that's the only ordering possible.

Bohm This higher order is not basically the order of space and time, but the order of space and time unfolds from it and folds back into it in the way we've been discussing.

Weber Our kind of space and time is one among perhaps infinitely many orderings possible in the universe yet we think that it's the only way and in fact the necessary condition for understanding. Kant almost said that and could not conceive of an alternative arrangement. The super-implicate order proposes an alternative to current narrow western epistemology.

Bohm Yes, it says that the information content out of which the implicate order unfolds is not determined in an order of space and time as we know it, but it contains that order within it.

Weber We might say this is the creative play (*lila* in Sanskrit) of the universe, where from its deep recesses it evolves different combinations.

Bohm Yes, and through that it's unfolding and developing and flowering (if you want to use that word as Krishnamurti does) and therefore evolution is fundamental. This involves both space and time. Time itself is an order of manifestation, you see. We are going to say that it is possible to have an implicate order with regard to time as well as to space, to say that in any given period of time, the whole of time may be enfolded. It's implied in the implicate order when you carry it through: the holomovement is the reality and what is going on in the full depth of that one moment of time contains information about all of it.

Weber You've said that the moment is timeless.

Bohm Yes, the moment is atemporal, the connection of moments is not in time but in the implicate order.

Weber Which you said is timeless.

Bohm Yes. So let me propose that also for consciousness; let me propose that consciousness is basically in the implicate order as all matter is and therefore it's not that consciousness is one thing and matter is another. Rather consciousness is a material process and consciousness is itself in the implicate order, as is all matter, and consciousness manifests in some explicate order, as does matter.

Weber The difference between what we call matter and consciousness would be the state of density or subtlety?

Bohm The state of subtlety. Consciousness is possibly a more subtle form of matter and movement, a more subtle aspect of the holomovement. In the nonmanifest order there is no separation in space and time. In ordinary matter this is so and it's even more so for this subtle matter which is consciousness. Therefore if we are separate it is because we are sticking largely to the manifest world as the basic reality where the whole point is to have separate units, relatively separate anyway, but interacting. In nonmanifest reality it's all interpenetrating, interconnected, one. So we say deep down the consciousness of mankind is one. This is a virtual certainty because even in the vacuum matter is one; and if we don't see this it's because we are blinding ourselves to it.

Weber Your implicate order philosophy treats space very differently from the usual way.

Bohm Yes. There are two views of space. One view is to maintain that the skin is the boundary of ourselves, that there's the space without and the space within. The space within is the separate self, obviously, and the space without is the space which separates the separate selves, right? And therefore to overcome the separation you must have a process of moving through that space, which takes time. Is that clear?

Weber That's how human beings have always thought of it.

Bohm That's right. But if we looked at it as a holomovement with this vast reserve of energy and empty space where matter itself is that small wave on empty space, then we should really say that the space as a whole is the ground of existence and we are in it. So the space doesn't separate us, it unites us. Therefore it's like saying that there are two separate points and a certain dotted line connecting them, which shows how we think they are related; or to say there is a *real line* and that *the points are abstractions* from that.

Weber Demarking the boundaries of the line.

Bohm Yes.

Weber You turn the whole thing around.

Bohm The line is the reality and the points are abstractions. In that sense we say that there are no separate people, you see, but that that idea is an abstraction which comes by taking certain features of the whole as abstracted and self-existent.

Weber So space is more fundamental and more real than the objects in it. Applying your theory to time, we would have to say that the *interval between* the moments is the real.

Bohm It could be considered to be that. But see, if we take the view that the space is what is real, then I think that we have to say that the *measure* of space is not what is real. The measure of space is what matter provides. So space goes beyond the measure of space.

It's the same with time. If we want to say that the interval is real, then the measure of time cannot be taken as fundamental. Therefore we are already outside of what we ordinarily would call time. But rather, if we have silence and "emptiness," it does not have the measure either of space or of time. Now in that silence there may appear something which is a little ripple which has that measure. But if we thought that the little ripple was all that there is and that the space between was nothing, of no significance, then we would have the usual view of fragmentation.

Weber   Taking what we call events as the points.

Bohm   Yes. Events are like the points.

Weber   But if you don't allow time to be measured by events, the line then . . .

Bohm   Then it's flowing movement, right?

Weber   Well, then in a way it's silence.

Bohm   It's just flow. If you look at nature and say, there's no event in nature, really, then it's just flowing. It's the mind that abstracts and puts an event in there.

Weber   But doesn't it follow that the flow or the silence cannot be broken up by any distinguishing characteristics?

Bohm   Yes. Except that's what thought puts in, the distinguishing characteristics. The distinguishing characteristics have their place in a certain limited domain of the explicate order and of the manifest.

Weber   Still, I think to some people all this is going to seem very strange. First of all, it challenges everything we've known or been taught. Second, it appears to be counter-intuitive, certainly to those who have been trained in modern science. Third, it may appear frightening or threatening. So let's spell it out. You're saying that the events are always distinguishable, they have characteristics, they are what we call happenings, and they're the ones we've seized upon as what transpires in the world, as the world's business, so to speak. Those – you're saying – are secondary, derivative, and less important than the absence of all that. And the absence of all that is emptiness, silence.

Bohm   It would be the holomovement, you see, the flowing movement. But it goes beyond that. We could say that even at this level of thought there is a way of looking at it in which emptiness is the plenum, right? I'm saying that what we call real things are actually tiny little ripples which have their place, but they have been usurping the whole, the place of the whole.

Weber By "emptiness" we don't mean a substantive emptiness like an "empty" box. We're talking about a plenum.

Bohm It's emptiness which is a plenum. Yes.

Weber An emptiness which is a plenum. What does that mean?

Bohm This is a well-known idea even in physics. If you take a crystal which is at absolute zero it does not scatter electrons. They go through it as if it were empty and as soon as you raise the temperature and [produce] inhomogeneities, they scatter. Now, if you used those electrons to observe the crystal (e.g. by focusing them with an electron lens to make an image), all you would see would be these little inhomogeneities and you would say they are what exists and the crystal is what does not exist. Right? I think this is a familiar idea, namely to say that what we see immediately is all there is or all that counts and that our ideas must simply correlate what we see immediately.

Weber From that, of course, it would follow that history and all multiplicity of objects and events are just ripples.

Bohm Yes. They're merely ripples and their meaning depends on understanding what underlies the ripples.

Weber And you're saying what underlies the ripples is the true and profound source.

Bohm Yes.

Weber And you've also said that man can connect with that emptiness.

[Bohm then makes the point that the human mind as it ordinarily functions cannot understand "emptiness," which lies beyond three-dimensional consciousness.]

Bohm It's not enough to say we are going to consider a consciousness which is more than this limited three-dimensional kind. The trouble is that we are still using the three-dimensional consciousness to guide us in that.

Weber To talk about it?

Bohm To talk about it. The point of meditation would be to stop doing that.

Weber What you have been saying sounds like mysticism – that we are grounded in something infinite. How does it differ from what the great mystics have said?

Bohm I don't know that there's necessarily any difference. What is mysticism? The word "mysticism" is based on the word "mystery," implying something hidden. Perhaps the ordinary mode of consciousness which elaborately obscures its mode of functioning from itself and engages in self-deception might more appropriately be called "mysticism." Or we could call it "obscurantism," and say there's an opposite mode that we could term "transparentism" (although I don't really like the suffix "ism" in any form).

Weber A transparence with respect to the whole.

Bohm Yes, as opposed to obscuring the whole.

Weber Kierkegaard had a wonderful phrase for that. He said true religion is "to be grounded transparently in the power that constitutes one."

Bohm Yes, that's exactly what it would mean.

Weber Speaking of mysticism, there is one important idea that I would like to discuss and understand and that is the idea of light. That is especially important to me because you are a physicist. Light has been used as *the* privileged metaphor in the language of mysticism and experimental religions, going back to the Greeks and the east. In all these, light is the symbol of our union with the divine. They talk about a light without shadow, an all-suffusing light, and it comes up as the central metaphor in near-death experiences. Do you have any hypothesis as to why light has been singled out as the privileged metaphor?

Bohm If you want to relate it to modern physics (light and more generally anything moving at the speed of light, which is called the null-velocity, meaning null distance), the connection might be as follows. As an object approaches the speed of light, according to relativity, its internal space and time change so that the clocks slow down relative to other speeds, and the distance is shortened. You would find that the two ends of the light ray would have no time between them and no distance, so they would represent immediate contact. (This was pointed out by G. N. Lewis, a physical chemist, in the 1920s.) You could also say that from the point of view of present field theory, the fundamental fields are those of very high energy in which mass can be neglected, which would be essentially moving at the speed of light. Mass is a phenomenon of connecting light rays which go back and forth, sort of freezing them into a pattern.

So matter, as it were, is condensed or frozen light. Light is not merely electromagnetic waves but in a sense other kinds of waves that go at that speed. Therefore all matter is a condensation of light into patterns moving back and forth at average speeds which are less than the speed of light. Even Einstein had some hint of that idea. You could say that when we come to light we are coming to the fundamental activity in which existence has its ground, or at least coming close to it.

Weber Why is speed the determinant?

Bohm Well, let's turn it around. If you look at Piaget and young children, movement is primary in perception. They see movement first and its unfoldment as time, and only perceive distance later. They have a tendency to say that if something went further it must have been going faster. They only learn later how to do it right. They are carrying some deeper perception into the ordinary explicate level, where it is inappropriate. In the deeper perception, movement is the primary reality in perception. The thing that is not moving is the result of the cancellation of movement. We say that there is no speed at all at light. To call it speed is merely using ordinary language. In itself, when it is self-referential, there's no time, no space, no speed.

Weber What is it?

Bohm It's just a primary conception. As you move faster and faster according to relativity your time rates slow down and the distance gets smaller, so as you approach very high speeds your own internal time and distance become less, and therefore if you were at the speed of light you could reach from one end of the universe to the other without changing your age at all.

Weber Isn't that saying that it's approaching a timeless state?

Bohm That's right. We're saying that existentially speaking or logically speaking, time originates out of the timeless.

Weber This is primary and time is derivative of it, cutting it down, freezing it, arresting it.

Bohm Yes, arresting it to a certain extent, not absolutely, but to a large extent.

Weber When mystics use the visualization of light they don't use it only as a metaphor, to them it seems to be a reality. Have they tapped into matter and energy at a level where time is absent?

Bohm It may well be. That's one way of looking at it. As I've suggested the mind has two-dimensional and three-dimensional modes of operation. It may be able to operate directly in the depths of the implicate order where this [timeless state] is the primary actuality. Then we could see the ordinary actuality as a secondary structure that emerges as an overtone on the primary structure. It's again the business of what is emphasized and what is secondary – the two kinds of music. The ordinary consciousness is one kind of music, and the other kind of consciousness is the other kind of music.

Weber The ordinary music can become noise, cacophony and disharmony. The music from the deep-structure cannot.

Bohm The ordinary music is harmonious only in a limited area.

Weber It's harmonious when it properly expresses this deeper harmony – Pythagoras' harmony of the spheres – but that other music is never disharmonious.

Bohm We might propose that. Let's say there are two poles where we can operate. We could operate from that extreme pole all the way to the ordinary pole, but we have accepted the distorted view that we can only operate at one pole, or very near it.

Weber We have already closed the gate when we needn't do so.

Bohm Yes.

Weber For the mystics there is always light. The primary clear light in the *Tibetan Book of the Dead* is the first thing the dying person is aware of. If he doesn't move towards it or away from it or feel awe or fear or manipulate it in any way as if it were outside himself, then he merges with it and is liberated, *enlightened*. Christ says: "I am the light," and so on. I've always asked myself, why light? You're saying that from the point of view of a physicist, it has to do with the absence of speed and the closeness of contact.

Bohm Light is what enfolds all the universe as well. For example, if you're looking at this room, the whole room is enfolded into the light which enters the pupil of your eye and unfolds into the image and into your brain. Light in its generalized sense (not just ordinary light) is the means by which the entire universe unfolds into itself.

Weber Is this a metaphor for you or an actual state?

Bohm It's an actuality. At least as far as physics is concerned.

Weber Light is energy, of course.

Bohm It's energy and it's also information – content, form and structure. It's the potential of everything.

Weber Physicists are not satisfied that they have understood light up to now because of the particle-wave paradox, right?

Bohm Yes, I think that to understand light we'll have to understand the structure underlying time and space more deeply. You can see that these issues are related in the sense that light transcends the present structure of time and space and we will never understand it properly in that present structure.

Weber How would implicate order philosophy handle light?

Bohm It could handle it more naturally, mathematically speaking, because it doesn't commit itself to the idea of separate points in space; but it may say that the underlying reality is something which is not localized, and light is also something which is not localized. One view says that light moves from one place to another through a series of positions, and the other view says it doesn't do that at all. Rather, light exists, it just simply *is*.

Weber It is at all points?

Bohm Points are defined by the intersections of different rays of light. That's the way we actually do it in perception. We infer a point from the fact that many light rays are coming from it, say a star or any point. In this view, points would be understood as the intersection of many light rays. The light is fundamental, the null ray. That's a technical term that shows the recognition of this fact in ordinary physics.

Weber It's where every particle of matter is in contact without the slightest gap between them.

Bohm Yes, it's possible to have that contact without a gap.

Weber So light is one continuous, unbroken, undivided whole?

Bohm You would have to look at it that way, yes; especially if you consider the quantum theory of it which says the action in it is undivided as well. What G. N. Lewis had in mind was to explain the quantum in that way. It was very mysterious to say that light is a wave which spreads continuously through space and yet that a single quantum of energy goes from one point to another. How could that happen? G. N. Lewis said this wave was some sort of an abstraction, and he said what actually happened in each ray was that there was an immediate contact from the source to the absorber. One understood the quantum in that way, that there was no spreading out of energy.

Weber It therefore takes no time, no transmission, no distance. There isn't any, is what you're saying.

Bohm That is the view I'm proposing. The ordinary view is another map of it. You can take many maps of the world; one of them is Mercator's projection, which is quite good near the equator but it says near the poles that the space is infinite. So maps can have the wrong structure. We can say that the ordinary space-time is a map which holds fairly well for ordinary speeds, but when you get to the speed of light it's as wrong in structure as Mercator's projection is at the pole.

Weber We say light is clarity, light illumines, light is energy, some mystics have said light is love, compassion, understanding, light can make whole or heal. If light is the background of everything, what would be its relationship to the foreground?

Bohm Light is this background which is all one but its information-content has the capacity for immense diversity. Light can carry information about the entire universe. The other point is that light, by interactions of different rays (as field theory in physics is investigating today), can produce particles and all the diverse structures of matter.

Weber You've stressed *information* and that has to do with *knowing* the universe.

Bohm A kind of knowing.

Weber The other aspect would have to do with its *being*. Maybe there's an undifferentiated realm of light and when it radiates itself as *being*, as particles, those might be its "shadows" or finite expression.

Bohm They are expressions but they are ripples on this vast ocean of light. This ocean of energy could be thought of as an ocean of light. But the information-content may be such as to predispose certain light rays to combine so that they move back and forth rather than moving straight ahead, and thus forming particles.

Weber Are those ripples, those particles, the silhouette of that light?

Bohm Implicit in the information-content of the light – you could say that. About silhouette, I don't know. Something would have to throw the shadow. What is going to do that? The light, as it were, determines itself to make particles.

Weber In order to do what?

Bohm I don't know. But we're proposing that this allows for a richer universe.

Weber To be consistent one might have to say that the light transforms aspects of itself into particles in order that those particles will reveal the light.

Bohm That's right, they will reveal the potential of the light in a new way. So the light and the particles together make a higher unity. Most physicists subtract off this infinity and say it doesn't count and what's left over are the particles, and they claim that these are all that count.

Weber But you're claiming that's incorrect and shallow because it's subtracting off the very thing in which these particles have their roots and being.

Bohm That's why I say present physics doesn't understand it, it's merely a system of computing and getting empirical results.

Weber We've given light a cosmological, a physical, and a metaphysical interpretation. What about the psychological and spiritual interpretation? Why do people who tap into that realm of light feel a rare peace and happiness even though light is considered neutral and value-free by physics?

Bohm The mind may have a structure similar to the universe, and in the underlying movement we call empty space there is actually a tremendous energy, a movement. The particular forms which appear in the mind may be analogous to the particles, and getting to the ground of the mind might be felt as light. The essential point is not that it's light but rather this free, penetrating movement of the whole.

Weber Somehow the energy it triggers in the experiencer is an integrated whole and that perhaps is what accounts for this profound sense of peace.

Bohm Yes. The analogy has often been made that even though the ocean is all stirred up and quite stormy on the surface, if you get to the bottom it is peaceful.

5 SOMA-SIGNIFICANCE AND THE ACTIVITY OF MEANING (1985)

In this chapter Bohm outlines the nature of soma-significant and signa-somatic activity. Here "soma" refers to the body, and by extension to any material structure or process, while "significance" refers to mind or meaning. These terms are meant to suggest complementary aspects of one indivisible process, rather than two qualitatively distinct domains. With this model Bohm furthers his argument that there is no essential difference between reciprocal processes in the objective world (Chapter 1) and reciprocal processes in the perception and cognition of human beings (Chapter 2), suggesting that active meaning is enfolded and unfolded throughout the whole of existence. Soma-significant and signa-somatic processes are thus seen as aspects of the dynamics of implicate and explicate orders.

Two examples indicate the scope of soma-significant processes. In the human realm, a somatic form, e.g., a traffic light, presents a significance to a driver. This significance is developed throughout increasingly subtle somatic structures – the visual system, the nervous system, and the brain of the driver (a soma-significant flow). These levels of somatic subtlety have corresponding meanings, and a cumulative significance for the driver – in this case, "stop." This significance becomes active, and the process then moves in an "outward" direction. The brain produces an intention to stop, which works its way out through increasingly "manifest" levels of soma – the nervous system and the musculature – resulting in stopping the car (a signa-somatic flow).

At the quantum level of matter, says Bohm, soma-significant processes

Extract from Chapter 3 of D. Bohm, *Unfolding Meaning: A Weekend of Dialogue*, ed. Donald Factor, Routledge and Kegan Paul, London ([1985], 1987).

also occur. In Bohm's version of quantum theory (Chapters 4 and 6), a "pilot wave" reads the somatic form of the environment and conveys this form to its accompanying particle (a soma-significant flow). The subtler somatic structure of the particle – which Bohm suggests is at least as complex as a radio receiver – develops a cumulative "orientational" significance from this information. When this significance is fully developed it also becomes outwardly active, giving rise to specific movements on the part of the particle (a signa-somatic flow).

For Bohm, the pilot wave model is not merely analogous to human soma-significance. He sees each of these examples as abstracted nodes in a continuum that includes the quantum level, the human domain, and the large-scale evolution of the cosmos. As a magnet divided into multiple parts will always exhibit positive and negative poles in each part, so also will any aspect of reality we select for examination show somatic and significant aspects. It is not possible to find an independent somatic phenomenon, or an independent significant phenomenon. Anywhere we make a cut in the fabric of reality, we will find this mutual interpenetration of soma and significance.

The implication of this perspective is central to Bohm's overall world view: meaning is not an exclusively human activity. We typically think of meaning as subjective attribution: "My wife means a lot to me." "That was a meaningful conversation." "What is the meaning of the universe?" The human mind is thus tacitly understood to be the exclusive source and repository for meaning. But Bohm's model turns this view on its head, seeing human meaning as a particular case of active, soma-significant meaning in the universe at large. From this perspective, we may have meaning for the universe. Further still, the universe may be meaningful to itself, with or without the presence of humans. A field of daisies, a cluster of galaxies, or the inner structure of an electron are understood as being actively engaged in soma-significant and signa-somatic processes. For Bohm the operative question and subsequent inquiry then becomes: How are our human meanings related to those of the universe as a whole?

I want to introduce a new notion of meaning which I call soma-significance, and also a notion of the relationship between the physical and the mental. This relationship has been widely considered under the name psycho-somatic. "Psyche" comes from a Greek word meaning mind or soul and "soma" means the body. If we generalize soma to mean the physical, the term psycho-somatic suggests two different kinds of entities, each existent in itself – but both in mutual interaction. In my view such a notion introduces a split, a fragmentation, between the physical and the mental that doesn't properly correspond to the actual state of affairs. Instead I want to suggest the introduction of a new term which I call "soma-significance." This emphasizes the unity of the two, and more generally, with meaning in all its implications and aspects. That is, "significance" goes on to "meaning," which is a more general word.

In this approach meaning is clearly being given a key role in the whole of existence. However any attempt at this point to define the meaning of meaning would evidently presuppose that we already know at least something of what meaning is, even if perhaps only non-verbally or subliminally. That is, when we talk we know what meaning is; we could not talk if we didn't. So I won't attempt to begin with an explicit definition of meaning, but rather, as it were, unfold the meaning of meaning as we go along, taking for granted that everyone has some intuitive sense of what meaning is.

The notion of soma-significance implies that soma (or the physical) and its significance (which is mental) are not in any sense separately existent, but rather that they are two aspects of one over-all reality. By an aspect we mean a view or a way of looking. That is to say, it is a form in which the whole of reality appears – it displays or unfolds – either in our perception or in our thinking. Clearly each aspect reflects and implies the other, so that the other shows in it. We describe these aspects using different words; nevertheless we imply that they are revealing the unknown whole of reality, as it were, from two different sides.

You can obtain a good illustration in physics for the unbroken wholeness underlying the aspects that are, nevertheless, distinguished, by contrasting the relationship of electrical poles or charges and magnetic poles. Electrical charges are regarded as separately existent and connected by a field; but magnetic poles are not that way. They are really one unbroken magnetic field. That is, if you take a magnet with a north and south pole, you may consider that there is a field going around the magnet from the north to the south pole. You may have seen this illustrated with iron filings (see Figure 5.1).

Now the point is that if you take this magnet and break it, you get two magnets, each of which has a north and a south pole (see Figure 5.2).

So you can see that there is actually no separate magnetic pole. In fact you may consider that even when it is not broken, every part is a superposition of north and south poles, and you may then understand the relationship as flowing.

With the aid of this concept of opposing pairs of magnetic poles, we

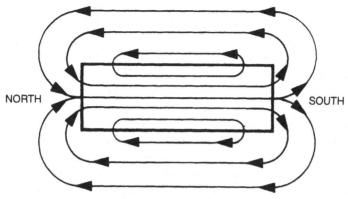

Figure 5.1

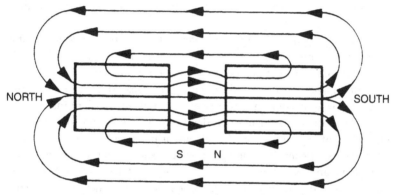

Figure 5.2

can contribute in a significant way to expressing and understanding the basic relationships in the overall magnetic field. I propose to look at soma-significance in a similar way. That is to say, I regard them as two aspects distinguished only in thought, which will help us to express and understand relationships in the "field" of reality as a whole.

To bring out how soma and significance are related, I might note that each particular kind of significance is based on some somatic order, arrangement, connection and organization of distinguishable elements – that is to say, structure. For example, the printed marks on this piece of paper carry a meaning which is apprehended by a reader. In a television set the movement of electrical signals communicated to an electron beam carries meaning to a viewer. Modern scientific studies indicate that such meanings are carried somatically by further physical, chemical and electrical processes into the brain and the rest of the nervous system where they are apprehended by ever higher intellectual and emotional levels of meaning.

As this takes place, these meanings, along with their somatic concomitants, become ever more *subtle*. The word "subtle" is derived from the Latin *sub-texere*, meaning woven from underneath, finely woven. The meaning is: rarified, delicate, highly refined, elusive, indefinable, intangible. The subtle may be contrasted with the manifest, which means literally, what can be held in the hand. My proposal then is that reality has two further key aspects, the subtle and the manifest, which are closely related to soma and significance. As I pointed out, each somatic form, such as a printed page, has a significance. This is clearly more subtle than the form itself. But in turn such a significance can be held in yet another somatic form – electrical, chemical and other activity in the brain and the rest of the nervous system – that is more subtle than the original form that gave rise to it.

This distinction of subtle and manifest is only relative, since what is manifest on one level may be subtle on another. Thus the relatively subtle somatic form of thought may have a meaning that can be grasped in still higher and more subtle somatic processes. And this may lead on further to a grasp of a vast totality of meanings in a flash of insight.

This sort of action may be described as the apprehension of the meaning of meanings, which may in principle go on to indefinitely deep and subtle levels of significance. For example in physics, reflection on the meanings of a wide range of experimental facts and theoretical problems and paradoxes eventually led Einstein to new insights concerning the meaning of space, time, and matter, which are at the foundation of the theory of relativity. Meanings are thus seen to be capable of being organized into ever more subtle and comprehensive over-all structures that imply, contain and enfold each other in ways that are capable of indefinite extension – that is, one meaning enfolds another, and so on. So you can see that the meaning of the implicate order must be closely related. The implicate order is a way of illustrating the way meaning is organized.

In terms of the notion of soma-significance there is no point to the attempt to reduce one level of subtlety in any structure completely to another. For example, if you meet a certain content on one level and then on another, the relationship between these levels is the essential content of yet another level. So it is clear that no ultimate reduction is possible. As the level under consideration is changed, the particular content of what is somatic (or manifest) and what is significant (or subtle) has always therefore to be changing. Nonetheless it is clear that it is necessary for both roles to be present in each concrete instance of experience. You see, it is like the magnetic poles. Wherever you cut the magnet you have a North and a South pole, and wherever you make a cut in experience and abstract something, and say, "This is the experience" (which is a bigger context) you have soma-significance. It would be impossible to have all the content on the side of soma or on that of significance.

I have emphasized so far, the significance of soma – that is, that each somatic configuration has a meaning – and that it is such meaning that is grasped at more subtle levels of soma. I call this the soma-significant relation, which is one side of the over-all process. I would now call attention to the inverse, signa-somatic relation. This is the other side of the same process in which every meaning at a given level is seen *actively* to affect the soma at a more manifest level. Consider for example, a shadow seen in a dark night. Now if it happens, because of the person's past experience, that this means an assailant, the adrenalin will flow, the heart will beat faster, the blood pressure will rise and he will be ready to fight, to run or to freeze. However if it means only a shadow, the response of the soma is very different. So quite generally the total physical response of the human being is profoundly affected by what physical forms mean to him. A change of meaning can totally change your response. This meaning will vary according to all sorts of things, such as your ability or background, conditioning, and so on.

This is different from psycho-somatic, because with psycho-somatic you say that mind affects matter as if they were two different substances – mind substance affects material substance. Now I am saying there is only one flow, and a change of meaning is a change in that flow. Therefore any change of meaning is a change of soma, and any change of soma is a change of meaning. So we don't have this distinction.

As a given meaning is carried into the somatic side, you can see that it continues to develop the original significance. If something means danger, then not only adrenalin, but a whole range of chemical substances will travel through the blood, and according to modern scientific discoveries, these act like "messengers" (carriers of meaning) from the brain to various parts of the body. That is, these chemicals instruct various parts of the body to act in certain ways. In addition there are electrical "signals" – they are not really signals – carried by the nerves, which function in a similar way. And this is a further unfoldment of the original significance into forms that are suitable for "instructing" the body to carry out the implications of what is meant.

From each level of somatic unfoldment of meaning there is then a further movement leading to activity on a yet more manifestly somatic level, until the action finally emerges as a physical movement of the body that affects the environment. So one can say that there is a two-way movement of energy in which each level of significance acts on the next more manifestly somatic level, and so on, while perception carries the meaning of the action back in the other direction (see Figure 5.3).

As in cutting a magnet it does not mean that these lines represent distinct levels; they are merely abstracted in our mind.

I want to emphasize here that nothing exists in this process except as a two-way movement, a flow of energy, in which meaning is carried inward and outward between the aspects of soma and significance, as well as between levels that are relatively subtle and those that are relatively manifest. It is this over-all structure of meaning (a part of which I've drawn in this diagram) that is grasped in every experience. We can see this by following the process in the two opposing directions. For example, as light strikes the retina of the eye, carrying meaning in the form of an image, the meaning is transformed into a chemical form by the rods and the cones. They in turn are transformed into electro-chemical movements in the nerves, and so on into the brain at higher and higher levels. Then in the other direction, higher meanings are carried electrically and chemically into the structures of reflexes and thus onward toward ever more manifestly somatic levels.

I have been discussing what you might call the normal soma-significant and signa-somatic process. Usually psycho-somatic processes are discussed in terms of some disorder, and you can see here that you

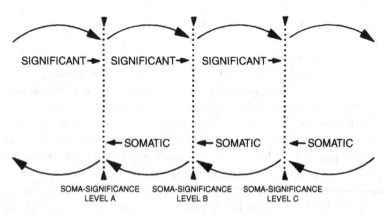

Figure 5.3

can also get signa-somatic disorder. For example, normally the heart will beat faster when something means danger. One realizes that that is the signa-somatic response to the meaning of danger. But it could also mean that something is wrong with the heart, in which case the danger will be indicated by the rate of the beating of the heart. In that case every time the heart beats faster it fills the person with more of the meaning of danger and causes the heart to beat faster still. So you get a runaway loop, and that could be an important component of neurotic disorders – the normal process gets caught in a loop that goes too far.

You can see that ultimately the soma-significant and signa-somatic process extends even into the environment. Meaning thus can be conveyed from one person to another and back through sound waves, through gestures carried by light, through books and newspapers, through telephone, radio, television and so on, linking up the whole society in one vast web of soma-significant and signa-somatic activity. You can say society is this thing; this activity is what makes society. Without it there would be no society. Therefore communication is this activity.

Similarly even simple physical action may be said to communicate motion and form to inanimate objects. Most of the material environment in which we live – houses, cities, factories, farms, highways, and so on – can be described as the somatic result of the meaning that material objects have had for human beings over the ages. Going on from there, even relationships with nature and with the cosmos flow out of what they mean to us. These meanings fundamentally affect our actions toward nature, and thus indirectly, the action of nature back on us is affected. Indeed as far as we know it and are aware of it and can act on it, the whole of nature, including our civilization which has evolved from nature and is still a part of nature, is one movement that is both soma-significant and signa-somatic.

Some of the simpler kinds of soma-significant and signa-somatic activity are just reflexes that are built into the nervous system, or instincts that express the accumulated experience of the species. But these go on to ever finer and more variable responses. Even the behaviour of creatures as simple as bees can be seen to be so organized in a very subtle way by a kind of meaning, in this case through a dance indicating the direction and distance of sources of nectar. Though they might not be conscious of it, there is a meaning going on. With the higher animals this operation of meaning is more evident, and in man it is possible to develop conscious awareness, and meaning is then most central and vital.

In these higher levels this soma-significant and signa-somatic activity shows up most directly. In fact the word "meaning" indicates not only the significance of something to us, but also our intention toward it. Thus "I mean to do something" means "I intend to do it." This double meaning of the word "meaning" is not just an accident of our language, but rather it implicitly contains an important insight into the structure of meaning.

To bring this out I would first note that an intention generally arises out of a previous perception of the meaning or significance of a certain total situation. This gives all of the relevant possibilities and implies reasons for choosing which of these is better. Ultimately this choice is determined by the totality of significance at that moment. The source of this activity includes not only perception and abstract or explicit knowledge, but what Polanyi calls *tacit knowledge* – that is, knowledge containing concrete skills and reactions that are not specifiable in language, such as riding a bicycle.

Ultimately it is the whole significance that gives rise to intention, which we sense as a feeling of being ready to act in a certain way. For example, if we see a situation meaning "the door is open," we can form the intention to walk through it, but if it means "the door is closed," we don't. But even the intention not to act is still an intention. The whole significance helps to determine it. The important point is that the intention is a kind of implicate order; the intention unfolds from the whole meaning. It doesn't just come out of nothing. Therefore a person cannot form intentions except on the basis of what the situation means to him, and if he misses the mark on what it means, he will form the wrong intentions.

Of course, most of the meaning is implicit. Indeed, whatever we say or do, we cannot possibly describe in detail more than a very small part of the total significance that we sense in any given moment. Moreover, when such significance gives rise to an intention, it too will be almost entirely implicit, at least at the beginning. For example, as I said, I have an intention to speak at this moment, and it is implicit what I am going to say; I don't know what I'm going to say exactly, but it comes out. Now the words are not chosen one by one, but rather are unfolded in some way.

Meaning and intention are therefore inseparably related as two sides or aspects of one activity. This is the same as we discussed with soma and significance, and the subtle and the manifest. We are saying that there is one whole of activity abstracted at a certain point conceptually – we make a cut in it – and we say it always has two sides. One of the two sides is meaning, and the other would be intention. But they don't exist separately.

Intentions are commonly thought to be conscious and deliberate. But you really have very little ability to choose your intentions. Deeper intentions generally arise out of the total significance in ways of which one is not aware, and over which one has little or no control. So you usually discover your intentions by observing your actions. These in fact often contain what are felt to be unintended consequences leading one to say, "I didn't mean to do that. I missed the mark." In action, what is actually implicit in what one means is thus more fully revealed. That is the importance of giving attention.

To learn the full meaning of our intentions in this way can very often be costly and destructive. What we can do instead is to display the intention along with its expected consequences through imagination, and in other ways. The word "display" means "unfold," but for the sake of revealing something other than the display itself. As such a display is perceived one can then find out whether or not one still intends going on with the original intention. If not, the intention is modified, and the modification is in turn displayed in a similar way. Thus to a certain extent, by means of trying it out in the imagination, you can avoid having to carry it out in reality and having to suffer the consequences, although that is rather limited.

So intention constantly changes in the act of perception of the fuller meaning. Even perception is included within this over-all activity. What one perceives is not the thing in itself, which is unknown or unknowable, but however deep or shallow one's perceptions, all one perceives is what it means at that moment, and then intention and action develop in accordance with this meaning.

The point is that as you act according to your intention, and as the perception comes in, there can arise an indefinite extension of inward signa-somatic and soma-significant activity. That is, you go to more and more subtle levels and the thing is, as it were, looking at itself at different levels ever more deeply.

Such activity is roughly what is meant by the mental side of experience. When something is going on that is not strongly coupled with the outer physical manifestation of some soma-significant and signa-somatic activity in which it is looking at itself, then we call that the mental side of experience. Now this is only a side. Once again I want to repeat that there is no separation between the mental and the physical. When it gets to the other side where it is primarily concerned with actions it just gets more physical.

Now we can look at this in terms of the implicate or enfolded order, for all these levels of meaning enfold each other and may have a significant bearing on each other. Within this context, meaning is a constantly extending and actualizing structure – it is never complete and fixed. At the limits of what has at any moment been comprehended there are always unclarities, unsatisfactory features, failures of intention to fit what is actually displayed or what is actually done. And the yet deeper intention is to be aware of these discrepancies and to allow the whole structure to change if necessary. This will lead to a movement in which there is the constant unfoldment of still more comprehensive meanings.

But of course each new meaning has some limited domain in which the actions flowing out of it may be expected to fit what actually happens. These limits may in principle be extended indefinitely through further perceptions of new meanings. But no matter how far this process goes there will still be limits of some kind, which will be indicated by the discovery of yet further discrepancies and disharmonies between our intentions, as based on these meanings, and the actual consequences that flow out of these intentions. At any stage the perception of new meanings may dissolve these discrepancies, but there will still continue to be a limit, so that the resulting knowledge is still incomplete.

What this implies is that meaning is capable of an indefinite extension to ever greater levels of subtlety as well as of comprehensiveness – in which there is a movement from the explicate toward the implicate. This can only take place however when new meanings are being perceived freshly from moment to moment. But if significance comes solely from memory and not from fresh perceptions it will be limited to some finite depth of subtlety and inwardness.

Memory, being some kind of recording, necessarily has a certain stable quality which cannot transform its structure in any fundamental way, and has only a limited capacity to adapt to new situations – for example, by forming new combinations of known principles, either through chance or through rules already established in memory. Memory is thus necessarily bounded both in scope and in the subtlety of its content. Any structure arising solely out of memory will be finite, and will be able to deal with some finite limited domain; but of course, to go beyond this, a fresh perception of new meanings is needed. And in fact, when you have a fresh perception you may also see new meanings of your memories. In other words, memory may cease to be so limited when there is fresh perception. To go on in this way to new meanings that are not arbitrarily limited requires a potentially infinite degree of inwardness and subtlety in our mental processes. And I am suggesting that these processes have access to an, in principle, unlimited depth in the implicate order.

Thus far I have suggested reasons why meaning is capable of infinite extension to ever greater levels of subtlety and refinement. However, it might seem at first sight that in the other direction – of the manifest and the somatic – there is a clear possibility of a limit in the sense that one might arrive at a "bottom level" of reality. This could be, for example, some set of elementary particles out of which everything would be constituted such as quarks, or perhaps yet smaller particles. Or in accordance with currently accepted views of modern physics it might be a fundamental field, or set of fields, that was the "bottom level." What is of crucial importance is that its meaning would be in principle *unambiguous*. In contrast, all higher order forms in this supposedly basic structure of matter are ambiguous – that is, their meaning is incomplete. There is an inherent ambiguity in any concrete meaning. That is to say, how the meanings arise and what they signify depends to a large extent on what a given situation means to us, and this may vary according to our interests and motivations, our background of knowledge, and so on. But if for example, there were a "bottom level" of reality, these meanings would be exactly what they were, and anybody who looked correctly could find them. They would be a reality that was just simply there, independent of what it meant to us.

Of course you also have to keep in mind that all scientific knowledge is limited and provisional so that we cannot be certain that what we think is the "bottom level" is actually so. For example, possibly something other than the present theories will come to reveal a "bottom level." But this uncertainty of knowledge cannot of itself prevent us from believing in the existence of some kind of "bottom level" if we wish to do so. It is not commonly realized however that the quantum theory implies that no such "bottom level" of unambiguous reality is possible.

Now this is a bit difficult to make clear in this short time, but Niels Bohr, one of the founders of modern physics, has made one of the most consistent interpretations of the quantum theory given thus far, and which has been accepted by most physicists (though few probably have studied it deeply enough to appreciate fully the revolutionary implications of what he has done). To understand this point, first we have to say that while the quantum theory contradicts the previously existent classical theory, it does not explain this theory's basic concepts as an approximation or a simplification of itself, but it has to presuppose the classical concepts at the same time that it has to contradict them. The paradox is resolved in Bohr's point of view by saying that the quantum theory introduces no new basic concepts at all. Rather what it does is to require that concepts such as position and momentum, which are in principle unambiguous in classical physics, must become ambiguous in quantum mechanics. But ambiguity is just a lack of well-defined meaning. So Bohr, at least tacitly, brings in the notion of meaning as crucial to the understanding of the content of the theory.

Now this is a radically new step, and he is doing this not just for its own sake, but he is forced to do something like this by the very form of the mathematics which so successfully predict the quantum properties of matter. This mathematics gives only statistical predictions. It not only fails to predict what will happen in a single measurement, it cannot even provide an unambiguous concept or picture of what sort of process is supposed to take place. So for Bohr the concepts are ambiguous, and the meaning of the concepts depends on the whole context of the experimental arrangement. The meaning of the result depends on the large scale behaviour which was supposed to be explained by the particles themselves. So in some sense you do not have a "bottom level" but rather you find that, to a certain extent, the meaning of these particles has the same sort of ambiguity that we find in mental phenomena when we are looking at meaning.

This kind of situation is what is pervasively characteristic of mind and meaning. Indeed the whole field of meaning can be described as subject to a distinction between content and context which is similar to that between soma and significance, and between subtle and manifest. Content and context are two aspects that are inevitably present in any attempt to discuss the meaning of a given situation. According to the dictionary, the content is the essential meaning – for example, the content of a book. But any specifiable content is abstracted from a wider context which is so closely connected with the content that the meaning of the former is not properly defined without the latter. However, the wider context may in turn be treated as a content in a yet broader context, and so on. The significance of any particular level of content is therefore critically dependent on its appropriate context, which may include indefinitely higher and more subtle levels of meaning – such as whether a given form seen in the night means a shadow or an assailant depends on what one has heard about prowlers, what one has had to eat and drink, and so on. So you see, this sort of context-dependence is just what is found in physics with regard to matter, as well as in considerations of mind or meaning.

Now I believe Bohr's interpretation of quantum theory is consistent, and he has produced a very deep insight at this point; but it is still not clear why matter should have this context-dependence. He just says that the quantum theory gives rise to it.

However, in terms of the implicate order, an alternative interpretation is possible in which one can ascribe to phenomena a deeper reality unfolding, which gives rise to them. This reality is not mechanical; rather its basic action and structure are understood through enfoldment and unfoldment. What is important here is that the law of the total implicate order determines certain sub-wholes which may be abstracted from it as having *relative* independence. The crucial point is that the activity of these sub-wholes is context-dependent, so that the larger content can organize the smaller context into one greater whole. The sub-wholes will then cease to be properly abstractable as independent and autonomous. The implicate order makes it possible to discuss the notion of reality in a way that does not require us to bring in the measuring apparatus, which Bohr does. He makes the context very much dependent on the apparatus; but he does so by making nature generally context-dependent. That is to say, the situation of any part of nature is context-dependent in a way that is similar to the way that meaning is dependent on its context – that is, as far as the laws of physics are concerned.

That would suggest that in a natural way one might extend some notion similar to meaning to the whole universe. It is implied that each feature of the universe is not only context-dependent fundamentally, but also that the grosser, manifest features depend on the subtler aspects in a way that is very analogous to soma-significant and signa-somatic activity. So something similar to meaning is to be found even in the somatic or physical side.

Now as I said, this holds for us both mentally and physically. It would suggest that everything, including ourselves, is a generalized kind of meaning. Now I am not thereby attributing consciousness to nature. You see, the meaning of the word "consciousness" is not terribly clear. In fact, without meaning I think that there would be no consciousness. The most essential feature of consciousness is consciousness of meaning. Consciousness is its content; its content is the meaning. Therefore it might be better to focus on meaning rather than consciousness. So I am not attributing consciousness as we know it to nature, but you might say that everything has a kind of mental side, rather like the magnetic poles. In inanimate matter the mental side is very small, but as we go deeper into things the mental side becomes more and more significant.

All of this implies that one can consistently understand the whole of nature in terms of a generalized kind of soma-significant and signa-somatic activity that is essentially independent of man, and that indeed it is more consistent to do this than to suppose that there is an unambiguous "bottom level" at which these considerations have no place. I would say that the crucial difference between this and a machine is that nature is infinite in its potential depths of subtlety and inwardness, while a machine is not. Although to a certain extent, a machine such as a computer has something similar. So it is in principle possible in this view to encompass both the outward universe of matter and the inward universe of mind.

In this approach, the three basic aspects arise:

Soma
Significance
Energy

To repeat, soma-significance means that the soma is significant to the higher or more subtle level. Signa-somatic means that that significance acts somatically toward a more manifest level.

Now I'm going to look at physical action in a similar way – to say that in the unfoldment of matter there is a kind of soma-significance; that the soma may be significant to a deeper level. So let's say that something unfolds and has a significance, and as a result something else unfolds.

In explaining this I should first discuss the work of the well-known psychologist Piaget, who has carefully observed and studied the growth of intelligent perception in infants and in young children. This led him to say that this perception flows out of what is in effect a deep initial intention to act toward the object. You can see the soma-significance coming in here. This action may initially be based partly on a kind of significance that objects have, which is grounded in the whole accumulated instinctive response to the experience of the species, and partly on a kind of significance that is grounded in his own past experience. Whatever its origin may be, Piaget says, what this action does is to incorporate or assimilate its object into a cycle of inward and outward activity. He moves out, he sees it, he acts on it and that changes his perception, and he acts again. His intention is implicitly in at least some conformity with what he expects the object to be, but it might be vague. The action comes back to the extent to which the object fits or doesn't fit his intention. Then this brings about a modified intention with correspondingly modified outward action. This process is continued until a satisfactory fit is obtained between intentions and their consequences, after which it may remain very stable until further discrepancies appear.

Piaget points out, however, that the initial intention need not be directed primarily toward incorporating the object into a cycle of activity in order to produce a desired result such as enjoyment or satisfaction. Instead it might be directed mainly at perception of the object. For example, the child may initiate movements aimed at exploring and observing the object, such as turning it around, bringing it closer to look at it, and so on. From such an intention it is possible for him to begin with all sorts of provisional feelings as to what the object might be, and to allow these to unfold into actions which come back as perceptions of fitting or non-fitting. This leads to a corresponding modification of the detailed content of the intention behind these movements until the outgoing actions and incoming perceptions are in accord. This is a very important development of intentional activity which makes possible an unending movement of learning and discovering what has not been known before. So we want to say that this soma-significant and signa-somatic activity, constantly going back and forth, is what is involved in learning. And we can say that this is going on, not only in regard to outward objects, but inwardly – that is, for example, with regard to thought. And there may be another level which picks up the meaning of the thought and takes an action toward that thought while thinking another thought to see if it is consistent. If it is not, then the intention changes until we get a consistent relationship between the thought which arises from the deeper intention and the thought that was first being looked at. You see, you may have a thought that you want to look at, and there may be a deeper intelligence which is able to grasp the meaning of that thought in a broader context and take an action toward it by, as it were, thinking again and seeing whether the thought which comes out is coherent with the thought with which you started. And if it's not, then you can start to change that action until it is. Or you can change the thought. Change can occur at various levels.

So all of these levels of meaning enfold each other and have a certain bearing on each other. This whole process is always soma-significant and signa-somatic, going to ever deeper levels. When I talk of these processes I don't only mean going outward into the manifest world, but also the deeper mental processes being explored by still more subtle mental processes. So you could say that the mind has available in principle an unlimited depth of subtlety, and learning can take place at all these levels.

Now what is important is not only what to think but how to think. But if we ask how we think, it may be just as difficult to answer as, how do you ride a bicycle? It is at the tacit level of knowing, or at the subtle level, that how to think takes place. You cannot say how to think but you can learn, as I have just been describing, through signa-somatic and soma-significant activity.

To sum up what I've just been saying, a somewhat similar view can be applied within matter in general. So one may think of the whole thing as one process – as an extended idea of meaning and an extended idea of soma. That is, meaning and matter may not have the same sort of consciousness that we have, but there is still a mental pole at every level of matter, and there is some kind of soma-significance. And eventually, if you go to infinite depths of matter, we may reach something very close to what you reach in the depths of mind. So if you consider it, we no longer have this division between mind and matter.

Now we have in this whole process these three aspects: soma and significance and an energy which carries the significance of soma to a subtler level and gives rise to a backward movement in which the significance acts on the soma. Modern physics has already shown that matter and energy are two aspects of one reality. Energy acts within matter, and even further, energy and matter can be converted into each other, as we all know.

From the point of view of the implicate order, energy and matter are imbued with a certain kind of significance which gives form to their over-all activity and to the matter which arises in that activity. The energy of mind and of the material substance of the brain are also imbued with a kind of significance which gives form to their over-all activity. So quite generally, energy enfolds matter and meaning, while matter enfolds energy and meaning.

You can see in Figure 5.4 how the middle term enfolds the other two.

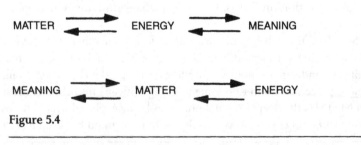

Figure 5.4

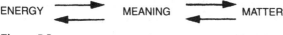

Figure 5.5

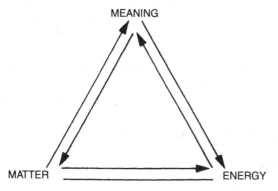

Figure 5.6

But also meaning enfolds both matter and energy. The way we find out about matter and energy is by seeing what they mean (see Figure 5.5).

So each of these basic notions enfolds the other two. It is through this mutual enfoldment that the whole notion obtains unity. So we can put all these relationships together (see Figure 5.6).

However in some sense the enfoldment by meaning seems to be more fundamental than the enfoldment of the other types, because we can discuss the meanings of meaning. In some sense meanings enfold meanings. But we cannot have the matter of matter, or the energy of energy. There seems to be no intrinsic enfoldment relation in matter-energy. Matter enfolds energy, and energy enfolds matter, according to this view, by way of significance. But meaning refers to itself directly, and this is in fact the basis of the possibility of that intelligence which can comprehend the whole, including itself. On the other hand, matter and energy obtain their self-reference only indirectly, firstly through meaning. That is, we can refer matter back to itself by first seeing what it means to us, and then going back. Or we can refer matter to energy, or energy to matter, by seeing what they mean. We refer them to each other reflexively, but only through their meaning.

Generally we have this problem of thought referring to something else, thus creating division and dualism. Even the thought that the universe is one unbroken whole in flowing movement refers to a universe which is one whole unbroken movement, and beside that there is the thought. So we therefore have two nevertheless. What we would like is a view in which the thought itself is part of the reality.

Usually we think of thought in correspondence with some object; the features of the thought correspond to some object. But as soon as you say a thought corresponds to an object, you immediately have, tacitly, a division between the object and the thought. In reality, we are saying that the thought is a part of the soma-significance and cannot be absolutely distinct from the object. Only in certain limited areas is the distinction useful or correct – that is, where the thought has a negligible effect on the object. This is the area of all practical activity, technology, and so on.

The modern mechanistic approach says that this area covers everything: but what I am saying is that it is a small area within a much vaster field. So we are not denying that kind of thought; we are saying it is only valid in a limited area.

The problem of conceiving of a universe that can refer consistently to itself has long been a difficult one that has not been resolved in a really adequate way. But the field of meaning can refer to itself, and of course, it also presupposes the context of the universe to which it also refers. Meaning, though, has nevertheless been regarded as peculiar to our own minds and not as a proper part or aspect of the objective universe. However if there is a generalized kind of meaning intrinsic to the universe, including our own bodies and minds, then the way may be opened to understanding the whole as self-referential through its "meaning for itself" – in other words, by whatever reality is. And the universe as we now conceive it may not be the whole thing.

The aspect of soma cannot be divided from the aspect of significance. Whatever meanings there may be "in our minds," these are, as we have seen, inseparable from the totality of our somatic structures and therefore from what we *are*. So what we are depends crucially on the total set of meanings that operates "within us." Any fundamental change in meaning is a change in being for us. Therefore any transformation of consciousness must be a transformation of meaning. Consciousness is its content – that is its meaning. In a way, we could say that we *are* the totality of our meanings.

If we trace some of these meanings to their origins, we find that most of them have come from society as a whole. Each person takes up his own particular combination of the general mixture that is available in a society. And so at least in this way, every person is different. Yet the underlying basis is characterized mainly by the fundamental similarity over the whole of mankind, while the differences are relatively second-ary. And insofar as man has the capacity to get beyond that, that also is common.

These meanings change as human beings live, work, communicate and interact. These changes are based for the most part on adaptation of existent meanings. But it has also been possible from time to time for new meanings to be perceived and realized – in other words, made real. Perceptions of this kind have generally occurred when someone became aware that certain sets of older meanings no longer made any sense. This may be understood as a vast extension of what happens in the development of intelligence in young children. That is, as they see some-thing about which they are puzzled, they have to see its meaning in a new way.

Now we can say that we are puzzled about the whole of life, and we have to see it with a new meaning. If you look at life as a whole it doesn't seem to make that much sense – the way we live, and so on. The childlike attitude would ask, "Well, what does it mean?" And some as yet incompletely formed notion of a new meaning that removes the contradictions in the older meanings may begin to penetrate a person's intentions. As I explained, the actions unfolding from the intentions would be displayed, for example, in the imagination, and the discrepan-cies between what is displayed and what is intended would lead to a change of intention aimed at decreasing this discrepancy, and so on. In this way a greater clarification of the meaning would occur along with a possibility of realizing it through a change in intention, because it is only when one's purpose or intention changes that a new meaning can be realized. Then, often in a flash that seems to take no time at all, a coherent new whole of meaning is formed, within which the older meanings may be comprehended as having a limited validity within their proper context.

Now if meaning is an intrinsic part of not only our reality but reality in general, then I would say that a perception of a new meaning consti-tutes a creative act. As their implications are unfolded, when people take them up, work with them, and so on, the new meanings that have been created make their corresponding contributions to this reality. And these are not only in the aspect of significance but also in the aspect of soma. That is, the situation changes physically as well as mentally.

Therefore each perception of a new meaning by human beings actually changes the over-all reality in which we live and have our exist-ence – sometimes in a far-reaching way. This implies that this reality is never complete. In the older view, however, meaning and reality were sharply separated. Reality was not supposed to be changed directly by perception of a new meaning. Rather it was thought that to do this was merely to obtain a better "view" of reality that was independent of what it meant to us, and then to do something about it. But once you actually see the new meaning and take hold of your intention, reality has changed. No further act is needed.

Seeing something intellectually or abstractly, though, will not change your intention. You may say that you need an act of will to change it, but I think that when you really see something deeply with great energy, no further act of will is needed. If you really see a new meaning to be true, then your intention will change – unless there is something blocking it, such as your conditioning, or the "program." And if something is blocking it, then the will is not going to help, because you don't know what the block is. Therefore you have to see the meaning of the block. So choice and will are of limited significance – valid in certain areas. But I think something deeper is needed if you are discussing the transformation of mind or consciousness or matter – they really all change together.

You see, the deep change of meaning is a change in the deep material structure of the brain as well, and this unfolds into further changes. Every time you think, the blood distribution all over the brain changes; every emotion changes it. Between thinking and the somatic activity there is also a tremendous connection with the heartbeat and the chemical constitution of the blood, and so on. The new meaning will produce different thought and therefore possibly an entirely different functioning of the brain.

We already know that certain meanings can greatly disturb the brain, but other meanings may organize it in new ways. And when the brain comes to a new state, new ideas become possible. But the new meaning is what organizes the new state. If the brain holds the old meanings, then it cannot change its state. The mental and the physical are one. A change in the mental is a change in the physical, and a change in the physical is a change in the mental. In fact, there has been some discussion of what is called subtle brain damage in animals in which no physical abnormality can be found; but some disturbance of function takes place when the animals are put under stress. So you see, we could say that living as we do, we probably have a great deal of subtle brain damage. In other words, the brain is damaged at a subtle level that might not show up at the cellular level, but deep in the implicate order. Eventually of course, it shows up in the cellular level too. So instead of saying that when we see a new meaning we make a choice and then act, we say that the perception and realization of the new meaning in our intention is already the change.

This point is crucially significant for understanding psychological and social change. For if meaning is something separate from human reality, then any change must be produced by an act of will or choice, guided perhaps by our new perception of meaning. But if meaning itself is a key part of reality, then once society, the individual and their relationships are seen to mean something different from what they did before, a fundamental change has *already* taken place. So social change requires a different, socially accepted meaning, such as in the change from feudalism to the forms that followed it, or from autocracy to democracy, or to communism, and so on. According to the meanings accepted, the entire society went.

These meanings may have been correct or incorrect. But once the meanings become fixed, the whole thing must gradually go wrong. Or to put it differently, what man does is an inevitable signa-somatic consequence of what the whole of his experience, inward and outward, means to him. For example, once the world came to mean a set of disjointed mechanical fragments, one of which was himself, people could not do other than begin to act accordingly and engage in the kind of ceaseless conflict that this meaning implies. The meaning of fragmentation includes conflict and self-centredness – in other words, not creative tension but meaningless conflict.

However if mankind could sustain a perception and realize this perception signifying that the world is an unbroken whole with a multiplicity of meanings, some of which are fitting and harmonious and some of which are not, a very different state of affairs could unfold. For then there could be an unending creative perception of new meanings that encompass the older ones in broader and more harmonious wholes which would unfold in a corresponding transformation of the over-all reality that was thus encompassed.

Here it is worth noting that our civilization has been suffering from what may be called a failure of meaning. Indeed from earliest times people have felt this as a kind of "meaninglessness" of life. Whether this is more prevalent today, I don't know, but people say it is. But in this sense, meaning also signifies value. That is to say, a meaningless life has no value; it is not worth living. But of course it is impossible for anything to be totally free of meaning. For as we have explained earlier, the notion of generalized soma-significance, regarded as valid for the whole of life, implies that each thing *is* its total meaning – which of course must include all of its relevant context. What I intend by "meaningless" therefore is that there *is* a meaning, but that it is inadequate because it is mechanical and constraining and is hence of little value and not creative. A change in this is possible only if new meaning is perceived that is not mechanical. Such a new meaning, sensed to have a high value, will arouse the energy needed to bring a whole new way of life into being. You see, only meaning can arouse energy.

At present people don't seem to have the energy to face this sea of troubles that threatens to overwhelm us, generally speaking. If we take a mechanical meaning, it tends to deaden the energy so that people remain indefinitely as they have been, or at best allows change in limited directions, such as the continuation of the development of technology, and so on. So I am saying that meaning is fundamental to what life actually *is*.

Now you can extend this to the cosmos as a whole. We can say that human meanings make a contribution to the cosmos, but we can also say that the cosmos may be ordered according to a kind of "objective" meaning. New meanings may emerge in this over-all order. That is, we may say that meaning penetrates the cosmos, or even what is beyond the cosmos. For example, there are current theories in physics and cosmology that imply that the universe emerged from the "big bang." In the earliest phase there were no electrons, protons, neutrons, or other basic structures. None of the laws that we know would have had any meaning. Even space and time in their present, well-defined forms would have had no meaning. All of this emerged from a very different state of affairs. The proposal is that, as happens with human beings, this emergence included a creative unfoldment of generalized meaning. Later, with the evolution of new forms of life, fundamentally new steps may have evolved in the creative unfoldment of further meanings. That is, we may say that some evolutionary processes occur which could be traced physically, but we cannot really understand them without looking at some deeper meaning which was responsible for the changes. The present view of the changes is that they were random, with selection of those traits that were suited for survival, but that does not explain the complex, subtle structures that actually occurred.

The question is how our own meanings are related to those of the universe as a whole. We could say that our action toward the whole universe is a result of what it means to us. Now since we are saying that everything acts according to a similar principle, we can say that the rest of the universe acts signa-somatically to us according to what we mean to it.

These meanings do not all fit harmoniously, but if we are perceptive of the disharmony, we may continually be bringing about an increase in harmony. That is to say, there is no final meaning or no final harmony, but a continual movement of creativity – or of destruction. In the long run, only those meanings which allow changes that tend to bring about accord between us and the rest of the universe will be possible. We can say that that is true for the universe as a whole, and that nature is experimenting with all sorts of meanings. Some of them will not be consistent, and they will not survive. So anything that has survived for quite a long time is bound to have a tremendous degree of coherence with the rest of the universe.

We are proposing that this holds for both living beings and for matter in general. We may say then that the harmony is never complete and cannot be so. Even now a further creation of meaning is going on in a process that includes mankind as part of itself. Not merely man's physical development but a constant creation of new meanings that is essential for the unfoldment of society and human nature itself. Even time and space are part of the total meaning and are subject to a continual evolution. As I indicated, at the beginning of the "big bang," time and space did not mean what they now mean. In this evolution, extended meaning as "intention" is the ultimate source of cause and effect, and more generally, of necessity – that which cannot be otherwise.

Rather than to ask what is the meaning of this universe, we would have to say that the universe *is* its meaning. As this changes, the universe changes along with all that is in it. What I mean by "the universe" is "the whole of reality" and what is beyond. And of course, we are referring not just to the meaning of the universe for us, but its meaning "for itself," or the meaning of the whole for itself.

Similarly there is no point in asking the meaning of life, as life too *is* its meaning, which is self-referential and capable of changing, basically, when this meaning changes through a creative perception of a new and more encompassing meaning.

You could also ask another question: What is the meaning of creativity itself? But as with all other fundamental questions we cannot give a final answer, but we have to constantly see afresh. For the present we can say that creativity is not only the fresh perception of new meanings, and the ultimate unfoldment of this perception within the manifest and the somatic, but I would say that it is ultimately the action of the *infinite* in the sphere of the finite – that is, this meaning goes to infinite depths.

What is finite is, of course, limited. These limits may be extended in any number of ways, but however far you go, they are still limited. What is limited in this way is not true creativity. At most it leads to a kind of mechanical rearrangement of the kinds of elements and constituents that are possible within those limits. One may think of anything finite as being suspended in a kind of deeper infinite context or background. Therefore the finite must ultimately be dependent on the infinite. And if it is open to the infinite then creativity can take place within it. So the infinite does not exclude the finite, but enfolds within it and includes and overlaps it. Every finite form is somewhat ambiguous because it depends on its context. This context goes on beyond all limits, and that is why creativity is possible. Things are never exactly what they mean; there is always some ambiguity.

6 THE CAUSAL-ONTOLOGICAL INTERPRETATION AND IMPLICATE ORDERS (1987)

In the following selections, Bohm gives concise descriptions of three aspects of his alternative to the standard interpretation of quantum mechanics, referred to here as the "causal interpretation" (later to become Bohm and Hiley's "ontological interpretation"). He goes on to explain in some detail the manner in which this interpretation is related to the implicate order.

To grasp the significance of the causal interpretation, it is useful to understand two aspects of the standard (Copenhagen) interpretation. The first of these is that particles, such as an electron in a laboratory, have only potential existence until they are observed. Once observed (e.g. with a measuring device), this potentiality "collapses" down into the concrete manifestation of the actual particle. Second, when not manifesting as a particle, the state of potentiality is represented by a mathematical wave form known as the "wave function." This wave function is understood to be mathematical only – there is no "real" wave there behind the numbers.

In the early causal interpretation, Bohm proposes that the particle is an objectively existing entity that does not depend on observation to bring it into existence. Further, the wave phenomenon also is understood to have objective existence – every objective particle has an objective wave which accompanies it. From this perspective there is no collapse from a (mathematical) wave state into a manifest particle. There is always the simultaneous objective existence of wave and particle, with no observer required for their actualization.

Extract from Chapter 2 of D. Bohm, *Science, Order, and Creativity* (with F. David Peat), Routledge ([1987], 2000).

Of central importance in the causal interpretation is the relationship between the particle and the wave. The wave is understood to carry complex but passive information about the form of the environment that surrounds the particle. Through the quantum potential – a feature unique to Bohm and Hiley's interpretation – the information contained in the wave is transmitted to the particle. This information – now active rather than passive – is "processed" by the particle and subsequently directs its movement. This relationship between wave and particle is analogous to radar waves directing a ship on automatic pilot – in both cases, a self-sufficient energy (the particle; the ship) is "informed" and directed by a wave whose form, rather than its intensity, is significant. This process can be understood as yet another variation of the "two-way" reciprocal relations discussed in Chapters 1 and 2, and of the soma-significant relations of Chapter 5.

The above features are characteristic of the wave-particle relation in an isolated, "one particle" system. Extended to a "many particle" system, the causal interpretation provides a framework for considering non-local connections between distant particles when they are in an "entangled" state; in such a state the whole system has primacy over the behavior of its parts. When the interpretation is applied to quantum field theory, further novel features emerge, particularly the super-quantum potential, which has the capacity to inform the sub-structures of the entire universe. Of equal significance, the super-quantum potential (equivalent to the super-implicate order as discussed in Chapter 4) coordinates the movement of quantum waves in a manner that allows for the very creation of "particle-like manifestations" in the first instance.

In this latter version of the causal interpretation there is a clear convergence with the perspective of the implicate order. The particle is no longer seen as a purely objective entity, but as a relatively autonomous and stable unfoldment from the implicate order, governed in its activity and creation by the super-quantum potential. Bohm thus outlines a coherent model in which the mechanically limited analogies of the glycerine and the hologram (Chapter 3) are given a more complete theoretical underpinning, while at the same time accounting for the organizational principles of whole and sub-whole in the universe at large.

There are several reasons for including a discussion of this theory within this chapter. To begin with, it provides a relatively intelligible and intuitively graspable account of how an actual quantum process may take place. Moreover it does not require a conceptual or formal separation between the quantum system and its surrounding "classical" apparatus. In other words there is no fundamental "incommensurability" between classical and quantum concepts and, therefore, a greater unity between the formal and informal languages used in its exposition. In addition, this theory has never before been presented in a non-technical way and it may be of interest to the reader to learn of a quite novel approach to the quantum theory.

Although the interpretation is termed causal, this should not be taken as implying a form of complete determinism. Indeed it will be shown that this interpretation opens the door for the creative operation of underlying, and yet subtler, levels of reality. The theory begins, in its initial form, by supposing the electron, or any other elementary particle, to be a certain kind of particle which follows a causally determined trajectory. (In the later, second quantized form of the theory, this direct particle picture is abandoned.) Unlike the familiar particles of Newtonian physics, the electron is never separated from a certain quantum field which fundamentally affects it, and exhibits certain novel features. This quantum field satisfies Schrödinger's equation, just as the electromagnetic field satisfies Maxwell's equation. It, too, is therefore causally determined.

Within Newtonian physics, a classical particle moves according to Newton's laws of motion and the forces that act on the particle are derived from a classical potential V. The basic proposal of the causal interpretation is that, in addition to this classical potential, there also acts a new potential, called the quantum potential Q. Indeed, all the new features of the quantum world are contained within the special features of this quantum potential. The essential difference between classical and quantum behavior, therefore, is the operation of this quantum potential. Indeed, the classical limit of behavior is precisely that for which the effects of Q become negligible.

(For the mathematically minded, the quantum potential is given by:

$$Q = \frac{-h^2}{2m} \frac{\nabla^2 |\psi|^2}{|\psi|^2}$$

where ψ is the quantum field or "wave function" derived from Schrödinger's equation, h is Planck's constant, and m is the mass of the electron or other particle. Clearly the quantum potential is determined by the quantum wave field, or wave function. But what is mathematically significant in the above equation is that this wave function is found in both the numerator and the denominator. The curious effects that spring from this relationship will be pointed out in the following paragraphs.)

At first sight, it may appear that to consider the electron as some kind of particle, causally effected by a quantum field, is to return to older, classical ideas which have clearly proved inadequate for understanding the quantum world. However, as the theory develops, this electron turns out not to be a simple, structureless particle but a highly complex entity that is effected by the quantum potential in an extremely subtle way. Indeed the quantum potential is responsible for some novel and highly striking features which imply qualitative new properties of matter that are not contained within the conventional quantum theory.

The fact that ψ is contained both in the numerator and the denominator for Q means that Q is unchanged when ψ is multiplied by an arbitrary constant. In other words, the quantum potential Q is independent of the strength, or intensity, of the quantum field but depends only on its form. This is a particularly surprising result. In the Newtonian world of pushes and pulls on, for example, a floating object, any effect is always more or less proportional to the strength or size of the wave. But with the quantum potential, the effect is the same for a very large or a very small wave and depends only on its overall shape.

By way of an illustration, think of a ship that sails on automatic pilot, guided by radio waves. The overall effect of the radio waves is independent of their strength and depends only on their form. The essential point is that the ship moves with its own energy but that the information within the radio waves is taken up and used to direct the much greater energy of the ship. In the causal interpretation, the electron moves under its own energy, but the information in the form of the quantum wave directs the energy of the electron. Clearly the term causal is now being used in a very new way from its more familiar sense.

The result is to introduce several new features into the movement of particles. First, it means that a particle that moves in empty space, with no classical forces acting on it whatsoever, still experiences the quantum potential and therefore need not travel uniformly in a straight line. This is a radical departure from Newtonian theory. The quantum potential itself is determined from the quantum wave ψ, which contains contributions from all other objects in the particle's environment. Since Q does not necessarily fall off with the intensity of the wave, this means that even distant features of the environment can effect the movement

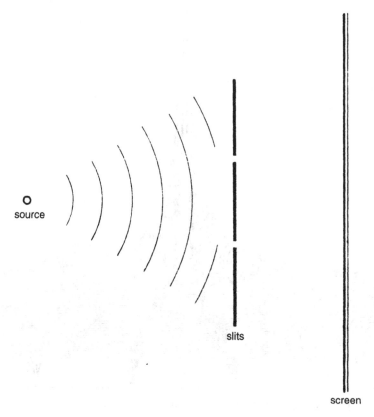

Figure 6.1 The Double Slit Experiment: an electron from the source encounters the double slits and ends up being registered on the screen behind. After very many of such individual events a pattern begins to build up on the screen. The conventional interpretation is that this interference pattern is evidence of the wavelike nature of the electron. In the causal interpretation, however, the pattern is a direct result of the complex quantum potential.

in a profound way. As an example, consider the famous double slit experiment. This is generally taken as the key piece of evidence of the wave-particle duality of quantum particles. When electrons are sent through the double slit, they exhibit a wavelike interference pattern on the other side which is quite "incommensurable" with the classical behavior of particles (see Figure 6.1). How does the explanation work in the causal interpretation?

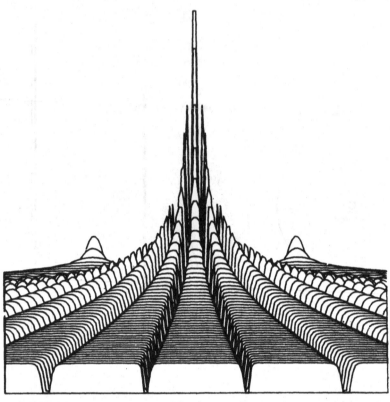

Figure 6.2 The quantum potential for the two-slit system.

The electron travels toward a screen containing two slits. Clearly it can go through only one slit or the other. But the quantum wave can pass through both. On the outgoing side of the slit system, the quantum waves interfere to produce a highly complex quantum potential which does not generally fall off with the distance from the slits. The potential is illustrated in Figure 6.2.

Note the deep valleys and broad plateaus. In those regions where the quantum potential changes rapidly, there is a strong force on the particle which is deflected, even though there is no classical force operating. The movement of the electron is therefore modified to produce the scattering pattern shown below. In this case, the wave-like properties do not arise in any essential duality of the quantum particle but from the complex effects of the quantum potential (see Figure 6.3).

The explanation of the quantum properties of the electron given above emphasized how the form of the quantum potential can dominate

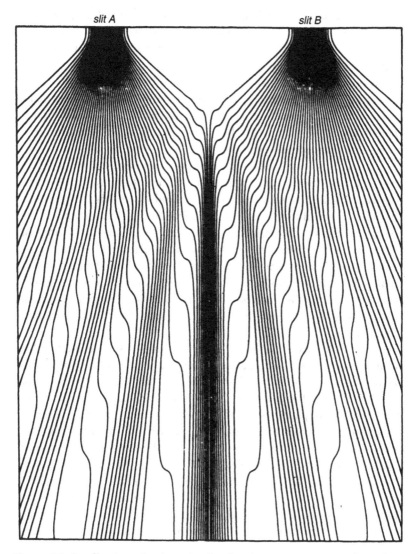

slit A   slit B

Figure 6.3 A collection of trajectories for the electron as it passes through the two-slit system.

behavior. In other words, information contained within the quantum potential will determine the outcome of a quantum process. Indeed it is useful to extend this idea to what could be called *active information*. The basic idea of active information is that a form, having very little energy, enters into and directs a much greater energy. This notion of an original energy form acting to "inform," or put form into, a much larger energy has significant applications in many areas beyond quantum theory.

Consider a radio wave, whose form carries a signal – the voice of an announcer, for example. The energy of the sound that is heard from the radio does not in fact come from this wave but from the batteries or power plug. This latter energy is essentially "unformed," but takes up its form from the information within the radio wave. This information is *potentially* active everywhere but only *actually* active when its form enters the electrical energy of the radio.

The analogy with the causal interpretation is clear. The quantum wave carries "information" and is therefore *potentially* active everywhere, but it is *actually* active only when and where this energy enters into the energy of the particle. But this implies that an electron, or any other elementary particle, has a complex and subtle inner structure that is at least comparable with that of a radio. Clearly this notion goes against the whole tradition of modern physics, which assumes that as matter is analyzed into smaller and smaller parts, its behavior grows more elementary. By contrast, the causal interpretation suggests that nature may be far more subtle and strange than was previously thought.

But this inner complexity of elementary matter is not as implausible as it may appear at first sight. For example, a large crowd of people can be treated by simple statistical laws, whereas individually their behavior is immensely subtler and more complex. Similarly, large masses of matter reduce to simple Newtonian behavior whereas atoms and molecules have a more complex inner structure. And what of the subatomic particles themselves? It is interesting to note that between the shortest distance now measurable in physics (10^{-16} cm) and the shortest distance in which current notions of space-time probably have meaning (10^{-33} cm), there is a vast range of scale in which an immense amount of yet undiscovered structure could be contained. Indeed this range is roughly equal to that which exists between our own size and that of the elementary particles.

A further feature of the causal interpretation is its account of what Bohr called the wholeness of the experimental situation. In, for example, the double slit experiment, each particle responds to information that comes from the entire environment. For while each particle goes through only one of the slits, its motion is fundamentally affected by information coming from both slits. More generally, distant events and structures can strongly affect a particle's trajectory so that any experiment must be considered as a whole. This [causal interpretation] gives a simple and tangible account of Bohr's wholeness, for since the effects of structures may not fall off with distance, all aspects of the experimental situation must be taken into account . . .

*

When one looked at the many-particle system, this new kind of wholeness became much more evident, for the quantum potential was now a function of the positions of all the particles which (as in the one-particle case) did not necessarily fall off with the distance. Thus, one could at least in principle have a strong and direct (non-local) connection between particles that are quite distant from each other. This sort of non-locality would, for example, give a simple and direct explanation of the paradox of Einstein, Podolsky, and Rosen, because in measuring some property of one of a pair of particles with correlated wave functions, one will alter the "non-local" quantum potential so that the other particle responds in a corresponding way.

Because the above response is instantaneous, however, it would seem at first sight to contradict the theory of relativity, which requires that no signals be transmitted faster than the speed of light. At the time of proposing these notions I regarded this as a serious difficulty, but I hoped that the problem would ultimately be resolved with the aid of further new orders. This indeed did happen later in connection with the application of the causal interpretation to the quantum mechanical field theory, but as this question is not relevant to the subject of the present paper, I shall not discuss it further here.[1] Meanwhile, however, I felt that the causal interpretation was affording valuable insight into a key difference between classical and quantum properties of matter. Classically, all forces are assumed to fall off eventually to zero, as particles separate, whereas in the quantum theory the quantum potential may still strongly connect particles that are even at macroscopic orders of distance from each other. In fact, it was just this feature of the quantum theory, as brought out in the causal interpretation, that later led Bell[2] to develop his theorem, demonstrating quite precisely and generally how quantum non-locality contrasts with classical notions of locality.

As important as this new feature of non-local connection is, however, the quantum potential implies a further move away from classical concepts that is yet more radical and striking. This is that the very form

Extract from Chapter 2 of B. J. Hiley and F. D. Peat (eds), *Quantum Implications: Essays in Honour of David Bohm*, Routledge, London (1987).

of the connection between particles depends on the wave function for the state of the whole. This wave function is determined by solving Schrödinger's equation for the entire system, and thus does not depend on the state of the parts. Such a behaviour is in contrast to that shown in classical physics, for which the interaction between the parts is a predetermined function, independent of the state of the whole. Thus, classically, the whole is merely the result of the parts and their preassigned interactions, so that the primary reality is the set of parts while the behaviour of the whole is derived entirely from those parts and their interactions. With the quantum potential, however, the whole has an independent and prior significance such that, indeed, the whole may be said to organize the activities of the parts. For example, in a superconducting state it may be seen that electrons are not scattered because, through the action of the quantum potential, the whole system is undergoing a coordinated movement more like a ballet dance than like a crowd of unorganized people. Clearly, such quantum wholeness of activity is closer to the organized unity of functioning of the parts of a living being than it is to the kind of unity that is obtained by putting together the parts of a machine . . .

During the 1960s, I began to direct my attention toward *order*, partly as a result of a long correspondence with an American artist, Charles Biederman, who was deeply concerned with this question. And then, through working with a student, Donald Schumacher, I became strongly interested in *language*. These two interests led to a paper[3] on order in physics and on its description through language. In this paper I compared and contrasted relativistic and quantum notions of order, leading to the conclusion that they contradicted each other and that new notions of order were needed.

Being thus alerted to the importance of order, I saw a program on BBC television showing a device in which an ink drop was spread out through a cylinder of glycerine and then brought back together again, to be reconstituted essentially as it was before. This immediately struck me as very relevant to the question of order, since, when the ink drop was spread out, it still had a "hidden" (i.e. non-manifest) order that was revealed when it was reconstituted. On the other hand, in our usual language, we would say that the ink was in a state of "disorder" when it was diffused through the glycerine. This led me to see that new notions of order must be involved here.

Shortly afterwards, I began to reflect on the hologram and to see that in it, the entire order of an object is contained in an interference pattern of light that does not appear to have such an order at all. Suddenly, I was struck by the similarity of the hologram and the behavior of the ink drop. I saw that what they had in common was that an order was *enfolded*; that is, in any small region of space there may be "information" which is the result of enfolding an extended order and which could then be unfolded into the original order (as the points of contact made by the folds in a sheet of paper may contain the essential relationships of the total pattern displayed when the sheet is unfolded).

Then, when I thought of the mathematical form of the quantum theory (with its matrix operations and Green's functions), I perceived that this too described just a movement of enfoldment and unfoldment of the wave function. So the thought occurred to me: perhaps the movement of enfoldment and unfoldment is universal, while the extended and separate forms that we commonly see in experience are relatively stable and independent patterns, maintained by a constant underlying movement of enfoldment and unfoldment. This latter I called the *holomovement*. The proposal was thus a reversal of the usual idea. Instead of supposing that extended matter and its movement are fundamental, while enfoldment and unfoldment are explained as a particular case of this, we are saying that the implicate order will have to contain within itself all possible features of the explicate order as potentialities, along with the principles determining which of these features shall become actual. The explicate order will in this way flow out of the implicate order through unfoldment, while in turn it "flows back" through further enfoldment. The implicate order thus plays a primary role, while the explicate order is secondary, in the sense that its main qualities and properties are ultimately derived in its relationship with the implicate order, of which it is indeed a special and distinguished case.

This approach implies, of course, that each separate and extended form in the explicate order is enfolded in the whole and that, in turn, the whole is enfolded in this form (though, of course, there is an asymmetry, in that the form enfolds the whole only in a limited and not completely defined way). The way in which the separate and extended form enfolds the whole is, however, not merely superficial or of secondary significance, but rather it is essential to what that form is and to how it acts, moves and behaves quite generally. So the whole is, in a deep sense, *internally* related to the parts. And, since the whole enfolds all the parts, these latter are also internally related, though in a weaker way than they are related to the whole.

I shall not go into great detail about the implicate order[4,5] here; I shall assume that the reader is somewhat familiar with this. What I want to emphasize is only that the implicate order provided an image, a kind of metaphor, for intuitively understanding the implication of wholeness which is the most important new feature of the quantum theory. Nevertheless, it must be pointed out that the specific analogies of the ink drop and the hologram are limited, and do not fully convey all that is meant by the implicate order. What is missing in these analogies is an inner principle of organization in the implicate order that determines which sub-wholes shall become actual and what will be their relatively independent and stable forms. Indeed, in both these models, the order enfolded in the whole is obtained from pre-existent, separate and extended elements (objects photographed in the hologram or ink drops injected into the glycerine). It is then merely unfolded to give something similar to these elements again. Nor is there any natural principle of stability in these elements; they may be totally altered or destroyed by minor further disturbances of the overall arrangement of the equipment.

Gradually, throughout the 1970s, I became more aware of the limitations of the hologram and ink droplet analogies to the implicate order. Meanwhile, I noticed that both the implicate order and the causal interpretations had emphasized this wholeness signified by quantum laws, though in apparently very different ways. So I wondered if these two rather different approaches were not related in some deep sense – especially because I had come at least to the essence of both notions at almost the same time. At first sight, the causal interpretation seemed to be a step backwards toward mechanism, since it introduced the notion of a particle acted on by a potential. Nevertheless, as I have already pointed out, its implication that the whole both determines its sub-wholes and organizes their activity clearly goes far beyond what appeared to be the original mechanical point of departure. Would it not be possible to drop this mechanical starting point altogether?

I saw that this could indeed be done by going on from the quantum mechanical particle theory to the quantum mechanical field theory. This is accomplished by starting with the classical notion of a continuous field (e.g., the electromagnetic) that is spread out through all space. One then applies the rules of the quantum theory to this field. The result is that the field will have discrete "quantized" values for certain properties, such as energy, momentum, and angular momentum. Such a field will act in many ways like a collection of particles, while at the same time it still has wave-like manifestations such as interference, diffraction, etc.

Of course, in the usual interpretation of the theory, there is no way to understand how this comes about. One can only use the mathematical formalism to calculate statistically the distribution of phenomena through which such a field reveals itself in our observations and experiments. But now one can extend this causal interpretation to the quantum field theory. Here, the actuality will be the entire field over the whole universe. Classically, this is determined as a continuous solution of some kind of field equation (e.g., Maxwell's equations for the electromagnetic field). But when we extend the notion of the causal interpretation to the field theory, we find that these equations are modified by the action of what I called a super-quantum potential. This is related to the activity of the entire field as the original quantum potential was to that of the particles. As a result, the field equations are modified in a way that makes them, in technical language, non-local and non-linear.

What this implies for the present context can be seen by considering that, classically, solutions of the field equations represent waves that spread out and diffuse independently. Thus, as I indicated earlier in connection with the hologram, there is no way to explain the origination of the waves that converge to a region where a particle-like manifestation is actually detected, nor is there any factor that could explain the stability and sustained existence of such a particle-like manifestation. However, this lack is just what is supplied by the super-quantum potential. Indeed, as can be shown by a detailed analysis,[6] the non-local features of this latter will introduce the required tendency of waves to converge at appropriate places, while the non-linearity will provide for the stability of recurrence of the whole process. And thus we come to a theory in which not only the activity of particle-like manifestations, but even their actualization, e.g. their creation, sustenance, and annihilation, is organized by the super-quantum potential.

The general picture that emerges out of this is of a wave that spreads out and converges again and again to show a kind of average particle-like behaviour, while the interference and diffraction properties are, of course, still maintained. All this flows out of the super-quantum potential, which depends in principle on the state of the whole universe. But if the "wave function of the universe" falls into a set of independent factors, at least approximately, a corresponding set of relatively autonomous and independent sub-units of field function will emerge. And, in fact, as in the case in the particle theory, the wave function will under normal conditions tend to factorize at the large-scale level in an entirely objective way that is not basically dependent on our knowledge or on our observations and measurements. So now we see quite generally that the whole universe not only determines and organizes its sub-wholes, but also that it gives form to what has until now been called the elementary particles out of which everything is supposed to be constituted. What we have here is a kind of universal process of constant creation and annihilation, determined through the super-quantum potential so as to give rise to a world of form and structure in which all manifest features are only relatively constant, recurrent and stable aspects of this whole.

To see how this is connected with the implicate order, we have only to note that the original holographic model was one in which the whole was constantly enfolded into and unfolded from each region of an electromagnetic field, through dynamical movement and development of the field according to the laws of classical field theory. But now, this whole field is no longer a self-contained totality; it depends crucially on the super-quantum potential. As we have seen, however, this in turn depends on the "wave function of the universe" in a way that is a generalization of how the quantum potential for particles depends on the wave function of a system of particles. But all such wave functions are forms of the implicate order (whether they refer to particles or to fields). Thus, the super-quantum potential expresses the activity of a new kind of implicate order. This implicate order is immensely more subtle than that of the original field, as well as more inclusive, in the sense that not only is the actual activity of the whole field enfolded in it, but also all its potentialities, along with the principles determining which of these shall become actual.

I was in this way led to call the original field the first implicate order, while the super-quantum potential was called the second implicate order (or the super-implicate order). In principle, of course, there could be a third, fourth, fifth implicate order, going on to infinity, and these would correspond to extensions of the laws of physics going beyond those of the current quantum theory, in a fundamental way. But for the present I want to consider only the second implicate order, and to emphasize that this stands in relationship to the first as a source of formative, organizing, and creative activity.

It should be clear that this notion now incorporates both of my earlier perceptions – the implicate order as a movement of outgoing and incoming waves, and of the causal interpretation of the quantum theory. So, although these two ideas seemed initially very different, they proved to be two aspects of one more comprehensive notion. This can be described as an overall implicate order, which may extend to an infinite number of levels and which objectively and self-actively differentiates and organizes itself into independent sub-wholes, while determining how these are interrelated to make up the whole.

Moreover, the principles of organization of such an implicate order can even define a unique explicate order, as a particular and distinguished sub-order, in which all the elements are relatively independent and externally related.[7] To put it differently, the explicate order itself may be obtainable from the implicate order as a special and determinate sub-order that is contained within it.

All that has been discussed here opens up the possibility of considering the cosmos as an unbroken whole through an overall implicate order. Of course, this possibility has been studied thus far in only a preliminary way, and a great deal more work is required to clarify and extend the notions that have been discussed in this paper.

Notes

1   For further discussion of this point, see Bohm, D., *Physical Review*, **85**, 180 (1952); see also Bohm and Hiley, B. J., *Foundations of Physics*, **14**, 255 (1984), where this question is discussed in more detail.

2   Bell, J., *Review of Modern Physics*, **38**, 447 (1966); see also *Foundations of Physics*, **12**, 989 (1982).

3   Bohm, D., *Foundations of Physics*, **1**, 359 (1971).

4   Bohm, D., *Foundations of Physics*, **3**, 139 (1973).

5   See Chapter 3 of this volume; also Bohm, D., *Wholeness and the Implicate Order*, Routledge and Kegan Paul, London (1980).

6   See note 1 (above).

7   Bohm, D., "Claremont Conference," in *Physics and the Ultimate Significance of Time*, ed. David R. Griffin, State University of New York Press, Albany (1986).

Part Two – Individual Orders

7 STRUCTURE-PROCESS AND THE EGO (1958–1967)

In this selection of letters to his brother-in-law Yitzhak Woolfson (addressed herein as "Isidore"), Bohm lays the foundation for his views of the ego, which he continued to develop and refine for some thirty years. The time-frame and content of the letters coincide significantly with Bohm's early meetings with the Indian philosopher J. Krishnamurti, whose influence in this area Bohm is quick to acknowledge. At the same time, Bohm brings his own perspective and unique investigative skills to the issue at hand, generating material essential to his subsequent considerations of soma-significance (Chapter 5) and proprioception (Chapters 8 and 12).

Very much like his critique of the notion of substantial "things" in Chapter 1, Bohm suggests that the ego is in actuality a dynamic "structure-process" which displays varying degrees of stability, regularity, and novelty, but on sustained examination is found to be without any inherent existence. The fact that the notion of a substantial ego has been reified over thousands of years – and never more so than in contemporary culture – is testament to the illusion-generating nature of the structure-process itself, rather than evidence that there is any actual entity which the term "ego" stands for.

At the core of this ego process, says Bohm, is an ordered array of unperceived contradictions, which result in endemic psychological and social confusion. Normalized through cultural and personal assimilation, these unperceived contradictions generate the conceptual superstructure for what we consider to be reality. Like the analogy of the single fish which appears as two separate entities (Chapter 3), the structure-process of the ego displays

Previously unpublished personal letters, courtesy of Yitzhak Woolfson.

experience in abstracted fragments: "I" and "me," thought and feeling, pleasure and pain, self and world, secure and insecure, and so on. We tenaciously identify with certain of these aspects, and strive to control or eliminate others. But if "thought," for example, struggles against "feeling" – while both are more deeply a single movement – we cannot help but exist in a state of confusion arising from a failure to recognize the contradictory nature of what is occurring.

Seamlessly woven into this reflexive structure of fragmentation, suggests Bohm, is the active operation of memory. While memory is essential for practical functioning, it has an insidious tendency to permeate and filter the whole of experience. Any prospect of authentic participation in the immediate present is thus generally overridden and replaced by a complex simulation, comprised primarily of "playbacks" of past experience. The fusion of such memory with the structural fragmentation outlined above yields a complete virtual world – one which seems to meet our criteria for "reality," but is nonetheless a tenuous construction, perpetually at risk of being undermined by its own inherent contradictions. Co-emergent with this world, and inseparable from it, is the "ego" – the sense of an inner entity which also seems stable and real, but is ultimately based on the same simulations as the world it appears to inhabit.

Bohm proposes that this entire process is sustained by a dysfunctional feedback loop between the thalamus (the reptilian aspect of the brain) and the cortex (the mammalian aspect of the brain). The cortex, with its immense capacity for memory storage and its ability to generate vivid images based on this memory, floods the thalamus with signals and images that overwhelm those coming from the natural world. The rudimentary fight–flight and pleasure–pain mechanisms with which the thalamus formerly engaged the natural world begin to run amok, triggered now by the milieu of cortical images, rather than by the direct stimuli of nature. In this way a loop is created, with the thalamus sending confused, overstimulating signals to the cortex, while the cortex generates an abnormal profusion of memories, images, and thoughts in an attempt to assuage the unnatural demands of the thalamus. The pervasive flow of neurophysiological energy generated by this looping serves to further substantiate a feeling of underlying "reality," thus bringing the process of fragmentation and simulation full circle.

Nonetheless, suggests Bohm, this feedback loop is not impenetrable. Through intentional recall of incidents where the sense of ego has been provoked or disturbed, there exists the possibility of seeing the entire structure-process at work, without being fully caught in its projections. In this way we may access and understand the concrete structure-process of the ego – the implicate order of consciousness as outlined in the remaining chapters of this *volume. It is direct engagement with the whole of this order – fragmentation, identification, the simulations of memory, and the somatic manifestations of these processes – that holds open the possibility of a creative movement outside the loop.*

Reflecting on his exchange of letters with Bohm, Yitzhak Woolfson comments:

The letters that appear here are part of a personal correspondence with my brother-in-law, David Bohm, between the years 1957 to 1980. It could be said that this correspondence began during the summer of 1956, because events that were taking place in Israel at that time affected the content of the letters I wrote. The early letters particularly reveal David's concern about the welfare of the family in Israel and his worry regarding the dangerous situation developing all over the world.

This correspondence has a special value for me because it is a sort of continuation of the many talks we had while walking together in England and in Israel, whenever it was possible for the family to be together. While staying with them in their home in Edgware, Dave and I would walk in the evenings through the quiet suburban streets around their house and talk. Dave did most of the talking about his work. He said that he was able to think more clearly while walking and his ideas and the explanation of them flowed from him with great energy. From time to time, as though worried that he might be losing me, he would ask if I were still with him, if I understood. When I asked for clarification, he would explain the point in different words and, with great patience, make sure that I too could see the concept clearly before continuing. Dave had a natural gift as a teacher.

In reading the letters I always hear Dave's quiet patient voice, choosing his words with great care to make sure that they indicated clearly and coherently the direction he was going. Dave showed me how to listen to the silence, which contains all sounds. How to understand about the silence of the mind, which is always there beyond the turbulence and confusion of the ego. And as in so much that he said and wrote, one can almost feel his conviction that in that silence when insight is uncluttered and true, the structure of thought will open like a flower in the clear light of the sun and reveal its inner depth.

These letters also show how Dave's concepts and ideas do not belong to any particular time or place. In the light of the difficulties we are presently witnessing in the efforts to make peace between nations, Dave's words in his letter of May 24, 1967 sound almost prophetic. The letters

are as pertinent today as they were when he began writing them and will
no doubt continue to be so until man can understand the need to change
his way of thinking radically.

Jerusalem, May 27, 2001

Sept. 9, 1958

Dear Isidore,

To continue the previous letter, what is now essential is that people must begin to think, to face the "real reality" and not just the superficial momentary aspects of reality that each meets, in his day to day experiences, within his narrow and limited sphere. To do this, man needs a general over-all philosophical point of view, which orients him in the chaos of shifting and unstable appearances that present themselves, when he focuses only on what is momentary and narrow. And I believe that my work in physics gives at least some elements of such a philosophy. For I am beginning to see that even in the apparently lifeless world of so-called "inert" matter, each thing, each particle (e.g., electron, proton, etc.) is not what it at first seems to be, i.e., a separate point in space, indifferent in its inner being to all the others, remaining always only just what it is, and interacting only externally with all the others.

Rather, each entity is continually being formed from the infinite background and falls back into the background, to be regenerated again and again (as long as it continues to exist). Thus each thing has its roots in the totality and falls back into the totality. Yet, it still remains a thing having a certain degree of independent being. And this is possible because each thing contains in itself, its own special image of the totality (cosmos) out of which it formed itself, and into which it is always dissolving (and re-forming). The apparent separateness of things as we see them immediately is that each thing has a certain degree of relative indifference to the others. But this indifference does not belong to it alone. For it is the cosmos itself which determines this indifference and which also determines the limits of this indifference.

If the above is true for the most elementary and inert kinds of things, it is much more true for more organized things, such as living beings, man and his consciousness and society. Each man draws his being from the totality and his effects fall back into the totality. His separateness, loneliness, indifference to the others are only relative, and determined by his relation to the totality (in this case, society). Change this relation and you bring out the deeper essential relations between man and man.

I will write more later.

Dave

September 25, 1962

Dear Isidore,

I have been seeing things a bit more clearly since you left. I would say that our concepts are like mirrors that we hold up to reality. If they are distorting mirrors, they may present many apparently different reflections of one thing. Thus, if I try to study my own ego, there appears the "me" with all its qualities, and an "I" that seems to be observing them. Yet we know logically that an "I" and "me" must be one entity. How then do we come to see them as two? I suggest that consciousness is a distorting mirror, which is able, in effect, to give two apparently different but related and interacting reflections of one process. In reality there is neither "I" nor "me," but the individual in his totality (individual = undivided). On the other hand, the ego process with the "I–me" division could be called the "dividual." In the individual, perception is "going on" without the need for a "perceiver" to do the job. Our language forces us to say that a subject is acting on an object. Thus, we say, "It is raining." But where is the "it" that is doing the raining? Similarly, we say, "I am observing."

Also, one can ask, "Is there really an 'I' that is 'doing' the observing or is there not just a process of observing that is going on?" When a person is serious about what he is doing, the ego falls away, and the individual as a whole is perceiving and acting. For instance, suppose that he is playing music. When he is finished, the ego process comes back into existence, and takes the credit. But in reality, the ego process never does anything at all, except to get in the way. Imagine trying to play music, while the ego is saying, "I am now playing music. Isn't it wonderful what I can do?"

So it seems to me that at all times, when the ego exists, the individual is in a state of confused perception. In this state he sees "the world," and also, he hears the words "this is 'I,'" along with a feeling of ownership or possession of a whole set of qualities, memories, urges, relations, desires, etc. This latter feeling can be called "identification." The individual also has the illusory perception of a process in which the centralized collection of qualities is initiating actions. But in reality, it is the individual as a whole who acts. The confusion is that the individual is seeing the process as if it were the separate ego that was acting, as in a moving picture we see the image of a person as if it were "doing" things.

I would say that as a man perceives, so he *is*. Here I include in perception, all of seeing, hearing, feeling, sensing, going on up to understanding, and the seeing of what is true and false. This latter is very important. If a man is confused in his vision of what is false and what is true, then nothing else that he does can mean much. Thus, if he wants to do good, he may nevertheless do evil, since he cannot see whether what he does is truly good or not. Probably even Hitler saw himself as doing good, but his vision was very confused. Similarly, an insane man may be responding in a natural way to his confused vision of the world. When a man sees differently (i.e., understands), then he *is* different. A man with a confused perception must act in a confused way, and therefore he *is* confused. As soon as his vision is really clear (not just in words, but in his whole being), then he turns away from confusion, and he is a different man. So the transformation of man must come through a new vision, a new understanding. Only the individual can do this. So it is the individual who can change, and not the collective.

The importance of perception by the whole man cannot be emphasized too much. Suppose that a man sees what he takes for sugar, but suddenly, he reads the label "Potassium Cyanide." His whole being immediately sees that this is poison, and he turns away from it without further ado, because he wants to live. His intellect, his emotions, his nerves and muscles, etc., are all aware that "this is poison," and each does its job in carrying out the appropriate action. Now, if we really had a corresponding total perception of the ego at work, we would see that it is as poisonous as cyanide. However, what may happen is that the intellect says, "This is poison," while the emotions, being more conservative and attached to memory, say, "No – it is sugar." So we are confused, we are in a state of contradiction. While we are in that state, all our thoughts and actions are confused, and each step only tangles us up worse. It is as if a scientist were to say, "$3 = 2$." Then he would feel uncomfortable because of the contradiction. Every idea that he introduced to resolve the contradiction would only make it worse, as long as he accepted the notion that "$3 = 2$."

When one is in a state of confusion, one can do nothing (as when one is lost in the woods, it is urgent to stop and try to understand rather than go around in circles). We cannot believe anything that is in our own minds, because it may only be an idea brought in to cover up our confusion. But there is one thing that one can see, and this is "I am confused." Here, one starts with the truth, and goes on from there. It is an objective fact that I am confused, as objective as "the temperature is now 65°F."

Then one must see the source of the confusion. This is of course often quite difficult. But here, it is helpful to ask the question, "Is there anything more important than seeing what is true and what is false?" If your mind puts forth some emotional demand as more important, then you will see, on asking this question, that here is one of the sources of your confusion. For it is plain to see that nothing can really be more important than to see what is true and what is false (not even the need to save your life, because if you mistake truth for falsity, you will act in a confused way, and will be more likely to lose your life than if you saw clearly).

It is clear now that it is no use to fight the ego, to "do" something "positive" about it. For this would only be a confused process, in which the ego tried to improve itself, not noticing that the ego process is the essence of the illness. When you understand confusion (i.e., see it deeply), then this perception will act of its own accord, and you will turn away from confusion, without further ado. The ego need do nothing at all. Indeed, if it acts, it must get in the way. If I confuse my image in a mirror with another man who is imitating me, then everything that I do to stop this man from imitating me will only confuse me more. When I understand that this is only a reflection of me in the mirror, then the whole problem disappears. As long as I do not understand the problem, it is insoluble, because it is based on confusion. As soon as I understand, there is no problem. And this is what happens with all the problems created by the ego process. They are all based on confusion, hence insoluble, until one understands. But when one understands there is no problem.

I would like to go a bit into the origin of the confusion that is responsible for the ego process. Now, an infant begins by not being able to recall to memory (in an internal image) an object that is absent. But he can still *recognize* it when he experiences it (this is often true even of an adult). How does it happen?

It seems to me that every experience leaves a kind of "negative trace" or imprint in the mind. When the experience is repeated, it fits this imprint as a key fits a lock. In this way one can recognize it. One can also produce an internal image in the imagination, which is recognized in the same way that direct perceptions are recognized, i.e., against the "negative trace." So memory is a positive internal imitation (in the imagination) of something that was once perceived, while recognition precedes imitation in the development of the infant, because it is basically a simpler procedure.

One can compare recognition to a set of grooves and scratches impressed by past experience on the mind, while memory is like the "play-back" of the record as internal images, sounds, etc.

It is important to notice that both recognition and memory involve the emotions as well as factual records of what happened outwardly. Thus, if the infant has a certain experience that is pleasant, his recognition traces start to demand a repetition. He tries to find a way to repeat it. But if they are unpleasant, he tries to find a way to avoid it. Here is the real beginning of the ego process. Evidently when the memory "play-back" develops later in the child, it too will be accompanied by emotional demands for or against the experience in question. Since thought is based on recognition and memory, it is clear that thought and feeling cannot be separated. They are two aspects of the one process, which is the response of recognition and memory to new perceptions.

Out of thought is then born desire, the urge to continue, to enhance, to possess, to make secure that which is pleasant and to guarantee the avoidance of what is unpleasant. Desire attaches itself to an object of the imagination, in order to attain permanence. But the object of desire is always changing. Firstly, the real object changes in one way, while the object imagined in desire changes in another way. We then discover when we get the object of our desire that it isn't what we expected; we soon encounter satiety and boredom. Other objects soon seem more attractive to desire. Besides, objects of desire change in unexpected ways, grow old, and even pass out of existence. So the attachment of desire to an object leads to contradiction (contradictory desires), and out of this comes confusion.

Religious people and moralists then tell us to suppress desire, shape it, control it, direct it to God or to the triumph of Communism. Psychologists and others tell us to sublimate it. But doing this only heightens the contradiction and confusion. Then comes fear that one will never achieve satisfaction of desire, a state of anxiety and despair, alternating with periods of hope, when there is the momentary belief that one can escape into a new job, a new religion, a new hobby, a new marriage, etc.

So we see that the ego process, with its attachment of desire to an object, is inherently in a state of confusion. What is the origin of this confusion? It is very simple. We mistake the demands made in the "play-back" of memory for true feelings. True feelings arise only in fresh perception of what is new. This perception is understanding on the intellectual side, and it has the wholeness of feeling sometimes called love on the emotional side. It can also be called creativity. But this creativity refers to creative living, and not just to the expression of creation in art, science, music, etc. It is essential to understand that the play-back of memory and the recognition "scratches" are not creative in this sense. They have their utility as factual memory to guide you in your life or your job (how to get home, etc). Memory is, like fire, "a good servant but a bad master." And as soon as you take the play-back of emotions seriously, you are the slave of memory, since your actions will then be only a response to these "memory scratches," and not to reality as it actually is from moment to moment. Since the "memory scratches" cannot fit reality (because reality is always changing), one comes into a state of contradiction between demands based on memory and reality, as well as between the different aspects of memory demands that contradict each other. So the ego is inherently in a state of confusion and contradiction.

What is to be done about all this? The answer is, as I said before, nothing at all. Whatever action is born of desire will also be self-contradictory (e.g., the desire to end desire which in fact only continues desire in another form). But the question is, "Why do anything at all about desire?" Desire seems to be necessary to mental life. It is like a many-colored flame, very beautiful and full of energy, always changing. When it is attached to something, it falls into contradiction and confusion – the flame turns into dense smoke. But if you understand the futility of doing anything at all about desire (satisfying it, attaching it to an object, shaping it, suppressing it, choosing "good" desires and getting rid of "bad" desires), then you will just turn away from these efforts, and let desire do what it will, to die as to unfold in its own natural way. Then there will be no contradiction. Desire does no harm if it is not attached. In other words, desire is something different, when you understand it. For recall, "As man sees, so he is." If you see desire in a new way, then desire is different in its operation in you.

If you don't let desire determine your actions, then what should do this? The answer is: The perception of what is true and what is false will operate by itself, if it is deep enough. For example, when one sees the poison, one simply leaves it alone, without bringing in a struggle between the desire on one side to live and on the other side to continue to take the poison. Also, when you see the truth – *that you are confused,* and the falsity of ideas that arise in the state of confusion – then this perception acts, and your mind is already starting to clear itself, without any effort by the ego to bring this about. You must ask yourself, "Is it possible for there to be such an extensive and deep perception of what is true and what is false, that the ego process as a whole will drop away like a dead leaf?" There is no way to answer this in words. It is foolish to try. The answer can only come by looking at the problem concretely.

Finally, it is interesting to compare all this with various forms of psychoanalysis, which also assert that self-understanding can lead to integration of the personality. The main difference is that they all urge us to adjust to the "normal" life in society. But this "norm" is confused and self-contradictory. So we are asked to adjust to confusion. Imagine a physicist who was asked to adjust to the assumption that "2 = 3." He would end up by going mad. Perhaps a similar fate awaits the one who tries to adjust to society. What is needed is to see through it as inherently confused.

Saral and I send you our love, also to Sheila and the children.

Yours, Dave

Nov. 12, 1962

Dear Isidore,

I was very glad to receive your letter, and to hear that all is well. I think that the problems you refer to are really very important ones today.

To see the totality of understanding, i.e., to understand the act of understanding itself, is indeed a difficult thing to do. Let us try to go into the question a bit.

First of all, it is clearly of no use to try to define the totality of understanding in terms of words. Rather, as we communicate, words are merely marks, showing the course of the process of awareness in the writer or the speaker. The hearer or the reader must, in effect, always be answering the question, "What could the writer or speaker be perceiving, thinking, feeling, that makes him put out these words?" And the answer will not be in words. Rather, it will come into being as perceived in the awareness process of the reader or hearer. Moreover, there are some questions that we can ask ourselves, whose answer is not in words, but only in the coming into awareness of what is referred to. Thus, when you ask, "What is the totality of understanding?", you must see that you do not and cannot "know" the answer, in terms of what can be recognized from the past. It must be fresh and new, what one has never seen. And even if one should see it some day, the memory of it will be false in the next moment, when in truth, the question must be answered by a fresh perception again.

What does it mean to see anything in its totality? As Krishnamurti points out, we can see a tree, or a river, first as a totality. We see the "treeness," the basic quality of all trees, which comes into our awareness as a sort of general structure and process; and then we particularize down to a given tree, then to a branch, or a leaf, if we wish. But if we started with the details, we could never get the totality by putting them together. It may be said that the part must be seen as abstracted from a totality. We can never abstract the whole from the parts, as this would be an absurdity. Nevertheless, it is a habit that we have, to try to do the impossible. Thus, we begin with various fields of specialization, and express the pious hope that some day these fields will automatically amalgamate to make human knowledge into a whole.

Now, understanding is the act of seeing experience as an integrated totality first. Then we may abstract down to a part. We do not accept experience as a bunch of separated fragments, but rather, we see it as a whole. This means that we must see ourselves too, since the "inner" and "outer" are inseparable, as aspects of experience. We experience the "outer" and "inner" worlds inseparably, on one field of total experiencing. Each one influences the interpretation of the other. So the whole truth cannot be understood by one who does not perceive his own motivations, along with the falsity of all motivation. For if you have a motive, then it is a certain result that you are mainly interested in, and not truth. There is no reason why your preconceived result should be fully compatible with the truth, and generally, it is not.

So to see the whole truth, you must not be in a state of conflict between "what is" and a motivation as to "what should be." What should be is always an illusion, which prevents you from looking at what is. When you see what is (psychologically speaking) it starts to change, and problems start to dissolve away, as one perceives that they all come from what should be – which begins by projecting what is illusory. Usually, our problems are insoluble because they result from contradictory demands of the ego process. As soon as one sees that these demands result from "the machine at work," one ceases to be fooled by them. As with the magician's trick that has been exposed, we are not deceived again.

So one has to see that the ego process is always engaged in setting up the illusion of "what should be," dressing it up in alluring colors, and pushing away "what is," hiding it in frightening and ugly disguises. But nothing can be done without our understanding what is, in its totality, at least as far as we are experiencing it. As soon as we push aside a certain part of what is in our experiencing, in favor of the illusion of what should be, we are in a state of contradiction, which leads immediately to an internal conflict. We then try to escape the conflict because it is very unpleasant, seeking to cover it up by introducing confused ideas and feelings, and by distracting our attention, drawing it to something else instead. But then our escapes, being confused, lead to even more and more conflicts, which in turn must be escaped. Thus, the whole process tangles up, in a sort of cancerous growth of contradiction, conflict, escape, and confusion, until it fills the whole mind. This is the typical state of most human beings in "civilized" society (and probably in "primitive" societies as well).

So if we wish to see the totality of understanding, we cannot be in a state of contradiction and conflict, as this destroys all understanding. Conflict fragments the mind, and is therefore incompatible with a state of understanding, in which the mind sees a totality, by functioning as a totality. People do not realize the importance of this. Some people imagine that one can have a true "intellectual" understanding, while the emotions are in a twisted and tangled state, while others (sentimental-ists) think that one can be full of love and good feelings, while the intellect is totally mixed up. Actually there is no thought without feel-ings (including motivation) and no feeling without thought. Thus, some artists think that they appreciate a picture by feeling and not by intellect also. But if this were the case, why should one picture have a different effect than another? Is it not evident that there is a complex mental process of seeing the picture? This process is probably too rapid for such a viewer to appreciate, so that all he sees is a sort of afterglow of emotion. But then he makes the mistake of supposing that this emotion is all there is. So he becomes confused about his way of viewing pictures.

Understanding evidently requires a state of truth in emotion, in intellect, and in the whole mind. It takes the mind in its totality to perceive the totality of anything. The fragmented mind inevitably sees in parts, while the person who begins with the part thereby fragments his mind. In a sense, to perceive the totality of understanding, one must be a totality. This requires that one see through the totality of illusion. Perhaps a man who is doing this will naturally, without any further action on his part, enter a state in which his mind is a totality. Therefore he will realize the totality of understanding, and in this way, he will perceive it.

Now, why is there illusion? The possibility of creating illusion is the precondition of intelligence. Thus, if we look into a mirror, we may first perceive a man. A partial perception (rays reflected in a certain pattern from the mirror) awakens a perception of the whole man in our minds. In this regard, we create our perceptions, which always look real at the moment that they are created (as they do also for the insane or drugged man, who perceives very convincing delusions). But then a healthy man is always trying also to perceive the truth and falsity of his perceptions. Thus, one very quickly sees that the man behind a mirror cannot be real because he is "imitating" us too closely. So in a fraction of a second, we no longer see another man. Instead, we see a mirror that is reflecting ourselves. We have seen the true meaning of the illusion. But some illusions are so good that we don't see through them. They may therefore lead us into confused and idiotic behavior.

It is clear then that the most important perception of all is the perception of what is true and what is false. Without this, all else becomes meaningless and confused. But this is just what society is always trying to destroy, by saying that certain things (religion, authority, family, nationalism, etc.) are too sacred or important to be questioned. As long as a person thinks that something else is more important than to see what is true and what is false, freshly from moment to moment, then that person is evidently very badly confused. Such confusion must lead to contradiction, conflict, and escape, with unlimited spreading of further confusion through his whole life. And a society constituted of such people leads to confusion to the power of 2000 million, which is a correct description of our world today.

The major source of illusion is the response of memory. We not only remember pictures and words, but our memories are also accompanied by active emotions. Thus, if I am angry at X, when I see him again (or a man who reminds me of him), then the memory is accompanied by a little nervous signal that arouses the reflex of anger. I suggest that you watch for it next time you are angry. The powerful emotions of anger, fear, envy, and many others can be seen mechanically to be following little signals, as if a switch were being pressed. So the illusion is that we are seeing actual experience, when the fact is that it is mostly a "replay" of past experience, adjusted somewhat to present circumstances. This replay is then recorded again, to add to the previous memory, thus cutting the "grooves" deeper. The more we experience in this way, the deeper we get stuck in illusion. And all the while, we are confused into thinking that we are basing our actions on "experience itself." The older a man is, the more "experienced" he is in this way, and the more sure he is that he "knows" what he is doing. So the essence of his life is now constituted of a series of illusions.

One sees through this illusion as one sees the illusory character of a mirror image – it lacks independence. As the image follows its object, so our false emotions follow these little signals. It is only necessary to be fully aware and attentive, and then the illusion loses its power. For no man will base his actions on an illusion that he has really seen through (i.e., if the whole of him has seen through it). Perhaps it is like a movie theatre. As long as your eyes are only on the screen, you "identify" with the characters, and feel that they are winning, losing, good, bad, etc. But on broadening your vision, you see that they are shadows on a screen. The same is true about the whole ego process.

Ego begins with the creation of illusion. How does this happen? Man evolved from an animal. His thalamus (central brain) is not so different from that of animals. Now, the animal's brain is more or less adequate for his normal surroundings. Thus, if he is afraid, he runs. If he is angry, he fights. But in man this thalamus is now surrounded by an intelligent cortex that can create illusions. These illusions can either be pleasant or frightening to the thalamus. When they are pleasant, the thalamus sends a signal to the cortex to produce some more of the same. When they are frightening, an urgent and disturbing signal is sent out, and the cortex is unable to function properly. Confusion results, and the cortex gets busy creating new illusions until the thalamus is satisfied. But reality is always bursting in, so that a man in the state of illusion is always being presented with crisis after crisis, and eventually tends to get into a state of chronic fear or anxiety.

It is no use blaming the poor thalamus, as it was never prepared for living in an environment consisting mainly of the cortex. Nor can the cortex help it, because it cannot function properly if the thalamus is always sending out urgent signals that mix it up. Thus, it no longer tries to see what is true and what is false.

Perhaps the above problem can be solved only if the whole mind understands what is happening, and sees through the mechanical character of the signals from the thalamus, as well as the illusory character of what the cortex "cooks up" to keep the thalamus quiet. To see this will amount to a psychological mutation in man. In other words, man is already a totality, physically speaking, in his mind. But because he does not realize this in his awareness, the thalamus and cortex each operate as if they were independent "minds," presented with problems from "outside." In this way, each confuses the other, in contrast to an animal without much cortex where the thalamus by itself functions sensibly. When man realizes his actual totality, then he will be a totally different being, since the thalamus and cortex will work together in an integrated whole, giving rise to a feeling of love, and an act of understanding. (This may perhaps be the totality of understanding.)

Saral and I send our regards to you, to Sheila and the girls.

Yours, Dave

P.S. Thanks for returning the manuscript on *Understanding in Science.*

Dec. 3, 1962

Dear Isidore,

Thank you very much for your last letter. There is little that I can add to what you said about the "I–me" problem, as well as about fear as the reaction to losing something with which one is identified as pleasant, desirable, necessary, etc. Perhaps the main point that I could stress is the illusory character of the division between "I" and "me." If you are sufficiently aware, you will sense a "signal" to which you react with fear, envy, desire, etc. That is to say, in the "play-back" of memory, there is a little signal, which means, "This is terribly important, urgent, and essential." It is this signal that creates the illusion of the ego process. For we react with emotions, just as if we were electronic machines. Each reaction is "recorded" in memory, thus strengthening the signal, effectively "cutting the grooves deeper."

Now once the emotion gets started, the signal is drowned out. But you can become aware of the signal, either in a real reaction of the ego process, or in an "experimental" reaction. By the latter, I refer to an experiment as follows: Begin with some ego characteristic that you note is common in your life. For example, one may have a sense that one needs the approval of other people. Just to see what happens, imagine to yourself that someone who is important to you is expressing disapproval of you (better yet, try to remember such a situation). You will probably discover every kind of emotional reaction, from anger and fear to the urge to defend yourself. If you can think of a problem on which you really did make a mistake, you will discover your mind fabricating every kind of confused justification, trying to prove that you really did right. Besides, there will be a cloud of fluctuating urges whose meaning is: "This is a dreary, boring, unpleasant, frightening, unimportant subject. There are many other things that should be done first, before coming to this." Before you know it, you are liable to find yourself thinking of something else.

The value of such an experiment is to reveal the *process-structure* of confusion, as it actually operates in the mind, below the level of rational thought. Freud has spoken of "repression" of unwanted desires, but I think that a better description is that they are smothered in a buzzing cloud of confused ideas and feelings, in which they can be mistaken for something that is less alarming. Even "deeply" repressed feelings frequently come out in momentary "jabs" and "jibs" that dig and needle us, setting us off on new trains of thought, action, and mood, without our realizing why we are doing it. The reason we don't realize this is that what we are really doing is quickly covered up by such a cloud of confusion.

The value of a detailed awareness of this process, without approval or disapproval, is that we can see that the whole process is an absolute sham and illusion. We do this, in effect, by working the "buttons" that control the creation of the illusion. We can see that when we purposely imagine a certain situation of disapproval, the "signal" is produced, and then the whole complex of emotions and blatantly false ideas follows. The signal may be felt as a peculiar nervous tension that is hard to describe in words (a sort of uncomfortable, tense, dissatisfied sensation, vibrating a bit in a peculiar way). You can learn to recognize it by seeing that the emotions and ideas always follow variations in the signals as if they were mechanical reactions to pressing a button.

If you are sufficiently aware of the mechanicalness of the process, recognizing the signal, and being intensely conscious of each little emotion in all the nervous tensions, how the thought process shamelessly invents thousands of fictions in order to assuage the urgent and unpleasant signal, then you will find the process changing. It will unfold its true character, "flowering," and then "dying away" like a leaf falling from a tree. But you must keep this awareness up, as the mind contains millions of such "weeds" and new ones are always being introduced by experience. I would like to compare the process described above to a modern "weed killing chemical" that makes weeds die by forcing them to grow so fast that they exhaust themselves.

Now very often in such investigations, you will find that the mind is trying to resist or disapprove of or get rid of something that is very unpleasant or frightening. But despite all efforts, it just won't go away. If you become intensely aware of each little nervous tension, pang of fear, feeling of annoyance, etc., as it fluctuates and changes, you will begin to sense a relationship, between these and various ideas and feelings flitting through the back of your mind. Slowly, the "hard-centered" something that you want to get rid of begins to unfold and reveal its structure. As it opens up, you always see something amazing. *You yourself* are sending in little signals of desire that are keeping the whole conflict going. What happens is that there is something you want – call it A. This something entails B. But you don't want B. Nevertheless, the ego process, being childish, tries to get rid of B while keeping A. Thus, it sends tremendous disapproval and negations against B. But this starts to get rid of A too. So it then sends in more urgent signals to hold onto A as well. This is all possible because, through confusion one loses sight of the relationship between A and B, as well as of the signal that is keeping A going and thus generating B. This buzzing cloud of confusion is generated by the mind, in response to another urgent signal, with which one reacts to the unpleasant fact that you can't have A without B. A great many conflicts have this childishly elementary structure.

In this connection, it is important to distinguish between the word and its meaning. A word is only a noise or a set of marks that evokes various memories, associations, desires, and so on. As long as you are only "thinking in words," you are really playing with a symbolic problem. After reasoning it out, one can feel satisfied that it is "solved." But actually, the emotional tangle and confusion is going on as before, except that one has added the further confusion that verbal reasoning has solved the problem. What is needed is that the words shall awaken the real feelings, the real process-structure of the ego, so that one can be aware of it in concrete reality. In this process, the words should drop away, as they tend to become an illusory substitute for real perception.

After going this far, one can observe a conflict of the type just described. Then you can say to yourself, "This conflict that is being observed has evidently only an illusory separation from the 'I' that seems to be observing it." Then you can ask yourself, "Is it possible to see that *I am* the buzzing cloud of confusion and conflict that is going on?" You may get an extraordinary sense of dissolving of all that is perceived. But I find that the process tends to be limited because a kind of fear and unwillingness to allow complete dissolution of the ego sets in.

It is interesting to speculate on the probable origin of the ego process in the human race. There is evidence that modern man, with essentially his present brain capacity, came on the scene not more than 30,000 years ago. At this stage, the poor creature was ignorant. The animal is ignorant too, but it does not possess a cortex that can remind it of unpleasant dangers, real and imaginary, as well as death, while at other times creating wonderful illusions of satisfactions, apparently to be obtained by certain actions, and so on. So the animal can sit peacefully until there is actual danger, in which case it either runs or fights. But our poor primal "Homo-sapiens" must experience fear of the known and the unknown as well as desire for satisfaction of a kind that never would enter the animal's brain. Being confused, he starts to invent imaginary "magical" means of dealing with dangers, and bringing him these satisfactions. At first, this is not so different from what a child does in his day-dreams. But eventually, man's imagination runs away with him. The magical forces that he has invented seem to escape him, and disclose themselves as even more dangerous than the unknown dangers that he first wanted to escape. So he must propitiate them. He confuses his mind even more, now being afraid even to look at what he has termed "taboo." But he doesn't see that it is all a game that he is playing with himself. How can he? After all, magical dangers are easily confused with real dangers that abound in his life, and there is no easy way at his disposal to study the problem properly.

Gradually man starts to accumulate tools, techniques, knowledge, language, weapons, etc. He develops agriculture, aggregates himself into stable communities, smelts metals, is able to guard against hunger by storing food, etc. This apparently happened in North Africa about 9000 years ago to 6000 years ago. V. Gordon Childe, in his *It Happened In History*, suggests that this was a great period of creative development in man.

Now up to this time, war had not developed, beyond occasional raids and quarrels. Slavery wasn't worth it, as a man consumed almost as much as he could produce. But with growing wealth, plunder became inviting. Weapons made raids practicable and slavery was now technically feasible. At some time, there began to occur to some people the brilliant idea that they could live off others. Thus started the modern age of war, plunder, slavery, and exploitation of man by men. Childe gives evidence that for thousands of years following this change, creativity almost dried up, as man began to look up to the "hero," the conqueror, the slave-holder, while technique and the arts were left to the despised slave. Even in Greek and Roman times, there was the same tendency, and only much later was there a change, allowing a self-respecting man seriously to interest himself in arts and techniques. Childe suggests that except for slavery, man could have reached modern technical levels long before the time of the Greeks.

It seems to me that the development of plunder, slavery, and exploitation as man's main mode of life determined the modern form of the ego process. Even when slavery was given up, exploitation remained the essential feature of man's relation to man, which it still is today. Once this mode was started, man was doomed to ever increasing confusion, for he had to justify his mode of life to himself. This is in fact impossible, except by continual recourse to confusion. For how else can you justify the arbitrary authority of some people over others? You can pretend that God or nature ordered it, that the others are inferior, that we are superior, etc. But once you start on this line, you can never allow yourself to think straight again, for fear that the truth will come out. You must glorify the "hero" who murders and plunders, while on the other hand, you tell the child that he must be honest, treat people fairly, and so on. Just this one point is enough to destroy the minds of most children. How can you square up the emotion of love and truth with that of plundering an enemy, stealing his wealth, murdering helpless people, and enslaving others? No wonder even brilliant people like Aristotle regarded slaves as basically inferior. How else could they stand life in their society?

The basic character of the ego process is to continue in the midst of superficial change. At its core is the desire to be satisfied. Since man began to project his satisfaction into an imagined future, he also needed to feel secure. For he is always seeing causes, real or imagined, that threaten his projected satisfactions and promise projected pains and dangers instead. The fear reflex is thus set in motion. This confuses the mind, which looks urgently for solutions. Usually these solutions put man in an ultimately more dangerous position than he was to begin with. For what else can be expected of a confused mind? Then his fear reflexes are activated again with still greater intensity. In this way, society has been developing into a situation of ever mounting fear. The movement is zigzag, with alleviations and improvements from time to time. But on the whole, each move of man to increase his security has brought him finally to greater insecurity than ever, until today, the "security" of atomic weapons threatens man with annihilation.

I would say that man is suddenly entering a situation where the whole idea of this mode of life is evidently absurd. It always was absurd, but many men had the illusion that they could get satisfaction out of it. It usually took a man 30 or 40 or 50 years to find out that this is false, and that life is "vanity" after all. But today, even the young people feel it, perhaps even more keenly than the older ones. Mankind has been in a chronic state of crisis for 6000 years or more. Now the crisis is acute, general, and inescapable. The old illusions don't seem to work very well any more, nor do the new illusions either. So mankind is presented with a unique opportunity to drop the ego process. This opportunity arises out of a unique danger. He may annihilate himself or degenerate to the level of a confused beast if he does not drop the ego process in a reasonable period of time.

Best regards to you and to Sheila and the girls. Saral will be there in a few days to talk directly with you.

As ever, Dave

Dec. 10, 1962

Dear Isidore

I am continuing the letter I sent you a few weeks ago.

The first thing I would like to do is to bring in the idea of structure-process. Structure is the static aspect of process (a kind of projection). Thus, every animal has a skeletal structure, which is the outcome of a long process of evolution of the species, as well as of growth of the individual from the embryo. This structure is in turn a basis for further movement and development in the process of the animal's life, and in the life of the species. So process and structure are two aspects of one movement and existence. Besides, structure is maintained by a balance of process at a lower level. Thus, the bone material is always wearing out and being replaced, but the bone-structure continues in the process, with a slow general change as the animal gets older.

All structures are based on a repetition of similar elements, in definite relationships. These elements are subject to variations so that there is not, in general, a perfect symmetry. Even in physics, with its crystal structure of ideally repetitious and symmetrical molecular cells, there is in reality a set of dislocations, which constitute variations, irregularities, changes in the crystal. Real crystals are always a repetition of similar elements with differences and variations.

The relationships of elements are important in a structure. Basically, they must fit together, one element against another, to form an integrated totality, that continues to be a totality as it functions, operates, and moves. This is evident for the skeleton of an animal. In a crystal, elements fit well enough to form a solid body, all of whose parts are integrated so that it behaves as a whole, and not as a collection of disjoint elements.

Each structure has a kind of order, a set of sequences of elements that are naturally most immediately related, as well as breaks as variations in this order. These variations open up a tremendous range of potentialities for the structures that can be built out of a limited number of kinds of similar elements. The patterns that are in the structures have a tendency to repetition, but also to breaks and changes, and to surprising new aspects.

Now, this structure is but an aspect of the over-all process of inner and outer movement, by which it develops, maintains itself, and eventually falls apart (or "dies"). Thus, the process has a set of related stages in a natural order, with breaks and variations opening up new potentialities. There is a tendency to a pattern of repetitious cycles (e.g., in the successive generations of animal life), but the pattern breaks, opening up new directions of living for the individual, and evolution for the species. The basic process is always creating, maintaining, and eventually destroying various aspects of a structure, ultimately replacing them with something new. On the other hand, the existence of structure expresses the fact that the process is not utterly chaotic, but can be understood in a natural way, if we can somehow begin with the basic principle of relationship, order, potentiality, repetition, and breaks in the process, and then go on to elaborate and articulate the structure in our minds. This is an important aspect of understanding a structure-process, rather than simply to deal with disjointed and arbitrary fragments "plastered" or "cemented" together in an ill-digested conglomeration.

The problem of structure is basic to my work in physics. In essence, I am trying to find the general principle of the process-structure that can abstract as time-space. If I can do this, then the laws of physics, the nature of the so-called "elementary" particles, their modes of interaction, and many other things will be understood, in the elaboration and articulation of this general principle.

Now, what I want to discuss here is the structure-process of the ego. We already saw that the ego is not a permanent entity, but a process, with continually changing and mutually contradictory aspects. But it evidently also has a structure, in the sense described above. There are basic elements in the process, related in certain ways, in a natural order, with breaks giving rise to many potentialities, with patterns of repetitious cycles, and changes into new structures.

Firstly, we must note that the process-structure level of the ego is not generally perceived with our usual training and conditioning, which leads one to regard the ego as an entity, or better, an "identity," with permanent or perhaps slowly changing features. In my previous letter, I indicated some experiments by which one might see this process-structure, below the illusory "entity-level" of the ego.

The illusion of the ego as an entity is based on confusion. Now, confusion may be explained roughly as *unperceived contradiction*. How is it possible for us not to perceive contradiction? I think that we do it by a process of identification and fragmentation. This goes in a number of stages. Let us take as an example a person who gets angry, and goes into a fit of resentment, rage, pique, self-justification, and what not. The stages are roughly these:

(a) At first, there is a "signal" producing a very powerful total reaction of anger and rage. This signal has the effective meaning that "This is terribly, overwhelmingly important. You can't treat me in this way. I am a special creature with rights of his own. I am not like other people, but my desires, my habits, my urges, my self-esteem must have special consideration and I shall engage in every manner of violent activity until I get what is due to me." This is the first stage of confusion. The person introduces a false fragmentation of the whole problem, into his rights and other people's rights, etc., along with a false identification of all his desires and urges, seeing them as if he were really wholeheartedly and single-mindedly devoted to the object of his anger. As yet, however, there is no conflict, because the "anger signal" produces such a powerful response that the person is not yet aware of the rest of his being, not all of which agrees with the implications of the emotion of anger.

(b) In the next stage, the person begins to glimpse that in his anger, he may destroy a great deal that he values, and wants to keep. So he is in a state of conflict. He wants to hold onto his anger, while rejecting certain inevitable consequences of this anger. This is a continuation of the childish "magical" notions, whereby one can exorcise what one doesn't want while keeping what one does want. In this way, he hopes to get rid of the painful conflict.

(c) He starts to look for solutions which give the appearance of keeping his anger while he gets rid of the bad consequences. Thus, his mind invents thousands of reasons justifying his anger as "natural," as "righteous indignation," and so forth. He says he will "teach the other fellow a lesson." He imagines that he is powerful enough to overcome all obstacles. Or he plots revenge in an imagined future. A buzzing cloud of confused ideas and feelings develops, in which he can get lost, mixing up the dangerous implications of his anger with something else that is less alarming. He has thus "escaped" his conflict. But now he must protect his escape. He must keep on proving to himself the false thesis that he is righteous, strong enough to have his own way, etc.

So here I have given a part of the process-structure of the ego. Fragmentation with identification lead to a narrowing of attention, on the basis of which the individual as a whole can throw himself into one aspect of the ego's desires. But soon, he glimpses contradictory aspects, which are not compatible with the realization of the implications of the aspect with which he is identified. He enters a painful, unpleasant conflict. Powerful signals confuse his mind, leading him to invent false and illusory escapes, eventually giving rise to a twisted, tangled, buzzing cloud of confused ideas and feelings in which the conflict is almost lost (except for vague and disquieting feelings of anxiety, guilt, fear, regret, annoyance, frustration, boredom, etc.).

Now, in this way the whole structure has been described in words and thoughts. But this is only a reflection, a "shadow image" of the real process-structure. One of the biggest confusions is that we identify the real problem with its "word image." One "solves" the "word image" problems, but one pays no attention to the unpleasant fact that the *real* problem is going on as before, only worse, because one has added to the previous general confusion the additional confusion between the word image of confusion and the real process-structure of confusion. So one is confused about confusion. In this way, by identifying oneself in an imaginary and illusory way with the word image in the process of thought, one in reality fragments oneself into the active but unseen "emotional" side, producing effective signals, along with an inactive but visible "intellectual" side, in which all is rationality, sweetness and light, and pure love for humanity.

A further confusion that can arise is the notion that there exists an entity "confusion," a sort of "beast" against which the ego is battling. In fact, it is the individual as a whole who is confused. Confusion has no more of a "positive" existence as a separate entity than do arrivals and departures of railway trains, apart from the process-structure of the railway system as a whole. The fact is that there is an individual who is carrying out an ego process. In this process, the whole mind produces signals that excite the reflexes of conflict and confusion, along with the illusion that there is an "observing, choosing ego," with a "will," who is battling against a confusion that has some origin external to it. So confusion is just an aspect of the individual carrying out his ego process. (Incidentally, I think it is instructive to say "the individual carrying out his ego process." This stresses that there is no separate ego process, but that it is more like an "act" or a compulsive ritual that the individual is in the habit of carrying out.)

Basically, the further process-structure of the ego is then that in the awareness, there is an illusion that there is an independently existing conflict, against which the "observing, controlling ego" is battling. In fact, if you will watch for the signals, you will see that the conflict is only a response to the signals, as the electric light responds to pressing the switch. In a similar way, the illusion of an observing, controlling ego is also projected into the awareness, in response to another aspect of the same set of signals, that sets the reflexes of conflict going. But one can become aware of the character of the whole business, by understanding its process-structure.

When this happens, the process-structure begins naturally, of its own accord, to fade away, without the need for any actions by the illusory ego. Indeed, the awareness never acts. Rather, the individual, with his body and mind, is in fact a totality, which responds naturally and of its own accord (at least when the nervous system and brain are not damaged) to the total extent of the awareness. If the awareness is confused, the response will be likewise confused. A clear awareness requires no further action, for the individual naturally acts clearly if he is clearly aware of all that is relevant in relation to what he is doing.

If there did not exist a natural way by which the mind could become aware of its own confusion, thus bringing the latter to an end, then humanity would have no way out. For it is evidently impossible that the ego process can bring its own confusion to an end, by an act of will. To try to do this is only to continue and enhance the confusion, which is in essence that there exists an ego that can make "choices" and exert its "will."

It is very important then to get below the usual level of thoughts, words, and feelings, coming to a *direct awareness* of the process-structure of the ego. This one obtains by learning the relationships of various feelings and thoughts (what is the signal, what is the reflex reaction to this signal), what is the order (how cause becomes effect and effect becomes cause), what is the break (a sudden needling, nagging, jabbing, nervous tension can switch you from one aspect of desire to a contradictory aspect), what are the repetitious cyclic patterns (going back to childhood), what are the breaks in the patterns (something formerly pleasant develops a painful conflict, etc.). You see the illusions behind our emotions and thoughts exposed, perceiving, for example, that one is engaged in a sham struggle against confusion, confined only to the verbal and intellectual level, while the real process of confusion is allowed to go on as before.

Soon you begin to wonder if words and thoughts really mean anything at all. Actually, they very often do not (the same holds for feelings). Our words function as a sort of "living card index system," such that when the right "button" is pressed, the memory projects pictures, sounds, smells, and feelings in such a profusion that we easily confuse them with real perceptions. These projections include "action" signals, and signals that lead us to mistake what is desired for what is true. So what happens is that we respond to fresh experience by trying to accommodate it into our "card index system," not noticing that we have replaced actuality with a fantasy having a superficial resemblance to what is really there.

Now, the proper function of words and thoughts is either to reflect what is actually perceived or to awaken an actual perception. Unfortunately, we often use words and thoughts as a substitute for perception in the manner described above (as a bureaucrat might tend to identify people with the data about them in his card index system). So the important point is to notice that there is a level below words and thoughts (a structure-process level) in which action, thought, and feeling are determined.

We must see this first, and then try to reflect it in words (if we wish to communicate what we see). So words are only an abstract reflection of the structure-process level. It is wrong to allow them to become the "buttons" that set up the signals based on memory, which lead to further thought and feeling, and to action. For this in reality leads only to confusion, since in fact action is always determined at the structure-process level. Hence, it is only an illusion if we imagine that our words and thoughts are determining action. What actually happens in this case is that remembered signals at the structure-process level are "re-played" and determine action in a way that usually escapes conscious notice. So we "say one thing and do another." This common habit is not the outcome of viciousness, but rather, of the very general kind of confusion described above.

The deepest aspect of the structure-process of the ego is the desire for what is pleasant, secure, self-expansive, and dislike of the unpleasant, the insecure, the uncertain, that which decreases the self. One would like to have a permanent state of pleasure, peace, gratification, interesting things going on, etc. All of this is united in an automatic reflex. As I indicated in earlier letters, there is probably a signal from the primitive "animal-like" brain in the center (the thalamus), going to the outer cortex, and tending to identify with the pleasant aspects of the ideas and images produced there, and against the unpleasant ones, thus starting a movement of fragmentation of the functioning of the cortex. But it is not really fragmented. Rather, the cortex is always actually a totality, and indeed thalamus and cortex together are always a totality. But because of lack of a broad and deep attention with understanding, the individual is not aware of this totality, but only of the aspects as fragments, within which are identified further aspects that are in reality quite abstract.

Now, the whole thought, feeling, and action of the individual responds to this whole awareness. Because the latter is wrong, the response is confused. Thus, if I imagine that the reflection of the world in a pool signifies an "upside-down world," and if I go to explore this upside-down world, I will get a nasty shock as I enter the water. Similarly, if I imagine that the pleasant aspects of the images and feelings produced by the cortex can be separated from the unpleasant ones, and permanently identified, "taped," and "secured," then I am in for some even more nasty shocks than in the case of the illusion of an upside-down world. For soon it will turn out that the more the thalamus sends in signals building up the pleasant aspects, the more the unpleasant aspects will also be built up. As the thalamus sends in signals to decrease or destroy the unpleasant aspects of the thoughts and feelings of the cortex, it will interfere with the pleasant aspects, because in reality the two are inseparably connected. Then I will have to strive all the harder in favor of the pleasant aspects. As the conflict builds up, there will be an extremely unpleasant sensation of effort frustrated; but here, the thalamus sends in signals that start the cortex cooking up escapes that alleviate the pain, thus creating still more confusion.

Now just as it is an illusion to separate the pleasant from the unpleasant, the secure from the insecure, so it is an illusion to separate the urges of the thalamus from the activities of the cortex. The thalamus is stimulated by the creations of the cortex, and these are in turn stimulated by the signals from the thalamus. In the cortex, the idea is produced that the signals that one feels are the operation of one's "very self," something that it would be senseless and meaningless to deny or question. And the reason for this is clear. For the thalamus cannot "force" the cortex to do anything. Rather, in the presence of these signals, the cortex responds naturally, and inevitably, of its own accord. Similarly, in the presence of cortical illusions, the thalamus very naturally starts to send out signals for and against various aspects of these illusions. So there is an appearance of complete spontaneity, as if the individual could not conceivably do otherwise. And indeed, without an integrated awareness, all of this behavior is natural.

But what is unnatural is the separation of thought and feeling, of the content of the mind from the thalamic urges that try to fragment this content, by identifying for and against various of its aspects. If this is the relationship projected into the awareness, then the natural response of the whole mind to the wrong "picture" will be to increase the confusion even more. In this way, a person can get lost in confusion. But as he understands and perceives the real process-structure of the illusion, it is this understanding that is present in the awareness. As a result, the whole mind responds naturally, by ceasing to stimulate the thalamic urges and by ceasing to create cortical illusions accompanied by an urgent signal meaning "this is the truth." So the mind will naturally start to clear up its own confusion, which latter will die away like a dead leaf. When this happens, the mind will function of its own accord, as a totality. The creations of the cortex will be accompanied by a perception of the truth and falsity, along with an understanding that pleasure and pain are not very relevant. The thalamus will send in signals to build up the true creations, and make the false creations die out. This is the intellectual aspect of the feeling of love, which attaches to nothing at all, for or against, and which wants only the truth.

If that happens, then fears will come to an end. For fear is a reflex that is set in movement when something that is identified as so necessary by the thalamic signals is seen to be threatened. The fear reflex causes the whole person to seek every possible means of protecting what is regarded as necessary, or failing this, to create illusions of escape from the approaching danger. Thus, fear confuses, corrodes and corrupts the mind.

It is worth experimenting with fear. Bring to mind any real fear that bothers you (death, illness, loneliness, etc.). Be intensely aware of all the pangs, nervous vibrations, variations of nervous tensions, the racing heart, and so forth. *This is the process-structure level of fear.* Be aware of the relationships of these pangs to thoughts that flit through the back of your mind. This will help expose the process-structure in all its detail. Be aware of the *order* of the fear-feelings, how one tends to give rise to another, of the *breaks* in the feeling, its recurrent repetitive *pattern*, and the *changes in this pattern*. Gradually, you will understand how mechanical it all is. Sometimes, there may be a literal "explosion" of fear sensations, creating a "hole" in the middle of the sensations. This "hole" expands, and the fear is dead. But as you return to the word "fear," to the memory and recognition of fear, the old signals get to work and it comes to life again. But after an hour or two of this, you will learn a great deal about what fear really is, how it is united in the mechanical response of protection of what is desired, and of how attachment to desire is also a mechanical process.

Now, this brings us to the problem of thought. If you watch yourself thinking, you will find that it does not go in a simple straightforward way. Suppose you have a problem, and the answer is not clear. You will find that the memory is being searched. But soon you may realize that the answer is not directly available. Perhaps it has to be reasoned out from what you already know. Watch this reasoning process. The mind is experimenting, creating trial solutions and seeing if they work, if they are logical, etc. But you will notice that the choice of these solutions is largely determined by emotional signals. Is this idea pleasant or profitable to you? Is it too disturbing? Does it reflect well on you? Does it do what you want? Does it create a feeling of security? Or does it make you uncertain, doubtful? Does it suggest that you did wrongly, that you are stupid, selfish, cowardly, or mistaken?

Long before the mind has time to apply "logical" tests, you will see that it rejects the unpleasant ideas, and builds up a buzzing cloud of confused feelings and further thoughts supporting the pleasant ideas. Thus, thought proceeds tortuously – and it must do so, because firstly, it is based on motivation. Any motivation is a thalamic urge, identifying for certain things and against others, quite indifferently to the question of truth or falsity, as well as logical coherence. Thus all motivation must limit the range of truth. No matter how wide this range is, it must be limited. Indeed, no thought is possible without some kind of motivation. Every motive is an arbitrary intrusion into the process, which will finally lead to contradiction (since as we have seen, there are always hidden connections between what we want and what we don't want, so that our motivation must finally make what we originally wanted come out as what we also do not want).

The confusing role of motivation can be seen in another way, as suggested by Krishnamurti. Every unsolved problem is felt by the individual as a challenge. He responds to the challenge out of his conditioning. In other words, his "active memories" determine what he will regard as the problem and what he regards as a satisfactory solution. For example, if I am a government official, I may be presented with the problem of how much money doctors should spend in the Health Service. If my conditioning tells me "to save money is the main point," I will cut down what is spent on drugs. That is my solution, my response to the challenge. But then, as a result many people may become seriously ill. There may be much misery, and even worse, money will be lost in the long run. My response to the challenge was therefore too narrow. It was determined by my motivation, which is the outcome of my conditioning. So my thought did not go "straight"; it proceeded tortuously to get a result that would satisfy my conditioning, rather than solve the real problem. And all thought has similar limitations, however broad, deep, natural, or precise it may be. It is only in the moment of understanding that these limitations cease to operate. But usually, if our ego will consent to get out of the way for a moment (after a long struggle that convinces us that the answer is not to be found by "talking it through"), it quickly returns to utilize the result of understanding in order to realize its old motivations once again. Our problem is whether we can have a sustained state of "non-thinking," i.e., of understanding. It is only this that can solve the problems of the modern world, which are no longer mainly the mechanical ones of finding means of surviving in comfort. These problems have in essence already been solved. So understanding is what is called for.

Now, thought is inherently involved in the process of time. One begins to think because of a challenge, a problem that one senses. The previous adjustments of the ego process are perceived to be inadequate, life batters down our old refuges, etc. So something must be done. The first stage is to perceive or sense a problem. Here, it is important to see the real problem. In science, for example, an experiment in effect "asks a question of nature." Half the battle is over when you ask the right questions. If you ask irrelevant questions, you will get answers that confuse you or mix you up. But the whole of life is that way.

Now, what determines how we see the problem? As I have indicated earlier, it is the response of your conditioning, the active memories playing back these emotions and stimulating confused ideas, which tell us that we must retain certain essential relationships, pleasures, and securities. So in this way, we respond to the challenge with a demand (often felt only implicitly rather than expressed explicitly). Then thought begins, as a process of seeking a means of satisfying this demand. This process goes on over a period of time, while the demand continues. First, the memory is searched for such a means. If this is of no avail, associative thinking is tried. "A" calls up "B" which calls up "C," etc. A more sophisticated person then tries logical thinking. He organizes, classifies all his knowledge, formulates relationships, and theoretically works out what is entailed by these formulations. If this doesn't work, he gives up for a while, but hopes someone else will suggest something, or that there will be an inspiration or a new discovery.

If the problem is really mechanical, these methods will probably eventually turn up a solution. But in life, the problem is seldom basically mechanical. Usually the problem is not external to the person, but really arises in the demands of the ego process themselves, which are self-contradictory, and therefore incapable of being satisfied.

Thought is incapable of understanding the demands that are its motive power, and that determine both the questions and the kind of answers to them that will be regarded as suitable. Even if the demands are questioned, this is done only on the basis of new demands (e.g., for peace, coherence, new kind of satisfaction). So if the problem is not mechanical, thought is necessarily confused, not being able to see that its basic assumptions are what create an insoluble problem. This can be seen only when the process of thought ceases, i.e., in understanding.

One can see that time is the factor that brings in confusion in thought. Out of the demand of the ego, one projects the essential features of what would constitute a solution. This projection is onto an aspect of the mind which we label "the future," and which evokes feelings of expectation, possibility, potentiality (being a part of our "feelings"). In reality, it is all of course in the present, but it is the future that we expect and hope for. This is the first confusion in the time process – to split a part of the mind off, and to call it "the future," with the implication that striving and effort are needed to bridge the gap between what is and this so-called future. In fact, there is no such gap, since this "future" exists now, but it is in reality only an illusory projection.

The next step is that the mind engages in a process that takes time. This is the second factor of confusion. For as time goes on, one continues to strive for what seems to be the same goal. But because of the inherently contradictory character of our demands, this goal will tend to change. Partly as a result of what is turned up in the process of thought itself, new aspects of our demands, previously hidden in buzzing clouds of confusion, will begin to emerge more clearly. Thus, even if by good logical reasoning, we produce a "solution" to what seems to be the initial demand, we find that it doesn't satisfy the demand in its more developed form. Yet, we think this demand is only a continuation of what it was. We still give it the same name. We still think that it is the "same" demand. Thus, our confusion is always on the increase. This is characteristic of the process of time. We identify our past and present goals, and in order to do this, we must fragment something else that is really one. For example, as a man gets older, his taste for the pleasures in life alters, but he does not notice this. So instead he may say, "Things aren't what they used to be." Or else he may say that the "eternal verities" are the same, but the younger generation is not as good as the older one was.

When is the thought process correct? It is correct when the conditions of the problem and the goals do not alter significantly while the thinking process is going on – either because of external factors, or because of the results of thought themselves. For example, if I want to get from A to B, I can try to find the best means with the aid of thought. This is possible only as long as A, B, the means available, and my goal do not significantly change while I am thinking. In practical life and in scientific research, many of our problems do have this character. But as we approach the frontiers of knowledge, things become more confused. It is not clear what the proper goal is. Our idea of the goal may alter, as we do our work. What is called for here is a creative process of understanding, rather than to begin with a demand that is the outcome of past experience and conditioning. Creativity does not accept the demands that motivate thinking; but it allows the question itself to be one of the things that must be understood. To be creative, therefore, one must not be responding to a challenge through one's conditioning. Rather, one must be in a state in which this conditioning itself has ceased to operate, because it too is understood.

This brings us to your question of "understanding of understanding." It seems to me that such an understanding must be the natural outcome of the proper function of the mind, which, when it is not confused, will probably understand understanding without any special action by the ego process. So the key problem that we can start on is to "understand non-understanding," that is, to see the totality of contradiction, conflict, escape, and confusion. This process, being mechanical, can be reflected into thought. When it is understood, the process will stop. Then the mind will be in a new state of totality, in which one can become aware of the totality of understanding.

By the time you get this, Saral will be in Israel. She would like to talk these things over with you.

Regards to Sheila and the girls, also to Saral, and to you.

As ever

Dave

P.S. I expect eventually to write an article on this topic; so could you please save these letters.

P.P.S. It has just occurred to me that one could consider the identification of the ego itself as part of the process of confusion, as described in this letter. One wants to feel that there is a permanent ego, and one "identifies" it with various feelings, desires, ambitions, possessions, achievements, relationships, etc. In reality, these are changing from moment to moment. But the ego cannot admit this, as it would destroy the "act" on which the whole process is based. So the ego identifies itself as "fixed and permanent" or else "slowly changing and developing." But because the ego is so different from moment to moment, its reaction to aspects of the world that are not essentially different will also have changed. To explain these, and thus to maintain the illusion of the permanence of the ego, one will have to fragment the world wrongly and artificially, introducing false differentiation (e.g., "Today is not as nice as yesterday" really means I don't feel as well about today as about yesterday). These false differentiations lead to false categories, and therefore to further false identifications, projected onto the structure of the world. Because I neglect my charges and identify myself as American, British, Israeli, Brazilian, etc., I will falsely ascribe great importance to nationality and other differences in people, which are really only a reflection of my own attitude, and I will falsely identify different residents of a given state. This projection of the ego process into one's perception of the world extends into every phase of life.

May 24, 1967

Dear Isidore

The news is very disturbing and we are very concerned as to what is happening in the Middle East. We do hope the crisis will pass very soon. Meanwhile we can only wait and see. The flames of nationalism are beginning to build up. Every Arab gets great pleasure out of identifying with the victory of the Arabs, and pain out of their "humiliation" by the Jews. So his mind is ready to accept any illusion, if he can only get pleasure rather than pain. The Jews are the same, at bottom. After all, the Jewish nation is also only an idea in the minds of various people, that gives them a satisfying and pleasing sense of identity and security. Any Jewish baby raised as an Arab would get great pleasure in Arab victories, and vice versa. It is all a matter of mechanical conditioning. The same is true of American, Russian, Chinese, or any other nationalism. Each nation is thought of as supreme. So if two nations disagree, in the end, there is nothing for it but war. Nationalism makes brutal and destructive wars inevitable. The idea of peace between nations is meaningless. The very existence of a nation implies a state of mind that makes war unavoidable in the long run. And politicians, along with their followers, are like drunken people, whose minds are befuddled with clouds of illusions. They don't really see what they are doing. The whole world depends on this "tight rope walk by drunken men." All we can hope for is that it won't explode in flames. Perhaps if it doesn't, people will some day see the absurdity of the whole structure, which will collapse like a bad dream.

Meanwhile, we have somehow to live, and not to get caught in confusion by identifying with what is happening. This is often a difficult and worrying thing. We do hope that all of you are all right in

Israel. Here, in England, people may worry more, because they don't know what is happening. Saral called up Betty, who is very worried about her mother, especially her reaction to leaving all of you in such a state.

If there is anything that we can do to help, please let us know. Don't hesitate to write, telegraph, or telephone.

The more one watches, the more one sees how meaningless the whole of human political action is. It is all based on a state of mind that is not only lost in illusion, but that is the generator of illusions. This is the state of mind that mistakes thought-induced pleasure for real joy and creativity. I have been looking at the way it works. It is roughly like this: All thought has a primitive or infantile component, in which a part of its own structure is a set of sensations of pleasure and pain, along with a tendency to present these as the basic substance of reality, at least at the psychological level. Now, all thought, including healthy thought, contains a tendency to formulate challenges and to seek to respond to them, by referring to further thought. Thus, if a machine is not working, one senses that something is wrong. This gives rise to a challenge. The mind begins to think. Perhaps one already knows the answer in memory. Perhaps someone else knows it or it is in a book. Perhaps one can figure it out logically from what one observes, etc. Such a challenge can give rise to an intense feeling of urgency, and this is only natural and proper, when there is an urgent problem to be solved.

However, when thought mistakes a part of its own structure for real pleasure and pain in the very "substance" of one's being, this too is inevitably sensed as an urgent problem. Thought will explore how to "solve" it (i.e., increase the pleasure and get rid of the pain). But in this exploration is inevitably a structure of confusion. For the "problem" is entirely due to thought itself. If thought stopped, the problem would vanish. And any thought is bound to have pleasure and pain components, therefore a similar problem. So as long as you are thinking, the "problem" is insoluble, and as soon as you stop, there is no problem.

The difficulty is that when thought gets an inkling of this fact, the "pleasure structure" begins to die out. Another level of thought, to which the brain is at the moment not paying attention, then goes into automatic and mechanical operation. The level of thought is like a machine, "set" to defend what appears to be the basic substance of one's being. So it causes thought to manipulate itself, not to reflect truth and fact, but rather, to preserve and enhance what appears to be the pleasure in the very center of one's being. It will do this by suppressing awareness of anything that would expose the emptiness of this mode of thought and therefore of the whole structure of pleasure itself. This suppression process involves not only feelings of dullness and deadness, but also rabbit-like darting from one subject to another, intense excitement that fills the brain and destroys clear discrimination, and the acceptance of every kind of false thought as true (i.e., fantasy, illusion, and delusion). So one sees that the illusion of a false pleasure-center in the mind is also the generator of an independently proliferating series of illusions. It is an "illusion-generating illusion." The key to sanity is to see through this basic illusion-generating illusion. It is no use merely to see through some of the particular illusions that arise in this process, as long as the "root" that generates all the illusions is not touched. But to see this root is very difficult, because the state of mind that wants to see it is already lost in illusion. So what it will actually see is an illusion about the root of all illusions. This will be worse than useless.

One reason this happens is that when there is internal conflict, thought projects the illusion of a "self" who is "observing" the conflict and trying to resolve it. Actually this "self" is only a cover-structure for the total process of generation of illusions. To see this, note that each idea "stands for" some concrete reality (e.g., the idea of the desk stands for the desk that can be touched). What does the idea of the "self" stand for? If you observe, you will see that it stands for what seems to be the "very substance of your being." But this in turn is nothing but the whole illusion-generating process. So it is necessary to realize that "I *am* the conflict, the process of generating illusions." Therefore, it is meaningless for me to try to "do" something about it. Whatever "I" do, the result will be to stir up the whole process even more.

Indeed, even when "I" seems to be "only looking" at the process, it is being stirred up. What is actually going on is that the brain is thinking about the process, while at the same time, its thought is governed by the pleasure principle, which puts pleasure first, ahead of factuality and logic. So the brain is throwing up a swarm of illusions about its conflicts. It is as if I were trying to look at something, while I stirred up a gigantic cloud of dust, so dense that I could not even see how my hands were stirring up the dust. I would say that I am "looking," trying to "penetrate" the dust. But to say this would be meaningless, as long as I did not understand that the dust was being stirred up entirely by my own action. The first step would be to stop doing anything and to let the dust settle. Then I would look without trouble. Similarly, if the brain can refrain from "trying" to resolve its conflicts, these will vanish of their own accord, spontaneously and naturally, leaving the "emptiness" in which clear perception takes place. Whatever one tries to "do" is based on the root of all illusion and is therefore just an extension of the very problem that one wants to get rid of.

Nevertheless, it is urgent that something be done. For we cannot go on as we are. It is just another illusion simply to cease to be aware of the problem, or not to pay serious attention to it. We must be intensely aware of how the mind is working, without attempting to do anything about it. This awareness is enough. Real awareness already is action, without the need for a "choice" by the "self" to do something. Thus, as soon as you are aware that something is poison, you have already stopped trying to eat it. No choices, decisions, or efforts are needed. But the "illusion-generating illusion" covers up and suppresses awareness of its poisonous character. To really see deeply the nature of this illusion is action enough. For in the light of this perception, it has to collapse. Whatever you do beyond this just stirs up the cloud of dust.

Even when one begins to see through the false role of the "self," there remain deep tacit assumptions about thought. We have been trained to believe that thought is always necessary. Of course, it is necessary in certain external problems. But in internal problems, it is the very source of the trouble. Nevertheless, we tacitly assume that the thought that is "looking" at an internal problem is different from and independent of the problem that it is "looking at." This is not true. Both the problem and the thought that "looks" at it have the same root in the generator of all illusions – i.e., the pleasure principle.

One can understand this better by noting that in addition to pleasure and pain, all thought contains a component consisting of active reflexes. In young children, this is the main component. Thus, children first learn to think of something by imitating it. This imitation requires a set of reflexes. Later, the activation of the corresponding reflexes can "stand for" the thing, without the physical act of imitation. Later still, images and words stand for the reflexes. But the reflexes are always there. The trouble is that they are confused with an active functioning reality, that would be at the core of one's being. This is because they are intensely active and deep within, therefore seeming to be vital, alive, and central.

A similar imitation occurs in emotional questions. Thus, children learn attitudes by imitating the emotional reflexes of other people, and later, these reflexes, dynamic and active, seem to be the very core of one's own being. One doesn't note that they are mechanically set into action by words and thoughts. (e.g., "Arab humiliation" leads to a set of reflexes in everyone, but especially in people who identify themselves as Arabs).

Similarly, when we try to think about our own internal problems (e.g., to get rid of the illusion-generating illusion), then the very thought with which we "recognize" the problem contains a set of active reflexes that *imitate* those out of which the problem is constituted. But this imitation of the problem is the same as the problem. It is dynamically active. Thus, the thought with which we think of the "generator of illusions" itself has the same structure as the generator of illusions, and is in fact only an extension of the latter. (Just as the thought of "Arab humiliation" is the *feeling* of being humiliated, and then applied to the word "Arab," so that if I am an Arab, this thought *is* my humiliation, with all its inevitable reactions that demand revenge.)

There is, however, an entirely different mental activity that does not imitate the structure of what is seen. Rather, it *reflects* this latter structure, but in itself, it has the structure that moves toward truth and factuality. Thus, when one understands brutality, one is not imitating a brutal structure in the mind. Rather, one's structure has to be based on love, truth, factuality. Within this is *reflected* the complete meaning and structure of brutality. We perhaps tacitly imagine that this is what we are doing when we think of brutality. But it is not so. Thus, it has been shown that one who watches a film of brutality sees it in terms of his own brutal feelings, which act dynamically to make him behave with brutality in other contexts. The trouble is that he sees brutality through the thought of brutality, which is based on a kind of internal imitation of the structure of brutality.

When one internally *imitates* an illusion-generating structure, one is thereby immediately lost in illusion, so that whatever he does is worse than useless. Therefore, what is called for is an ending of the response of thought, which is too mechanical. Rather, what is needed is response from the emptiness, which sees the structure of illusion generation, without imitating this structure.

Saral and I send all our love and best wishes to you, to Sheila, and the children. We do hope that all turns out well.

Dave

8 UNFOLDING THE SELF-WORLD IMAGE (1987)

In the following conversation, Bohm explores the manner in which implicit values and assumptions actively constitute both our image of our self and our image of the world. Normally, assumptions and values do not present themselves as such in our experience; rather, they appear simply as "the way things are" and "the way things must be." Thus unexamined, values and assumptions tend to generate conflict that is fueled by great "soma-significant" energy (Chapter 5), particularly when issues of supreme value are at stake.

The logical solution to this problem – to acknowledge the limitations of our assumptions and modify them accordingly – usually involves a structure of experience in which "I" am examining some part of "me" (in this case my values and assumptions). The difficulty with this approach, suggests Bohm, is that the "I" that ostensibly observes the values is itself subtly infected with those very values. Any change that occurs through this process is likely to be topical, modifying but leaving intact the essential nature of the values in question. Consequently, ethical injunctions, arising from a reservoir of cultural mores and manifesting through the "I," may reconfigure or suppress behavior, but rarely change it at a fundamental level.

As an alternative to this reflexively ingrained mode of self-observation, Bohm points out that we have recourse to the body as an immediate display of the actual movement of values and assumptions. It is through the body that we can experience values and assumptions as concrete processes, rather than as purely abstract ideas. In this way we discover, as a very intimate example

Previously unpublished interview, courtesy of Saral Bohm and Lee Nichol.

of soma-significance, that the ideational aspect of a value or assumption is inextricably linked to neurophysiological activity. From this perspective the information displayed in the body can be considered a bridge between the collective "implicate" nature of society and the individual "explicate" nature of the self.

This approach to experience – in which the symbiotic nature of idea and energy is suspended and displayed throughout the entire organism, rather than just in "the mind" – has the potential to engender proprioception. Typically, proprioception refers to the body's capacity to instantly comprehend and orient its own movement in a coherent manner. However, says Bohm, our thought process currently lacks the capacity to grasp the whole of its own activity, largely because it does not recognize its inseparability from the energy patterns that motivate us on a daily basis. But through suspension and display of values and assumptions, we may come to the threshold of a new mode of proprioceptive awareness in which the conflicts of socio-cultural injunction and the limitations of "I" observing "me" are resolved at a fundamental level.

Nichol In attempting to understand the nature of the self, it seems that one can spend a great deal of time and energy, yet still make fundamental mistakes with regard to observation.

Bohm Yes. You see the whole field is very deceptive. Things are not what they appear to be. The structures are a lot different from what they seem. For example, one of the basic assumptions that we make is that one can look at the mind as if one were a separate observer, looking at something different, as I, for example, can look at the chair and see that my thought is one thing and the chair is another. The chair is independent of my thought, and my thought can move independently of the chair. We may make a similar assumption as we look at our own internal processes, but this is not true. Our thought profoundly affects the emotion and the whole state of the body, which in turn profoundly affects thought in a cycle, a feedback loop that tends to build up. This is one of the basic mistakes. If you thus start with a false assumption, your whole enquiry may make things worse, and add more complications to those already there. There are many such false assumptions that are operating within our sociocultural context.

You see, if the assumption of the separation of observer and observed were correct (which it isn't), it would make sense to project, to find out what is the problem and try to bring about some desired result as a goal. In such an approach, which is suitable, for example, in practical affairs, you may change your goal through further insight, but the basic idea of having some kind of a goal to direct you is always there. On the other hand, within the mind, this approach may be totally out of place because there is no separation of the kind that has been assumed. The goal you project is therefore fantasy, with arbitrary features of certain ideas that you are simply trying to impose on top of the confusion that's already there, about which you're actually doing nothing.

Nichol It seems that part of the difficulty is that we may read this or hear it, and in some ways it seems quite clear. Then we assume that we can move on to more important issues – but without having really gotten to the bottom of this basic question of how we observe ourselves.

Bohm Yes, it's not so easy to clear it up, you see, because we're caught up in it. One can say that one of the problems is, that we may have insight into this issue on a certain level, but that then there is still the problem of distraction. In this connection, I have a friend who was studying young children. There has been a belief, based on the work of Piaget, that children learn certain concepts such as conservation of water, at a certain age. But my friend has shown that such learning has to do with the function of distracting factors. If you can reduce the distracting factors, they can learn it much earlier. And if you increase the distracting factors, there may be delays. Or to put it differently, attention is required to learn, and distracting factors may draw the attention elsewhere. Similarly, at an intellectual level, you may see fairly clearly that the problem that we are talking about here is that of the observer and the observed, but when the time comes to look in another context, there are a lot of distracting factors. One of these is the ability of the mind to create very powerful, vivid, convincing images that are experienced as real, especially when they move very fast. Thus, if we take a television set and there is a telephone bell ringing, when we look into the image and see a telephone, we experience that telephone ringing in the image though there is no telephone, nothing there except spots of light. But on the other hand, if it doesn't look consistent – for example, if nobody answers it – we may think it's the telephone in the next room and experience it that way. So the way we experience depends on *attribution*.

A basic property of thought is to attribute a quality or a property to something. And then it's experienced as intrinsic to that thing, right? So I suggest that once you have the assumption of the observer and the observed, the mind can create an image of an observer looking at the observed, as you could have in the television set. You could have some man looking at something and you could say there's the observer, and there's the observed – but nothing is going on at all of that nature. And similarly, in the mind, there will seem to be the observer and the observed, and various little things indicating that combination. Thought attributes the whole of the process to the observer who is looking at the observed, and who says that thought comes out of the thinker. What actually happens, however, is that thought creates the image of the thinker, and then it attributes its origin to that image. Thought then behaves as if it were being produced by a thinker, but in fact, thought is producing an image which it calls the thinker and attributes itself to that. The thinker and the thought, and the observed and the observer are just different phases of one thing, one process. And therefore, as a person is thinking, very often tacitly and implicitly without knowing that he's thinking, all of this is attributed to a thinker, which gives it great authority.

Nichol   It seems that this separation is well-hidden.

Bohm   What is covered up is the true nature of the whole process. Actually there is no real separation, but the assumed separation is attributed to an image, and the resulting experience is regarded as proof that there is a real separation. That is to say, the image is experienced as if it were real, and that is taken as proof that the assumption is correct. This is part of the way in which the real nature of the process is covered up.

Nichol   All that you're describing is generally an unconscious process.

Bohm   We'll call it unconscious, implicit, tacit. The thought behind it is implicit.

Nichol   If this process of obscuration is implicit or unconscious, it seems that it would take something more than conscious, analytic thinking to reveal the actual dynamics.

Bohm   Yes. You may say consciously and rationally and logically this is what's the case, but if your whole feeling and whole experience and sensation are telling you otherwise, you really can't be deeply convinced by it, right?

Nichol   So there are two things going on – an intellectual recognition that something may be operating in one way, but at the same time, a deeper set of sensations and experiences apparently indicating something very different.

Bohm We wouldn't necessarily say deeper, but different. It is a set of experiences that don't agree with your intellectual conclusions, even though your intellectual conclusions are probably right; you've probably had a real intellectual insight at that level. So we mustn't decry the intellect or say it is never of any value in this context.

Nichol Instead of viewing the contradiction that you've just described as a further difficulty, perhaps that contradiction, properly attended to, could actually lead to a deeper understanding of the whole process.

Bohm Yes, you have to give attention to this contradiction – that's quite right. And the question, then, is how. For this whole process of covering up and deception is going on. There's a constant "show" being put on, implying that all this is real, and that the intellectual stuff is not real. For example, the person may well say, "I'm not an intellectual, that's just a lot of ideas. My real *gut feeling* is that it's the other way." And, "I don't go in for this intellectualism," so I ignore all that you say, right? What I wanted to say is that this gut feeling is what is deceptive. There are *true* deep feelings, you know, you may get all sorts of responses if somebody dies that you're close to, or if you look at nature, seeing the beauty and so on. But then I say there are also feelings which *appear* to be deep feelings, but are not, because they are produced by thought.

Nichol But they have all the attributes of such feelings.

Bohm They don't have *all* of the attributes or else we could never get out of it. But they have enough attributes to get by, to be accepted by us as real. The point is, now, to be able to see that this is what's going on – that we are producing feelings out of thought. Everybody knows you can whip up feelings by certain shouts and cries and clamors and marches and songs, political rallies, etc. It's well known that feelings can in this way be whipped up, essentially by actions directed by thought, so that such a response need not be a surprise. What about this sort of feeling as compared with deep feelings? *At the moment that it is happening a person might not be able to tell the difference.* You have a crowd shouting and screaming and a great leader in front of them shouting and screaming and driving and urging them on, and so on. So that establishes the principle that feelings can be produced artificially. But what I was talking about is much more common than this. It doesn't require a demagogue or some unusual set of shouts, screams, and cries to do it. Rather, one simply has to notice that the meaning of a thought tends to be carried out in terms of feelings all over the body.

In order to demonstrate this, you may take the case of getting angry. This is a feeling that is not as difficult to look at as say fear or pleasure – deceptive feelings of pleasure – which you know too can be produced by thought, a seductive thought. You see, a person may first get an outburst of anger and then cool down – it simmers down, but it's still there. You may put it in abeyance because something more important comes up, but it's still there ready to come up. My suggestion is to call it up on purpose by trying to find the words that express the reason for being angry. Thus, you may say, "I'm angry, and I have *good reason* because he did this and that and that." You will find that you are getting still angrier. Usually you'll say, "I shouldn't get angry, so I'd better stop this." But now we're going to use this on purpose, not for the sake of getting angry, because we're going to suspend the angry feelings, neither by stopping them, nor letting them come out. Is that clear what I mean?

Nichol Yes, but there are some difficulties with suspending.

Bohm Well you see, it's not being done right in the heat of your original outburst of anger, but still, you're not calling it up to get rid of the angry feelings. Your first impulse might be to try and go out and insult the person and do something, and in earlier times you might even have hit the person. And now you don't do any of those things, but let the feelings come up and watch what's going on. We're regarding it as a sort of test display of the process, you understand? So then you'll see these angry feelings which will produce tension in the solar plexus and the belly and the chest, and affect your breathing, and heartbeat, and all sorts of things. You'll be able to see a sort of movement of responses all over the body, such as a tension of the jaw, in the neck.

Nichol Even if one waits a bit beyond the heat of the moment, there still comes up a very strong resistance to acknowledging that one is actually in this state.

Bohm Yes, that's part of our socio-cultural conditioning, which says that you shouldn't be angry. And not only that, you yourself have seen by clear thought that it's leading you astray. You see, both reason and society . . . everything is telling you that you shouldn't be angry. Now there's a serious mistake in there. Of course, it's right that you shouldn't be angry, if only because it is very destructive to your deeper interests. But the attempt to say you just shouldn't be angry is simply not affecting the anger, it's just trying to impose another pattern on top of the anger. This will come out as we go along, but the first point is to realize that such resistance is false and that this falseness will come out as we go through this process and pay attention to it.

Nichol The falseness of the socio-cultural as well as the personal judgment.

Bohm Yes. This is very tricky, because in some ways the judgment appears to be right. But there's a fundamental, deeper falseness in it. So we also have to give attention to our tendency to say, "I shouldn't be angry, I must stop being angry," and we will see that this too has to be suspended. In this process one will begin to get certain feelings, at first perhaps very faintly because of all the resistance, and later more strongly – you'll see the play of these feelings over the body, because the action is being suspended. If you actually *did* something, you would no longer notice the feelings. If you went out and hit somebody or punched him in the nose, or insulted him, or otherwise tried to get redress for your anger, you might momentarily feel a lot better, because the tension would go away (until the other person retaliated in a similar way). But now, when action is suspended, you can see that the words are calling up the feelings, and you'll be able to get a sense that there's some sort of mechanical connection between the words and the feelings.

For example, you may find that the words may be, "He shouldn't have done this; he shouldn't treat me that way; he hasn't due regard for me, he's always doing that; he's never taking my rights into account. It's not the first time." So you may notice the feelings coming up rather mechanically, and that those feelings are producing mechanical pressure, making it very hard to look at those thoughts and see whether they're right or not.

Nichol Now it seems that there's a very thin line here, because when you do what you suggest, if it's really activating these responses you're talking about, they don't *feel* mechanical.

Bohm No, but you can see a certain mechanical quality in the sense that the word is followed by the feeling. And you'll see there is a little something also in that pressure of the feeling to avoid examining the meaning of the words, to avoid seeing whether you really have a good reason to be angry. You see, if it were really a straightforward process, there would be no resistance to examining it. Now you can begin to suspect that it looks a little mechanical. Here you can use a certain amount of knowledge which has come from biofeedback. You have a device, the so-called lie detector, with which you attach an electrode to your finger, and this measures the activity of the autonomic nervous system. When the autonomic nervous system is aroused, you'll get the solar plexus and heartbeat and the adrenaline and all those things acting. When somebody says something disturbing, or you think something disturbing that arouses you in this way, then roughly three seconds later the needle jerks. If it's not very disturbing, you may be hardly aware that anything has happened, yet the needle jerks. So it does look very mechanical when you look at it that way.

This process takes three seconds. We could say that your thought is in the pipeline for three seconds, but you don't pay attention to this. Then suddenly the emotional response appears. It suddenly appears in this way as if it were spontaneous. However, there's been another thought in the background all the time saying that everything which appears suddenly like that is deep gut feeling, so that it's really very important. That produces more thought which goes into the pipeline, and three seconds later there comes another jerk, and the whole process thus builds up. The thought that this is a deep gut feeling is now taken as further proof that you have good reason to be angry. See, the original proof was that he's always doing this, right? Now you have an additional proof – the deep gut feeling – saying, "I have a deep feeling which is instinctive. I've been badly treated."

In the case of fear, that's even more clear. You can similarly produce a response of fear by thinking of danger, and a short time later, you have this sinking sensation in the solar plexus. You now say I have an instinctive feeling of danger. The animal would get just that feeling as the first sense of danger, right? Or you yourself might be in a very dangerous situation and get it. So that could be a real warning of danger. But the assumption is that it's always a warning of danger. *This ignores the fact that it could be an entirely false warning arranged by thought mechanically.*

Nichol Why would you think there is resistance to looking at the mechanical nature of this process?

Bohm Because it has gotten tied up with the self-world image. One feels uneasy about saying that one's deep gut feelings may have no meaning because it begins to threaten the notion of the self-identity. For you identify yourself, among other things, with those deep gut feelings. So if you begin to think these deep gut feelings may have no meaning, and you have depended on them for the foundation of a lot of your life, you begin to worry about your whole self, right? There's a thought behind it that's ready to defend the self by not allowing this to be seen. It's really defending the self-image. We don't know what the self is, nobody has ever managed to look at the self, but what we have is a kind of an image of a self with an image of a world in which it lives. This image creates a wide range of neurophysiological effects, implying that this is all a reality of very great significance. We have already discussed some of these effects. And if this image is altered, the whole neurophysiological system goes into chaos, so that there's then a response from the body and from the brain to do something to restore equilibrium. The most immediate way to do that would be to produce thoughts which would change those responses toward equilibrium. But then that would be a mechanical way of thinking, which is false. So you get mechanical feelings and mechanical thoughts working on each other.

Nichol It seems that this apparently simple notion of observation, pursued in the way that you've described it, will at least initially lead one into difficulty, and not necessarily clarity. For the act of looking at the connections in the way that you've indicated will eventually lead to this very point of questioning the meaning of one's deepest feelings.

Bohm Yes, and also one's deepest thoughts.

Nichol This is not a particularly comfortable position to find oneself in. Perhaps this is one reason that observation never penetrates beyond a certain point.

Bohm Yes, I think that there's a kind of defense which is based on the assumption that whenever this whole system starts getting too disturbed, it's best to keep away from whatever is disturbing it. The whole body reacts instinctively that way to pain, moving away from the pain, and then that same reaction is carried up into the higher functions of the brain by some movement of thought, away from the issue which is disturbing it. It moves in such a way as to ease the system. And that's not an intelligent way for thought to operate.

I think the idea is to find a certain skill of pushing this to the point where you can observe, and yet not to push too much, because that's really more mechanical action again. You need insight, you see, and the whole point of suspension is merely to get insight, not to produce predetermined results. Only the insight can change you. The insight that this process is mechanical will change you. It will decrease the importance of the process in your mind, and therefore, the whole thing will change.

But there's always a danger that you haven't gone far enough in doing this, because there's more to the process. There's all sorts of deeper things you haven't touched yet, and you are beginning to shake them, too. But all I'm saying is: Don't go too fast. Start with anger, where people generally realize that the thing is destructive, so that you are able to work on it. It doesn't shake you too much to discover these things about anger. You might then work on fear, because fear has a similar structure. And so do desire and pleasure – they all have a similar structure. In fact, desire comes from projecting in the imagination, the thing you want, and anticipating the satisfaction of pleasure, or whatever. And fear is the same thing except you anticipate all the pain and trouble that's coming. So between desire and fear, there's very little difference; it's just that you anticipate something nice or something bad. Anticipation is a function you need, but in this context it's begun to go wrong, because you're anticipating the internal state of the mind and not realizing it's just an image.

In exploring this, a person will find that as certain issues come up in relationship, that they're distracting. In this way, he loses sight of the insight because of powerful distractions. Then he needs to get insight into that distraction in a similar way. What Krishnamurti once said was "there is no distraction" – every distraction is just a part of the process, which helps to reveal the process. We call it distraction, and the assumption that it is a distraction *makes* it a distraction. Now this misleads you. If it's a "distraction" you're going to say, "Well, my job is to get back on line." But your job is not to get back "on line." You see, *the "line" is the distraction.* Your job is to look at the distraction, not to say, "I was looking at fear before, and now I'm looking at something else. I'd better get back to fear." But rather, you now say, "I'd better get on to the fact that I'm distracted, and find out the thoughts that are distracting."

Nichol In that respect, everything is a basis for observation.

Bohm Everything that happens is part of the process. There really is no genuine distraction in this process. Every one of those distractions is just part of the cover-up, you see. If there were no cover-up, the process couldn't exist, right? I mean, it's too absurd to go on with if it's not covered up and given a false interpretation.

Now all this anger and fear and pleasure and desire is part of the constitution of the self and its relation to the world. These are part of the values which move you. "Value" has the same root as valor and valiant. It means strength. And values, or things of high value, give great strength to what we do, and give it high priority. Now we have a vast set of values which thus moves us. Some respond to one situation and some to another. We are moved by the values much faster than we can think. You see, if somebody is prejudiced, he's got a value judgment which he may not be conscious of – for example, that people of a certain group are bad. Now he experiences this not as a thought, but rather as an apparent perception of the badness which is projected into a particular person who is being perceived at a given moment – the same as the telephone in the television set.

Nichol   So it has the full appearance of reality.

Bohm   Yes. All sorts of value judgments of that kind affect your perception and your intentions at the same time. You see, people have the notion of freedom of choice and freedom of will. But these values operate very fast, and people think they have chosen, but they haven't.

Nichol   How are these values related to the self?

Bohm   The self is determined by the values with which it's identified, for example the supreme value of your religion or your country or your money or fame or power or ambition or your family, or whatever it is. Or your body, your security, your comfort. Whatever has highest value will override the other values. And this generally leads to contradiction, you see. Thus you may have the value of honesty and truth, and so on. But if your value is also comfort and security of the self, or if your country comes first no matter what else is at stake, then when the time comes, such values may take over. And though you profess the right values of honesty and truth, they're not really the dominant values. So the self is determined, in a way, by the whole set of values which are as much socio-cultural as individual.

Nichol   Is this all there is to the self?

Bohm   Well, it is a dominant feature. If the values were not there, the self would collapse, would have no energy. It would be like something which is inflated. When someone removes the air, it collapses.

Nichol   When you say a dominant feature, do you mean, as well, an essential feature?

Bohm   Yes, it's a moving essential power. There may be some other assumptions behind it, but these values are the moving essential power without which the self would have no power. It would be just an image. You may ask, "How could an image ever get power?" Thus the telephone in the television set never does anything except produce a pattern of light. Why should the image of the self have such power? Because there are tremendous values which are attributed to it. Whatever has value, the whole system must try to act on it. That is an absolute necessity and the way it works.

Nichol So we can suspend these values . . .

Bohm Well, not so easily. They come very fast, you see. We have value judgments, such as in prejudice or prejudgment. Suppose you're very prejudiced against a certain group of people and you immediately react against a particular member of this group. The way it goes is like this. There is a thought in the background, an implicit thought, which you don't know about: all people of this group are inferior. You may never have overtly thought it, because you picked it up non-verbally, implicitly, by the way people behaved toward that group. Now you come along and say that this is a person of that group, therefore, he's inferior. As I've already pointed out, you don't actually think that, it just comes out immediately as a perception of inferiority, apparently. Our *immediate* experience of that person is of inferiority, seen as inhering in that person. So that's a value judgement. The value judgment operates implicitly. The implicit is more active than the explicit. With an explicit thought, it's going to take time before you act on it. Therefore you can examine it. But normally you can't really examine an implicit thought.

You see, the implicit thought is organized in the sense that there is in it a certain kind of order. The order is that one thought entrains the other, and that sometimes these lines of entrainment meet and entangle, forming a kind of web. Each person is caught in that web. The web is as much sociocultural as it is individual, if not more. For example, as a child, almost everybody may feel weak or isolated or not properly supported by his parents or his environment, and he also feels that he can only gain their love by fitting in with their assumptions. So one of his basic assumptions may be that "whatever they assume, I must assume too, or else I'll be out." He doesn't have the feeling that he can stand being rejected in this way. If he had felt, for some reason, very strong within himself, then he could have said, "Okay, I'm out." He may, for example, have felt that at least at home he will be all right, so that he can have the strength not to accept the assumptions that are false from the other children or the rest of society. But in the home, it's the same. He must accept their assumptions or else he'll be in terrible trouble. So wherever he looks, that's the way it is.

Nichol It seems that if one took every singular value, every singular feeling and tried to deduce from that a root cause, that may not work.

Bohm That may not work, and yet it may be useful to get familiar with that structure. You're not trying to deduce the cause, but just as with anger, to see how the structure works. Can we suspend the activity of the structure sufficiently to see it working? This will be far more convincing to the system than just talking about it. Rather, see how it works non-verbally as well as verbally, as we saw anger working. The value system works in a rather similar way. The difficulty is that it is based on implicit assumptions. Actually the anger was basically implicit, too, but it's a little bit easier to get at. Thus you say, "I have good reason to be angry" – that was a value judgement, because the word "good" means a value judgement. What was the "goodness" of the good reason, for example, "He's always doing this"? There are implicit assumptions that somebody always treats me in such and such a way – and that is a good reason to be angry. This person is doing that and therefore I have a good reason to be angry at him. And that works immediately. So a value judgment is involved in getting angry or in becoming frightened. It is similar when we say, "I have good reason to be afraid or good reason to be pleased." The point of this is that we didn't really get to the bottom of the anger when we saw the process of anger going between the head and the gut feelings, for we didn't see the value judgement that fuels the whole process. As soon as the "goodness" of the good reason is gone, the process collapses.

There are all sorts of implicit assumptions of this nature. The thing is to go further and just experiment with trying to verbalize, in order to make explicit the implicit assumptions and see how they affect the process. You'll see, for example, that a certain assumption already implicitly contains anger and all the other reactions. Moreover it contains the assumption that there is "me," who is experiencing whatever is going on, and it attributes all of the activity to "me," right? *Now if you suspend that whole process and keep repeating the verbal expression of the value judgment for a while, that begins to be more visible as a non-verbal process, which involves both thoughts and feelings.*

Nichol It seems that the common feature in each of these experiments is the picture of "me."

Bohm Yes, the image of "me" divides into the observer and the observed. Or you could say "I" and "me." "I" am the subject, the active subject who has the intention and the perception, and "me" is that same entity considered as the object. There was a fellow I knew who believed in solipsism. and he argued for its validity. His basic statement was that "I" become "me." You understand what that means? Whatever was in the subject which he regarded as sort of a creative source crystallizes into the "me." So the assumption was that the subject is the unlimited creative source, and that this created the "me" as the object. The implication was that you could always do this in any way whatsoever, but actually you can't. The actual fact is that the "me" will be found to control the "I." Thus, if you have created anger, in the next step you are compelled to justify it. You're then no longer free to say, "Okay, I'm not going to be angry." In this way the "me" constricts the "I," and the whole process goes in a cycle.

Nichol You're suggesting that this "me" is in fact nothing more than a picture.

Bohm Yes, and the "I" too. I'm not saying that there's nothing left; I'm saying what you *experience* as the "me" and the "I" is a picture. Perhaps there's some sort of "me" or some sort of true being which we don't actually perceive.

Nichol What we normally think of as "me" may simply be the sum total of a certain number of values.

Bohm Values and assumptions along with a general picture that goes with them, a picture of something having supreme value as the "me." You see, the "me" and the "I" are taken as of essentially unlimited value. Therefore they will tend to override all other values. The "me" may be identified with money or power, or the country or the religion, but it's still the same process. And whatever is attributed to the "me" takes unlimited value. It's *mine*, you see. Suppose you're looking at some land. You may first say it's land, but then suppose you suddenly say, "It's mine." It then takes on a tremendous value and the meaning is totally different. Or if you say, "It's his but it ought to be mine," still again, it's different. *Notice the tremendous power in that judgment, in which a very special kind of value is given the land when it is sensed as mine or yours or his.*

Nichol Well, the difficulty seems to be in coming to terms with whether or not this image of the "me" is true. That is to say, is the "me" something substantial, tangible, that lives and breathes here and now and possibly whose soul will continue after death and so on, or is the "me" not of that nature?

Bohm That's the question. Perhaps a human being is of the nature of the "me" as we know it, or he or she may have another nature. If a true self does exist, it's surely hidden by, or made inactive by, this "me" process that seems to fill the whole system. It's like the lights of the city which shine brighter than the stars, so that you don't see the universe. That holds also, for the world-image as well as the self-image. Is the world just as we experience it, or is reality something different? That we don't know.

Nichol   Are you suggesting that the self, and the world that we experience, are not different?

Bohm   I am suggesting that they are not different. It's one image. You see, it's the image of the self, and of the world in which the self lives. No self could exist without a world in which to live, and therefore, the image of the world is arranged according to the needs of the self. And thereafter, the self has to try to adjust somehow to this world.

Nichol   Perhaps the notion of absolute value is tied to the notion of a solid, fixed self.

Bohm   Yes, it's identified. "Identity" means being always the same. The self is assumed to have an identity. Certain features are thought to remain essentially the same, though it may change in other features that are superficial. But there's an essence that remains always the same. And we would like it to go on forever, after death. Moreover, it is implied that it has always gone on. Thus people think of reincarnation, and so on. It would be hard to understand what it could mean for the self suddenly to emerge into existence at birth. How could you say exactly when? Indeed, a common concept of the self is that it is eternal in its essence, though its superficial features change. This implies that it is unlimited in its value. And you see, for all we know there could be such a self, but what we are now experiencing as the self is an image. We won't say anything about whether that notion of the self is right or wrong since we have no knowledge of what the real self would be, but we can see that what we are now experiencing as the self is an image mistaken for reality. And the activity of the organism guided by that image gives it an apparent power in reality. Just as the telephone bell coming out of the image makes the image seem very active.

Nichol   As an image, then, the "me" has no substantial reality.

Bohm   Well, it has a kind of substantial reality in the sense that because the image has been established, the whole nervous system is affected. This involves a certain kind of reality, *but no reality independent of the image.*

Nichol   Somehow this must be put to the test.

Bohm   That's why I suggested these experiments of watching anger and watching the value judgment originate and trying to produce these results by thinking the thought again, and by making it explicit. You see, in order to test this, you've got to see the connection between thought and the rest of the activity. That's the key task. But if thought is implicit you don't see it, or else even when you see it, you don't see the connection to the rest. And then you're going to make this mistake again and again. So even in the attempt to test it, you'll make the same mistake.

So you keep at it, you sustain the work. What you are really aiming for is an analogue to what is called *proprioception* in the body. The word "proprioception" has two parts. "Proprio" means "self" in Latin and "ception" is like perception. So it means self-perception. Now that's a technical term used by people discussing the body, physiology, to describe the fact that the body knows immediately its own being; it knows its own movement; so it can tell right away its movement without thinking, and can distinguish it from movements that originate independently. That is necessary for survival. Now the mind doesn't seem to have this. Thus, we can think something and suddenly there appears a gut feeling but we don't see that the thought produced the gut feeling. If your hand suddenly moved, and you didn't know you moved it, you would be in a bad way. For example you might hit somebody in the nose and say, "I didn't know I did it, it happened by itself."

Nichol   If one enters into these experiments with the wrong kind of observation, most likely they would be fruitless.

Bohm   What would be the wrong kind?

Nichol   Observing with the intention of modifying.

Bohm   Yes, the right intention is to reveal that which is, rather than to modify it. Thus, if you said, "I am angry," then you're implicitly separating yourself that way from the anger. It will then follow immediately that it makes sense to want to change the anger, by saying for example, "I shouldn't be angry." This is a contradiction, because you also have another implicit thought which is, "I have good reason to be angry." So if you superimpose the two thoughts, "I have good reason to be angry," and the other, "I don't want to be angry," or "I shouldn't be angry," that's crazy. We must make it very clear why we shouldn't want to modify anger because in many other contexts modification is a perfectly rational activity. But here it is not rational because you have the implicit thought, "I have good reason to be as I am." And at the same time you say, "I want to change." Evidently that doesn't make sense. But the trouble is that the thought that I have good reason to be as I am is basically implicit, you don't see you've got it.

So it seems to make sense to say, "I'm going to modify all this," because you see the whole result coming out without seeing the implicit thought that is constantly maintaining the present state of affairs. All this is a result of the lack of proprioception. Thus, once you know that you moved your hand, you can stop moving it if it's doing the wrong thing. But if your hand moved without your knowing it, say because the nerves telling that your hand is moving are gone, you could not initiate an intelligent action by trying to control your hand.

Nichol Allowing this contradiction to expose itself seems to be necessary.

Bohm Yes. Otherwise there could be panic because "I must hold onto what I am" would be one implicit thought, and another thought would be "I must change it." This could create a great fear. I might think that if I don't hold on to what I am, I'll go to pieces, and this would be one implicit thought. The other thought would be "I must change what I am." Then you have the thought, "I may go to pieces." These thoughts make it impossible to take any meaningful action.

Nichol Both thoughts, "I must stay the way that I am" and "I must change the way that I am" lead to further confusion.

Bohm Yes. Especially because both are there together and they are both conflicting with each other. You may not know it, but you already have the opposing implicit thought. When you say, "I must change," you would never even have that thought, except you've also got the implicit thought, "I must not change. I have good reason to remain as I am." So it's urgent to find these implicit thoughts. It's very important to make them explicit, to find out what your thoughts are by putting them in words and seeing how they work. If the thoughts are not put into words, you generally won't know you've got them. Maybe eventually you'll get so subtle you won't need the words. But in the beginning, the mind is not able to do that. To put the same point again in order to emphasize it, you have a thought and you don't know you've got it because it's implicit. So it apparently makes sense to put in another thought saying, "I should change," because you don't know you've got the thought, "I have good reason to remain as I am. It's absolutely necessary to remain as I am."

Therefore you don't realize the contradiction because one of the thoughts, at least, and very often both, are implicit. *That's the lack of proprioception.* The implicit thought does not appear in consciousness at all. Only by its effect does it show itself, unless you make it conscious by expressing it somehow explicitly to yourself or to others.

The point is that you've got to see this contradictory movement there and express it, but not for the purpose of doing anything about it, because that would be just the same mistake again. Thus another implicit thought would be, "I can avoid this contradiction." But rather than resolving in this way to keep this contradictory thought out, I need simply to be aware of the thought and all of its effects, all through the body and so on. However, the only way I can see to be aware of it is first to make it explicit. There may come a time when there is another way, but I am trying to say this is a way to begin to loosen up the mind. And the word is thus being used, neither for analysis, nor for its content, but simply to make the process visible. Not that you believe or disbelieve the word – the meaning of the word is not really at stake. That's not of any basic importance – it's really what the word is *doing* that's interesting in this context. The word is part of a non-verbal process. Fundamentally, the actual activity of the word is non-verbal. It's a real activity based on all sorts of nervous processes, sounds, and so forth. As for the meaning, we want to see what the meaning *does*, not so much what the meaning *is*.

Nichol   The word itself is just the tip of the iceberg.

Bohm   Yes, but the point is, at least to have a tip. You see, if somebody could chop off all the tips of icebergs, then ships would be sinking all the time. We try to get rid of the problem by chopping off the tip, and it's all there beneath. Then we collide with it.

9 FREEDOM AND THE VALUE OF THE INDIVIDUAL
(1986)

In this essay Bohm addresses a series of issues at the heart of the Western cultural value system. Foremost among these are the questions of free choice, free will, and the nature of the individual. It is Bohm's view that in general we do not exhibit genuine free will, and thus do not rise to the original definition of "individual" – one who is undivided.

A systemic limitation on free will springs from a fundamental misunderstanding of knowledge itself. Pervasive and active in our experience, knowledge is typically a mechanical projection from the past, whether from three seconds ago, or from a thousand years ago. As such, suggests Bohm, it is incapable of meeting the rich and complex nature of the living present – the very milieu in which true free will would necessarily act. Ignorance of how knowledge functions leads us to believe that we can freely choose our will, while in fact the parameters of this will have already been set in place by the crystallized predispositions of accumulated knowledge.

Bohm proposes that it is only by fully comprehending the activity of knowledge that we have any real prospect of expressing free will. This comprehension cannot be strictly abstract or intellectual, but is acquired through direct engagement with the concrete implicate order of knowledge. Here, the movement of past knowledge may be seen at its generative source, from a perspective other than that of its own self-referencing content.

In Bohm's view, consciousness that is dominated by knowledge echoes the conservative nature of matter, where stability, repetition, and relative

Extract from Appendix to Chapter 13 of D. Griffin (ed.), *Physics and the Ultimate Significance of Time: Bohm, Prigogine and Process Philosophy*, State University of New York Press, Albany (1986).

invariance are at a premium. By loosening these conservative restrictions, we may begin to discern our inherent link to the holomovement – *the unknown timeless present, the qualitative infinity of nature at the experiential level. In this way we embody true individuality, manifesting our creative potential through participation in – rather than objectification of – a total field of experience.*

Freedom has been commonly identified (especially in the West) with *free will* or with the closely associated notion of *freedom of choice*. In these terms, the basic question is: Is will actually free, or are our actions determined by something else (such as our hereditary constitution, our conditioning, our culture, our dependence on the opinions of other people, etc.)? Alternatively, can we or can we not choose freely among whatever courses of action may be possible?

Such a way of putting the question presupposes that the mind is always able to know what are the various alternative possibilities and which of these is the best. Evidently, however, if one does not have correct knowledge of the consequences of one's actions, freedom of will and choice have little or no meaning. It must be admitted that, in most of human life, lack of knowledge of what will actually flow out of one's choice prevails.

In order to deal with this, we try constantly to improve our knowledge. But as we have seen, reality is infinite both in its depth and in its extension. Although relatively independent contexts do exist, such independence is always limited. Very often the question of what these limits are is obscured not only by ignorance but also by the sheer complexity of the entangled web of interrelationships on which the consequences of our decisions may depend. It often does not seem to be at all likely that we will be able by increasing our knowledge alone even to keep up with this ever-expanding and agglomerating mass of interdependent processes, especially in the field of human relations (consider, for example, what is happening today in politics and in economics).

There has been an enormous expansion of scientific knowledge, especially in the last century, along with a flood of many other kinds of knowledge (this has been called the "information explosion"). But has all this knowledge contributed to our general freedom in any significant way? Or has it not, in many ways, led to yet further unresolvable entanglements in the problem of trying to establish orderly and harmonious human relations? One can mention here the obvious example of how knowledge of nuclear physics has brought us to the need to make all sorts of decisions in situations in which there is little or no reliable information about how human beings will behave (or, even about all the significant physical consequences of our nuclear devices). Without such knowledge no sensible choice is possible. But even if we consider medical knowledge, which is on the whole beneficial (except for the growing number of cases of "iatrogenic" disease), this too is leading to situations in which we have to decide many questions without an adequate basis for making such decisions (e.g., shall terminally ill people be kept alive against their expressed wishes?). In such contexts what do free will and free choice actually mean? This is one of the questions that I hope can be discussed.

Thus far, we have considered lack of knowledge of what may be called the "external world" as a serious limitation on *meaningful* freedom. But there is a much more serious limitation, a lack of what may be called "self-knowledge." Schopenhauer has, at least implicitly, already called attention to this area when he said that though we may perhaps be free to choose as we will, we are not free to will *the content of the will*. Evidently, this content is a key factor determining what sort of person one actually is, and yet it appears somehow to be "given." The significance of this question becomes especially clear when we note that people so often seem to be unable actually to do the good things that they have (apparently freely) resolved to do. No person can be said to be free who is for reasons of internal confusion unable consistently to carry out his or her chosen aims and purposes, for evidently such a person is driven by inner compulsions of which he or she is unaware. This inner lack of freedom is far more serious than a lack arising from external constraints or a lack of adequate knowledge of external circumstances.

The problem has often been approached with the aid of moralistic injunctions, telling people to "pull themselves together" and to choose, once and for all, what is right and good. But to those who are unaware of what is actually determining the content of their wills (which includes, in many cases, a content that divides and weakens the will), such advice has little meaning. Moreover, the content is ultimately based on overall self-world views that the individuals have usually not chosen for themselves. These include general notions not only of the sort of world we live in, but also of models of what constitutes a normal right-thinking good human being, of how such human beings are to be related, and of what are their duties and obligations, etc. This all-pervasive web of shared thoughts and feelings, propagated not only explicitly but also by tacit and subtle clues picked up since the time of one's birth, operates in most people as an almost overwhelmingly powerful limit on freedom, of which they are essentially ignorant. Indeed, this web not only determines what will generally be thought or felt to be the right choice. Much more important is that it determines what is regarded as the correct range of alternative possibilities in any actual situation. If something is not considered a real possibility, there is no chance at all that it will appear among one's choices. Is there any meaning to freedom of will when the content of this will is thus determined by false knowledge of what is possible, false knowledge that we do not even know we possess (or, more accurately, that possesses us)?

The problem is seen to be even sharper if one considers that the question of choosing what is right and good so often arises in circumstances in which one's desire is in some other direction. Indeed, if something is clearly seen to be right and good, and if one has no desire to do otherwise, it hardly seems that any particular act of choice is ever needed. Will one then not spontaneously have the urge to act according to what one has perceived to be right and good? But often, as has been mentioned above, one finds that one has an irresistible desire in some other direction. One has by no means chosen this desire. Rather, it also arises from the totality of remembrances, reactions, and "knowledge" accumulated over the past, which responds to present "needs" as overwhelmingly urgent in ways of which one is not aware.

The attempt of will to struggle against such desire has no meaning, for this sort of desire contains in it a movement of self-deception, along with a further movement aiming to conceal this self-deception and to conceal the fact that concealment is taking place. Thus, one will often accept as true any false thought that makes one feel better (or more secure) or that makes one believe that the object of one's desire can be realized. This is, for example, the basis of the activity of the confidence trickster, who paints a false picture of satisfying greed that the victim cannot resist accepting as true. As long as one is ignorant of how this sort of self-deceptive desire operates, what can it mean to talk of freedom of any kind?

It appears, then, that the principal barrier to freedom is ignorance, mainly of "oneself" and secondarily of the "external world." This ignorance is also the main barrier to true *individuality*. For any human being who is governed by opinions and models unconsciously picked up from the society is not really an individual. Rather, as has been made very clear, especially by Krishnamurti, such a person is a *particular manifestation* of the collective consciousness of humankind. He or she may have special peculiarities, but these too are drawn from the collective pool of thoughts and feelings (here we may usefully consider the word *idiosyncracy*, which, in its Greek root, means "private mixture"). A genuine individual could only be one who was actually free from ignorance of his or her attachment to the collective consciousness. Individuality and true freedom go together and ignorance (or lack of awareness) is the principal enemy of both.

It is important to note that the main kind of ignorance that destroys freedom and prevents true individuality is ignorance of the *activity of the past*. As has been brought out earlier, although the past is gone, it nevertheless continues to exist and to be active in the present, as a nested structure of enfoldments, going into even the distant past, which are carried along (with modifications) from one moment to the next. (I have treated these as projections of various kinds.) Here I have been indicating how this activity of the past can interfere with freedom and individuality as long as one is not aware of this past.

The past is also absolutely necessary in its proper area (as, for example, it contains essential knowledge of all sorts), but when the past operates outside awareness it gets caught up in absurdities of every kind and becomes something like the sorcerer's apprentice, which just keeps on functioning mechanically and unintelligently, to bring about destructive consequences one does not really want.

This brings us to the question: can the past contain adequate knowledge of its *actual activity in the present*? As I have already pointed out, the content of knowledge (which is necessarily of the past) cannot catch up with the immediate and actual present, which is always the unknown. Since the activity of the past is actually taking place in the present, this too is inherently unknown. Thus, knowledge cannot "know" what it is actually doing right now. As an example, one may "know" from hearsay or from general conclusions drawn from particular experiences that people of a certain race are inferior. When this "knowledge" responds to a particular member of this race, one's immediate perception is shaped, colored, and twisted so as to present that person as inferior. One is not aware of just how all this is actually taking place, as it happens very rapidly and in very subtle ways. Moreover, the whole process is accompanied by a great deal of distortion and self-deception. For example, as one perceives the "inferiority" of the other person, thus implying one's own "superiority," one experiences a short, sharp burst of intense pleasure. To sustain the pleasure the mind continues with further false thoughts along this line while concealing from itself the fact that it is doing do. Clearly, because the response from accumulated knowledge always lags behind actuality and because, in cases such as this, the mind is caught up in feeling the need to distort, it is not possible through such knowledge alone thoroughly to free the mind of such prejudices.

How, then, is it possible for there to be the self-awareness that is required for true freedom? Along the lines of what has been said in this [book], I propose that self-awareness requires that consciousness sink into its implicate (and now mainly unconscious) order. It may then be possible to be directly aware, in the present, of the actual activity of past knowledge, and especially of that knowledge which is not only false but which also reacts in such a way as to resist exposure of its falsity. Then the mind may be free of its bondage to the active confusion that is enfolded in its past. Without freedom of this kind, there is little meaning even in raising the question as to whether human beings are free, in the deeper sense of being capable of a creative act that is not determined mechanically by unknown conditions in the untraceably complex interconnections and unplumbable depths of the overall reality in which we are embedded.

This leads us to a more fundamental question of what the relationship is between the truly free human being and the totality that goes beyond the explicate order, beyond the implicate order, and beyond time and space. To pose this question is of course itself subject to questioning, since one may ask whether one has properly laid the ground for doing so by giving serious and sustained attention to the actual activity of past knowledge in one's own mind. Is one sufficiently free of this past meaningfully to inquire into the infinite totality? Any real inquiry of this kind must have such attention in it from the very beginning, or else one's mind may be so bound by preconceptions and desires enfolded in one's past that true inquiry at this depth is not actually possible.

Nevertheless, there may be a possibility of usefully engaging, at least to some extent, in a meaningful discussion of this question, even though one inevitably begins from the common state of collective consciousness, which is not properly aware of the actual activity in the present of that false and self-deceptive "knowledge" which is part of its base.

In doing this, the first question to be discussed is: just what is it in our past that binds us, misleads us, deceives us, and thus prevents true freedom? I suggest that what limits us is the attempt to identify oneself with a certain part of one's past that is regarded as essential to what one is. As we have seen, we *are* primarily the present, which is the unknown.

The past is, in its actuality, merely a part of this present that is also active in the present, and therefore it too is the unknown. All that we know is the *content* (i.e., the meaning) of the past that is gone (and this resembles the actual present only for those contexts in which changes are slow and regular enough so that the difference between past and present makes no significant difference).

If we *are* the unknown, which is the present, then time can be seen in its proper meaning only in the context of that which is beyond time (i.e., the holomovement or eternity). Any attempt to treat the whole meaning of existence in terms of time alone will lead to arbitrary and chaotic limitation of this existence, which then takes on the quality of being rather mechanical. If we are to be creative rather than mechanical, our consciousness has to be primarily in the movement beyond time. Implicitly, this is well known to us. No one will be creative who does not have an intense interest in what one is doing. With such an interest, one can see that one will be at most only dimly conscious of the passage of time. That is to say, though physical time still goes on, consciousness is not organized mainly in the order of psychological time; rather, it acts from the holomovement. On the other hand, if the mind is constantly seeking the goal of finishing its task and reaching its aim (so that it is organized in terms of psychological time) it will lack the real interest needed for true creativity.

Given that a human being may be creative when his or her consciousness arises directly from the "timeless" holomovement, we come to another question: is the creative human being merely an instrument or a projection of the creative action of totality? Or does one act from one's own being independently? I suggest that this is a wrong question, as it presupposes a separation of the human being from the totality, which I have denied at the very outset of this inquiry. A better question is: can we be free to participate in the creativity of the totality at a level appropriate to our own potential?

The need for this question becomes clear if we note that ultimately *everything is* participating creatively in the action of the totality. For matter in its grosser levels, this creative participation consists of continuing to re-create its past forms, with modifications, in a way that is approximately mechanical. (This was implied in the statement that each moment is created with its past as a *projection* containing a further projection of the past of previous moments.) Such creation of a sustained but ever-changing existence of matter at the grosser (mechanical) levels opens the way for the action of higher levels of creativity, such as life and mind.

But once these higher levels are possible, why are they not always fully and harmoniously realized? I have proposed that, at least in part, this is because of ignorance. Such ignorance leads the mind to continue its past, mechanically, through identification rather as if it were a form of matter at a grosser level. The mind is trying in a confused way to realize the kind of creativity appropriate to such grosser levels of matter. In doing so, it is clearly unable to realize the kind of creativity appropriate to its own level.

Ending this state of ignorance may then open a new possibility for the mind to be creative at its own level. When it does this, it is still participating in the universal creativity, but now it is realizing its proper potential.

I suggest that this is the essence of freedom, to realize one's true potential, whatever the source of the potential may be. It is unimportant whether it is grounded in the whole or in some part (e.g., the individual human being). And, indeed, as has been said earlier, the attempt even to raise the question of whether creativity originates in the totality or in the individual presupposes a kind of separateness of the two that we have already denied. So I propose that the question of freedom has to be looked at in a different way.

This new way flows out of giving sustained and serious attention to how unfreedom arises basically from identification with the past, in which the mind commits itself to act as if it were determined mechanically in the ways in which grosser levels of matter are determined. We have to use the past, but to determine *what we are* from it is the mistake. To do this implies that such grosser mechanical existence in time has supreme value and that the main function of the mind is to sustain this sort of existence by continuing the past with modifications. The clear perception that we are the unknown, which is beyond time, allows the mind to give time its proper value, which is limited and not supreme. This is what makes freedom possible, in the sense of realizing our true potential for participating harmoniously in universal creativity, a creativity that also includes the past and future in their proper roles.

Part Three – Collective Orders

10 KNOWLEDGE AS ENDARKENMENT (1980)

The following small-group discussion highlights Bohm's view of knowledge as the central factor in the "endarkenment" of human consciousness, exploring at length the overview of knowledge first outlined in Chapter 9. Bohm claims that an active, self-sustaining pool of human knowledge – accumulated and refined through millennia – is thoroughly infected with misinformation, *thus polluting human experience at its generative source. In going to this generative source, the problems of humanity may be resolved in radically new ways. Attending to particular problems, while clearly necessary, will not significantly affect the conflicts and challenges we face in the world today.*

To illustrate the meaning of a generative source, consider the term "roses." At one level this is a sheerly abstract category which, in our concepts, distinguishes roses from rhinoceroses, tangerines, and so forth. Bohm designates such a category as the abstract general. *But there is another aspect –* the concrete general *– which indicates a living, generative process. This concrete general is the actual emergence of rose after rose after rose, and is not just "in our concepts." In this context, any single rose is the* concrete particular, *a temporary manifestation derived from the more fundamental generative process. Our attention is usually split between the concrete particular rose and the abstract category "roses"; we rarely attend to the concrete generative process from which all roses emerge.*

Similarly, when we think of "knowledge" as residing in minds, books, computers, and so on, this is the abstract general aspect of knowledge. We thus categorically distinguish "knowledge" from oceans, trucks, or roses. But knowledge also has its concrete general aspect – a living pool of collective

Previously unpublished seminar, courtesy of Saral Bohm.

meaning that gives rise to our perceptions, emotions, dispositions, thoughts, and actions. Bohm asserts that this pool of generative knowledge produces all forms of particular knowledge, including the experience of an apparently concrete particular ego. Such a perspective inverts the common view wherein the self or ego acquires, possesses, and applies knowledge.

These category confusions between ego and knowledge, between abstract and concrete, and between general and particular, are but a few examples of the misinformation coded into the generic activity of knowledge. And since all variety of such information permeates human experience, the appearance of genuinely creative knowledge is sporadic at best. What is thus called for, says Bohm, is sustained investigation into the concrete general movement of knowledge, rather than the reflexive application of strictly abstract knowledge.

Bohm   We've said the world is in this terrible state. I think you can see that everyone is following his own ego, his own individual or collective ego, and that's why nothing can be done. You cannot get people to agree; it's obvious what should be done and yet nothing can be done. For twenty years people have known there would be an oil shortage. For seven years it was extremely obvious, but in this seven years no one was able to get together and do anything. They couldn't agree. Different people had different ideas, or even the same person had contradictory ideas. People would like to say there's plenty of energy. To many people, there is no oil shortage and they believe we can go on with our customary driving habits. So clearly there is self-deception or distortion. People don't see the fact because they would prefer not to. That is, it would make people too uncomfortable – or at least they think it would become too uncomfortable – if they saw this fact about the oil shortage, or about anything: about the fact that the world is falling apart economically or about the fact that we are ready to thoroughly annihilate the world. Every alternative is dangerous.

Atomic energy would be quite safe if everybody were very rational, but since people are not very rational, it's not very safe. Even a small mistake can cause a plant to blow up and devastate an area the size of Pennsylvania, and make it radioactive for thousands of years. If there were a nuclear war every one of these radioactive plants could be counted on to blow up. But nobody thinks about that. On the one hand they're preparing for nuclear war, and on the other they're building nuclear plants; these are two inconsistent approaches. So in one compartment we have a nuclear-plant-building requirement, and in the other compartment the preparation for nuclear war. They're not allowed to meet, you see. If they did meet, there would be too much of a disturbance. Somebody facing this would feel that he might go to pieces.

Now, what is the origin of all this? We've said the "self," but that doesn't get us very far, because the self has been with us for as long as we know. Nobody knows what to do with it. People may say, "Let's get rid of the ego," but then it's the ego saying, "Let's get rid of the ego." Therefore, it doesn't mean anything. So is there something deeper, some source, some generating process that is really responsible for all this? That's the question I want to consider.

I want to say that the word "general" means "to gather together everything that is important, to put things together abstractly into a general class." But it also comes from the root "to generate." You see, the "genus" of a species is that which shares a common generating process. People feel more closely related when they say that they've been generated from a common source. The word "relative" means "common source of generation." Now if you understand the process of generation, you can make things like agriculture possible. But to do that you must study the general abstractly and gather the relevant facts, and that leads you to a perception of the generating source of the whole thing. Is that clear? Similarly, in chemistry you find all the elements and you classify them abstractly into categories and so on, and that shows how to generate all sorts of chemical compounds. Or in physics, we can see that from the general laws of physics we have come to the generative source of energy, and so on.

Similarly, perhaps, in the mind we tend to look at particular problems one after another. You may have trouble being angry, being fearful, being jealous; we have political problems, economic problems, military problems, social problems – you can go on endlessly. You won't learn very much if you deal with them one after another. The next one will surprise you, and you will be caught in it once again. From a particular problem treated *as* particular you cannot learn very much. But if you can see these things as general – as not only belonging to you, general to you, but general to all mankind over all history or even before history began, perhaps even including the animals in some cases, then this may lead you to the generative source of the difficulties.

To know the generation of plants they did not know all the details about plants. It was careful observation that showed people that the seeds are what control the growth of plants. It didn't mean they knew every detail about every plant. It requires careful observation, intelligent observation to see what to abstract in the abstract general. You see, the general is both abstract and concrete – it's abstract when you form a class and it's concrete when you see the generating process itself. *Now the important point is to see what to abstract that will lead you to the concrete general.* With plants, it's the seeds; in chemistry it's understanding the elements; in physics they hope to understand the elementary particles to see how matter is generated.

In society people thought you could treat each person as an atom, and in that way you would understand how society is generated. That's been a common idea that obviously breaks down. A person is not an atom. He has a tremendous amount inside of him – he cannot be treated as an atom. So that theory of generation is wrong. It doesn't mean that you'll have to know everything about everything, but you'll have to be observant to find out what is the important point. If you have particular problems you may get overwhelmed by one after another – that's just what's happening to us. There are so many particular problems that you feel they can never be solved. In fact, they can't.

By seeing the general source of all this, we may find things simplified. It would take some time to explain how to get over this, but I think that Krishnamurti has been suggesting that *knowledge* is the general source of our difficulty. He calls it "thought" at times, but it would perhaps be better to call it "knowledge," which includes thought, but includes a bit more. It's dominated by thought, but knowledge goes further than thought, further than abstract thought. You see, when we talk about thought, the tendency is to think of something rather abstract. Knowledge may be much more concrete. The Latin languages have two words for knowledge. One is the abstract, as in the French "savoir"; the other is "connaître," which is the concrete knowing. The English "recognize" is that same root; when you recognize something you don't have time to think.

Also, skill is knowledge. You use skill in driving your car; that is, you got the knowledge and it became part of you. Many other things are knowledge. Knowledge acts through various dispositions of the body. If you're walking down stairs, your body is set to walk in a certain way. I remember I was walking in the dark, not expecting that the stairs would end, and the body was set wrongly. In other words, it was the knowledge that these were stairs that produced the set of the body. The knowledge that there are no more stairs means changing that set. And you have other knowledge of that kind – if you know that some person is your enemy, you will be disposed to him in a certain way. He will see that you are his enemy, and he will make the same disposition. The thing will be set. So knowledge is involved in enmity, right? If you didn't know he was your enemy, I don't know how it would work.

I read a science fiction story a long time ago – in the Thirties – where a scientist invented a machine that would remove people's memories altogether, immediately, all over the world. Hitler was talking and he suddenly forgot he was Hitler. People had to rediscover how to do everything. It shows that all those political problems were in the form of knowledge. These people *knew* they were Nazis and they *knew* what they had to do. Other people *know* they are communists and this and that. So because of what people know, not only abstractly but concretely, they're faced with all these problems. It seems silly to have problems based on what you know. That is, knowledge includes not only information but misinformation; it also includes confused information and it includes nonsense. It's mixed with all sorts of useful and correct things. Even an idea which is correct in one context becomes nonsense in another. You can't so easily fix it.

So, you could say that knowledge is not just something in the library that you can look up any time you want. It's not just sitting there waiting for you to refer to it. That's one picture of knowledge – that it is entirely abstract, sitting there in the computer waiting for you to use it, and then you *choose* to use it when you want to and give it up when you don't want to. But that doesn't work, you see. If you *know* this fellow is your enemy, you can't give it up. If you *know* you're in danger, you can't give it up. Suppose we take people getting angry at each other. You can see that knowledge is involved, because somebody can say, "I was just sitting here peacefully and he attacked me," or, "He's always doing this; he does it to annoy me." That knowledge will produce anger, right? From there on, your thought is no longer clear, because once the anger has been created, then your thought is directed toward justifying the anger. You'll only look at the evidence that justifies your side and not the other side – or you'll even invent evidence. Also, you may finally say, "I shouldn't be angry," but that's rather silly, because one part of your knowledge says, "I should be angry," and the other part says, "I mustn't be angry," and you can't stop it, right? Why can't I just wipe out the knowledge that says I should be angry? Then I don't have to fight with the other knowledge that says I shouldn't be angry. But when you carry on this fight, you just get more confused and worn out. The brain cells perhaps start to break down.

Participant Also, it can be projected into the future as fear: it may happen again.

Bohm Yes, I *know* it may happen again. Or else two people come together and somebody says, "I know I'm right." Now that's not only abstract knowledge, it's concrete knowledge. You get that feeling of rightness concretely, which happens immediately. The other person says, "No, I know I'm right." Or else the other person says, "Yes, you're right, I'll just do what you say." Either of those responses produce trouble. That feeling that *I know* – you base everything on it, right? But that feeling "I know" may be properly founded, or it may not be.

Now how does this come about? That is, you feel you *know* when there's nothing underneath it at all. But still you're quite confident and you get a very strong feeling that you know. You see, the feeling that "I know" is immediate, concrete; it is not just an abstraction. If it were only an abstraction, there wouldn't be any trouble with it. You could just say, "How do I know?" You would soon find out whether you know or don't know. These people in Iran *know* that they're right; the people in Washington also *know* that they're right, so there's no way to meet. Now, if you analyze it you could say each one shows what the other has been doing wrong. And if you just listen to one side you would say, "Yes, that's terrible, they shouldn't have done that." Then you go to the other side and listen and you're apt to say the same thing. But they can't drop all this. Both sides say, "Yes, we will stick to this no matter what happens. It may lead to nuclear annihilation, but still we'll stick to it. We can't do otherwise." There seems to be something very powerful there, and yet if that science fiction machine I described were to operate, the thing would evaporate, right? There wouldn't be any problem. The ayatollah would forget who he was, so where would the problem be?

So evidently one important form of knowledge is that we know what we are, we know who we are. It's very important, right? But how do we know? Do we really know? Therefore, knowledge, as I say, is not just this abstract thing; it is a whole process that is autonomous and moves on its own. *We don't choose to apply it; it applies us.* The knowledge of who we are, what we are, and what we've got to do determines this whole future, and we're going to be driven by it to prepare for nuclear war or whatever. Perhaps not – perhaps we'll do something else according to further knowledge. But it's entirely out of our hands. We have no control over this knowledge; it controls us.

Participant The distinction you began by making was that first we may think of knowledge as something in a book, but you're pointing out that it's the way we set our bodies, it's who we think we are.

Bohm It's also our emotions. Knowledge produces emotions, like anger, fear, pleasure, pain.

Participant But when you say that it drives us rather than we drive it, it's getting less clear to me who the "we" is that is different from this knowledge.

Bohm As human beings we are driven by knowledge; we are dominated by knowledge, let's put it that way. We may imagine that we control this knowledge, that we can select it and use it according to our convenience. And in some areas we can. But fundamentally, it drives us. There are some areas of technical knowledge where we might select, but that only makes it more dangerous. You see, in a certain limited area we can be in control and be very rational, but this rationality is in the service of wild irrationality. The aims towards which this knowledge is used are determined entirely by the irrational parts of the knowledge. The more rational you are, the more dangerous it gets.

Participant That pressure of the past, basically.

Bohm Well, whatever it is. We say the pressure from knowing who you are and what you are. That is the past, but why should the past press on you? Where is it? It isn't making sense. At first you may want to put it, "Who are we?" That's the way people think: "I am such-and-such a person, I am religious, I am this, I am that, I belong to this group, I belong to that group." That's the way people start out.

Participant Is this knowledge synonymous with the term "conditioning"?

Bohm It includes conditioning, yes. We can say conditioning, like knowledge, has two sides. Knowledge may be useful, and it's necessary for us to exist. Then there's another kind which is driving us to disaster. So we're not understanding knowledge, you see. That is, we know all sorts of things – not "we," but *knowledge* knows all sorts of things. We're saying that knowledge is moving autonomously – it passes from one person to another. There is a whole pool of knowledge for the whole human race, like different computers that share a pool of knowledge. There's one pool of knowledge that's been going on many thousands of years, developing and so on, and this knowledge is full of all sorts of content. It has gone on to great achievements of technological and scientific content, but at the same time it is leading us to disaster. This knowledge knows all those things, but it doesn't know what it is doing. This knowledge knows itself wrongly: *it knows itself as doing nothing.* But what this knowledge also knows is that there is somebody else there called "me" or "us" who are using it. So this knowledge says, "I am not responsible. I'm just here for you to use."

Participant   But actually it's driving.

Bohm   It's driving. It says, "You are doing it."

Participant   But actually it's us.

Bohm   But who are the "us"?

Participant   The "I" that started the whole thing.

Bohm   But was there an "I" that starts it? This knowledge tells us that an "I" started it, but how do we know? Maybe knowledge produced the "I." If there were no knowledge, where would this ego be?

Participant   But isn't that like the chicken and the egg?

Bohm   Well, not necessarily. It might be conceivable that man would be free of this ego and still have knowledge, right?

Participant   Right, but the kind of knowledge you're talking about is not factual. It is an interpretation of . . .

Bohm   But interpretation arises in all good knowledge too. You can't do science without a certain amount ot interpretation. Knowledge is there, and somehow it went wrong. One could put it like this. Perhaps in the distant past, man took a wrong turn and began to develop knowledge in the wrong way. We could speculate that it then got worse and worse.

Participant   This knowledge that you're talking about – the "wrong turn" – is that something we've absorbed or something we conjure up?

Bohm   We haven't found out yet. We don't know. It's important in this inquiry not to assume what you don't know. One of the things that knowledge has been doing wrong is taking for absolute truth things it has never known.

Participant   Couldn't we also say, as you suggested, that somewhere in the past this has gone wrong? Maybe one could also put it the other way around: we are taking the wrong turn all the time. So we have to find out where we are taking this wrong turn all the time.

Bohm   Yes, it's both. That is, possibly in the past we made a wrong turn, and that wrong turn was such as to be continually repeated, and we are continually doing the wrong turn. If we could discover that wrong turn, then it might change.

Now let's say a few things about the structure of knowledge. First of all, we have the abstract knowledge. Then we have something more concrete, which we call "imagination," that displays this abstract knowledge in a more concrete form. You can imagine not only what things look like, but what they feel like – what it feels like to be such-and-such a person and so on. It produces in your nervous system a set of excitations similar to what might be produced by an actual occasion, but different because it's being produced in the memory, like a tape recording. In addition to imagination, our knowledge includes direct sense impressions and feelings and so on. Let's distinguish between the two forms of knowledge, which we could call "immediate" and "non-immediate," which is mediated. If we take thinking, we say, first of all, we have the immediate necessity. You begin with your sense experience, which we can call "immediate," right? Then you start thinking, and that introduces mediation between you and that fact. It tells you this is a table, you can use it to write on, it's made of wood – you know, thousands of items of information which will be useful and necessary.

That's the way it begins. It appears there are two sides to know-ledge: the immediate, which is concrete, and the mediated, which is abstract. Now "abstract" means "to take out." The power of thought is that it can abstract; it abstracts what's important. If you had to do every part of this in concrete detail, you couldn't handle it. By means of thought, people have found out how to abstract that which is important in various cases, put it together and make it general rather than particular, so it would apply in a tremendous range of cases. So a general concept such as "table," you know what it means, you know all sorts of tables. You know many kinds of tables you've never seen before; you could even say there's a table mountain, and so on. That's the general concept, you see, the general notion, which is abstract. That abstraction abstracts what is important and shows the general significance. The word "concrete" comes from the Latin "concrecere," which means "to grow together; all grown together." It's like a jungle that has so much in it that you would never find your way through it. Nonetheless, the abstract must come from the concrete. That is, if you generate abstractions out of pure imagination, they don't mean very much, as a rule. Ultimately, they must be connected with the concrete.

So, the concrete is one side, the abstract is another. They don't exist separately; it's an abstraction to use those two words. It's like the two sides of a coin – the two sides of a coin have no existence as such. They are entirely an abstraction, but nevertheless they have a mean-ing. You could never have the sides of the coins standing by them-selves, you know, with the coin somewhere else. So the concrete also is not separated from the abstract, but both the concrete and the abstract are abstractions. They are also both concrete, as we'll see later.

The same is true of what is mediated and what is immediate. We begin by saying, "This is immediate," and we form an abstraction. The abstraction process takes time by *thinking*. It takes you time to make that abstraction and apply it, right? But now that past process – this is where conditioning comes in – that past process of abstract thought being put down in memory and generalized now comes out in the concrete reality as the way you see it and the way you react to it. We said that in the case of skill, the abstract knowledge of driving a car becomes concrete. You don't think about it; you move right away. All that you've seen about road signs and so on comes up immediately – you don't have to think, right? Similarly, you know that this is a table, you feel it immediately. You don't think, "This is a table, therefore I can write on it," but you immediately start to write, if that's what you want to do. Similarly you say, "That's my enemy." That's immediate, right? You don't say, "I've been thinking that's my enemy – that's an abstraction," but rather you *concretely feel* him immediately as the enemy. Therefore the way you feel, the way you see, the way you move, the way you act concretely now, contains the effect of past abstractions.

Participant   What I have a picture of is a sequence in time which begins with the concrete, the sensation, the immediate thing, the conditioned abstraction which operates . . .

Bohm   It may be unconditioned or conditioned, you see; it might be a creative acting. Now, part of our knowledge – we'll call it "misinformation," which mankind seems to have accepted – is that the concrete is quite different from the abstract, divided from it; also, that the immediate is quite divided from mediation. We say, "There is non-thought and thought." I have my "immediate" experiences and I say that thinking is something entirely different. Now that leads to confusion. Confusion arises when you mix one thing up with another, or when you take two things and call them one or one thing and call it two. It's a failure of your categories, of your classification.

Participant   But the funny thing about this is that once I set up this structure, I find evidence all the time that this structure is correct. So I cannot really check it anymore.

Bohm   Well, that's the difficulty. We haven't found out yet why the mind starts to distort around this structure and produce only evidence that it's correct, and failing to notice evidence that it's not.

Participant It's also important to point out what's wrong with abstractions.

Bohm I say they are very good.

Participant Yes, they are very good in certain ways. For instance, when one wants to read a road sign, it's very good to remember the principle feature of a certain road sign being triangular, for instance. It has then a very good function as being an instant response, because you know it. In the psychological field this becomes very dangerous.

Bohm That's right. You could begin by thinking about the outward and the inward. Outwardly, we deal with the material world, and though we can become confused about it, that process of the abstract and the concrete, or the immediate and the mediate, is absolutely necessary. It's the way by which we have learned how to deal with the world, right? Without it we would be lost. Even animals do it; they make abstractions of what's general. They know what's good for them and what's not.

Then the thing comes inward, because we are able, through the imagination, to produce inward experiences similar to what might be produced by an outward fact. This can cause confusion, because we will fail to see the difference between an experience produced by memory and an experience coming from an outward fact. In fact, they are generally fused and they should be fused. They merge together; the outward and the inward merge in the experience, and that's the way it has to be. I mean, there's no other way. But still, there is the danger of confusion. They can never be separated, and yet somehow we must be able to keep clear these two sources. The whole mechanism of the brain, whatever it may be, does this. The recording combined with the senses produce consciousness, an experience of consciousness in which these two are fused inseparably. You cannot tell at that immediate moment what came from what.

The important point is, that is part of the function of knowledge. Knowledge is not just what's in the library or on the tape recording. Knowledge is all that. If you didn't have that knowledge, it would be of no significance. It would be a sheer abstraction that you could keep in the library; it would never have any significance.

So knowledge is this process of the abstract and the concrete, the mediate and the immediate, the outward and the inward. It's all one process, *but knowledge does not know this,* generally speaking. Knowledge knows itself as *not* being this one process, but as being split up into the abstract and the concrete, the mediate and the immediate, and so on. That's the way people think of it, right? Since people have thought of it that way, that thought is going to be recorded, and that thought will react, immediately, and people will then experience the outward as the mere outward and the inward as the mere inward. Therefore the experience will be wrong.

Now people say, "Whatever I experience is the source of what I know." So knowledge gets caught in a trap, because it produces experience, and then knowledge says, "I must depend on experience to know what I know." But knowledge *produces* an experience and then proves itself to be correct. Knowledge produces that sense of truth. It produces a thought which produces the sense of truth, and a feeling which produces that sense of concrete, immediate experience. So knowledge has lost track of the fact that it's doing that. You see, if knowledge knew that it was doing that, it wouldn't make a fool of itself.

Participant Is there some sort of corrective device, a mechanism . . .

Bohm With knowledge, no mechanism is able to do anything. The only mechanism – which is not a mechanism – would be to see what's happening. If you don't see what's happening, then knowledge can do nothing. Knowledge is in the dark.

Participant Is knowledge capable of learning?

Bohm Of course it's capable of learning; it's learned all these things. It's capable of learning, it's capable of mislearning, it's capable of all sorts of things.

Participant Unlearning?

Bohm That's a question we must come to. It can unlearn some things, but it finds it very hard to unlearn the structure we're discussing, as we'll see. It gets caught. Once knowledge took this wrong turn, it got caught in something it doesn't know how to get out of. Now knowledge says, "Whatever I do I must base on knowledge." That's the way knowledge has been thinking and working. Knowledge says, "I have nothing to go on but what I know. I don't know any way out of this. I only know what I know, and I'm in the dark about these things."

Participant Is knowledge, as you use the term, distinguishable from thought?

Bohm Thought is part of knowledge.

Participant But is there knowledge outside of thought?

Bohm Yes, the unconscious knowledge by which you do all sorts of things, react to all sorts of things. It's been based on thought, and it's been built up by thought, but it happens without your thinking. You're not conscious of thinking. You may be thinking unconsciously, but it won't help you to say that.

Participant   But essentially you're saying it all comes out of the same movement.

Bohm   Yes, it's one movement: knowledge, which includes thought feeling, desire, will, physical reaction, tensions, and all sorts of things. You know how to fool yourself as well as knowing how to do all sorts of useful things. It's one movement. It cannot be separated any more than the concrete can be separated from the abstract.

I'm trying to draw a map here by saying that knowledge has this structure. The reason I'm doing this is because knowledge already knows it has a structure, and what it knows is *misinformation*. Knowledge knows that the concrete is different from the abstract. That the inside is different from the outside, that the immediate is different from the mediated, and so on. And knowledge not only *knows* it, but *experiences* it that way. It says, "I know what I experience and I experience what I know." Is that clear? Therefore, knowledge already knows this and the mere abstract statement otherwise – that knowledge doesn't know – isn't good enough for me: "I need concrete proof that it's different, otherwise I won't change. You don't want me to take any airy-fairy abstraction and just follow that." Knowledge says, "I'm good and hard-headed and I stick to the concrete" – except when there's a concrete fact that tells us we must go back to the abstract.

Participant   What are we doing as we are listening to this? What is knowledge doing as it listens to this conversation about knowledge?

Bohm   That's the question. Knowledge is listening. One part of knowledge says, "Yes, very interesting, quite logical," and the other part says, "This is nothing but an abstraction; I can't count on that."

Participant   Why is it that we've created this division, this concrete/abstract, inward/outward division? Maybe if we could observe the point at which the mind sets that up . . .

Bohm   Well, it's necessary to do this, you see. Look, man began to think. There are psychologists like Piaget who claims that sometimes, or even frequently, young children just beginning to think, having strong imaginations, cannot distinguish their thoughts from real things. They expect other people will see their thoughts standing in the middle of the room, like they do. Then later on they learn it's otherwise, so how are they going to explain it? They say, "There's a real thing out there and it's concrete, and in here is some sort of abstraction, which we call thought." The child has got to do that. I mean, nobody can see your thoughts; we can all see the table, so we say that it is concrete, factual, and actual. And we say, besides that,

that we have thoughts, which are quite abstract and somewhere in your head. Not necessarily in the head, though. You see, the ancient Greeks said that the heart was the seat of the mind; they deduced because of the many folds in the brain that it was a good organ for cooling the blood. Hence the phrase, "a warm heart and a cool head." Anyway, they probably thought that thoughts were taking place somewhere around the heart. Knowledge doesn't know where it's taking place, right?

Participant What's the seat of the awareness that can be aware and still not be caught in this trap?

Bohm Well, we'll say, "Is knowledge all there is?" is the question you are asking. Knowledge says, "Knowledge is all I know." But then the next question is, is knowledge all there is? You can see rationally that it can't be all there is. The whole thing would be meaningless if knowledge were all there is. There must be something beyond knowledge, some kind of reality, at least, some kind of truth. From the senses we often get information that shows us our knowledge is wrong, and we're usually ready to drop it, outwardly. Now, can we get information from *within* showing us our knowledge is wrong? That's the question.

We're asking whether new information can come. Knowledge assumes that its information is coming basically from somewhere beyond knowledge, from some reality, let us say, beyond knowledge. In the case of the senses we seem to get quite a bit of evidence of that. When we discover things that contradict our knowledge, we drop it. Things happen that are entirely surprising, that we don't expect, and are quite contradictory to our knowledge. Therefore we can see a reality which appears to be independent of our knowledge, or at least substantially so, although we can affect it by our knowledge.

Participant In the physical world there are illusions. For instance, there might be a painting of a violin, and when people reach for it, they can't get it. So their senses tell them that in fact it's not a real violin, it's a painting. Your interpretation then has to be changed on the basis of your senses. But the experience you have of anger or another inward experience, there's not a comparable feedback.

Bohm Somehow we don't have it, but that doesn't mean we can't have it. By and large, people haven't had it; they get no feedback inwardly, and they have no way of knowing that this whole construction of experience is entirely a construction of the imagination. It looks very real, it feels very real. All the signs of reality are carefully imitated: solid, strong, fast, necessary, and so on. Whatever test you have for reality is produced inside. You see, this table resists being changed; and your feeling of anger resists being changed. This table remains stable; your feeling of anger does also. Any test for reality you propose is satisfied by your feeling of anger. But I propose that this satisfaction is part of the illusion, that the illusion is so constructed as to satisfy all your tests. Now, knowledge knows how to do that because knowledge is what knows all the tests in the first place, so it can also know what tests have got to be satisfied and know how to satisfy them. You can't fool knowledge . . .

Participant It's got all the cards.

Bohm It's got more cards than you've got!

In any event, we could say that the brain has to deal with itself differently than it deals with, say, a chair. But in fact mankind has fallen into the trap of implicitly treating the brain in the same way that it treats the chair.

Participant It's quite interesting, first of all, to realize the flow between the inward and the outward. This might be a starting point for a new observation.

Bohm Yes, I think that would be a way. Now, I think we should clarify a little bit on what we mean by "inward." In the beginning, "inward" might mean just what's inside the skin, but then we find that's not exactly what we mean, because the stomach is inside the skin, but it can be treated outwardly by pills or operations. When you open up the body you see that it's just the same as the outward; it just happens to be inside. If something is inside a box, that doesn't mean it's inward. That's not what we mean by "inward," right? Rather, "inward" has a deeper meaning in the sense of going deep into the depths of consciousness, where you can't say where it is.

Participant Inside of *me* – isn't that the notion?

Bohm Yes, inside me, but what do I mean by "me"? You see, I might mean the body sometimes, or the soul or the spirit, but it means something at the very depth of things, at the very innermost source, rather than just something inside a box or inside a certain region of space which is covered by the skin. But even what happens inside the skin is highly affected by what goes on outside. Not only temperature and pressure and all that, but other people affect you; they give you a pain in the neck, or they stir up your blood – all sorts of things happen. Somebody did an experiment with radioactive tracers in the blood, and he watched the distribution of blood in the brain. Every thought produced a radical redistribution of blood in the brain. So in that sense even the physical outward and the mental inward are not separated; what happens in the one happens in the other. If you change the distribution of the blood in the brain, it will change the whole way the brain works. If you change the way the brain is working, it will change the blood. So they are really one, but still we're getting a notion of what we mean by "inward."

Now, if you are in a close personal relationship – or even with nature, seeing a beautiful scene outwardly – you can see that the beauty of that takes place inwardly. Or seeing something ugly outwardly, the ugliness is taking place inwardly, probably both physically and mentally.

The whole tradition of the human race has been that this is not so; our language is built that way, and we experience it that way. Knowledge, which goes by experience, logic, and reason, has no way out of this. It says, "I can only start from what I experience." But now we say experience cannot be counted on, and furthermore not even reason, because reason gets distorted when you're angry – your reason cannot be trusted. So neither experience nor reason can be trusted in this area. Even when scientists get angry at each other, their reason can no longer be trusted, nor their choice of facts.

We have to ask: How are we going to get at this and perhaps something beyond all this? That's really one of the questions we must keep coming back to. One of the points, then, is to look in more detail at how experience is formed. We accept experience by saying, "That's experience – nothing more can be said about it. That's where we start." But now we're saying it can't be that way; we know experience is mediated by the past. I've said this abstractly, right? You can see by reason that it must be so. Reason can work from the abstraction and draw necessary conclusions. *But then the rest of the mind doesn't pay any attention to reason unless it has the concrete along with the abstract.* In a way that makes sense, because reason could produce all sorts of nonsense if it starts from something wrong. For that reason alone, the mind is reluctant to go any further, and perhaps for other reasons which are even stronger. So you sort of get stuck there.

Participant When you say the outward and the inward are the same, is there a difference between saying that, and saying the outward affects the inward?

Bohm Yes. Anybody can see the outward and inward affect each other, but ultimately there can be no distinction, because the consciousness of the outward is inward, right? Whatever you see outwardly is being projected from the inward movement of the brain.

Participant   And the inward affects the outward as well?

Bohm   Yes, it obviously affects the outward, but your "outward" *is* the "inward." You see, we may say there is some abstract outward which nobody knows about, which is just outward. It's pure reality, right? But any reality that man knows is inward as well as outward. The reality that man doesn't know might as well not even be there as far as he's concerned, unless he has some sign of it. Therefore, the reality that man experiences is both outward and inward at the same time, inseparably.

Participant   Do you mean, in as much as he is storing his outward experience inwardly?

Bohm   No, the very experience is an inward one. You see, if you cut the nerves going to the brain, there's no experience. It may be taking place outwardly, but if you put an anesthetic in there, nothing happens inwardly and you say no experience has taken place, right? Experience is an inseparable fusion of the outward and inward. If there's no consciousness, there's no knowledge. The word "conscious" means "knowledge"; it's "conscience"; "ponsciare" means "to know." "Conscience" would mean "to know it all together." In ancient times, when the word was formed, it did mean what everybody knew, all together. Consciousness was inherently belonging to everybody. Since then we have gotten the idea of private consciousness, that each person is a separate individual whose consciousness is entirely separate. And that's the way he experiences it because that's what he knows. On the other hand you can ask, Is it truly private? Isn't it formed from the general consciousness, which passes on the pool of information? I could refer to the Greek word "idiosyncrasy"; its root means "private mixture." The idiosyncrasies of a person are his particular mixture of the general ingredients.

That's the suggestion: that this consciousness is general, and by looking at the general consciousness we get to the generating root of consciousness. If we try to look at our own consciousness, we're looking way up on a top branch somewhere, and we won't be anywhere near the root.

Participant   Usually we think of each individual having a mind and his own thoughts and so forth, and the sum of those thoughts making up a consciousness. But you're really suggesting flipping it around, saying, in fact, that there's a general consciousness which contains the appearance of these individual . . .

Bohm Yes, the manifestation as particular individuals. It's like saying, basically, that all trees of a certain kind are one tree, really. Like the Eskimos used to think there was only one seal that they were continually hunting, and they prayed that the seal should reappear.

Participant And is that consciousness also thoughts, feelings, and emotions?

Bohm That's right. We want to go into that – it's thought, feeling, desire, will, and so on. Now we have to go into this carefully, because we have said that our ordinary way of experiencing things finds that thought is different from feeling, they're both different from will (possibly desire is more feeling, but they're all different from will), and then physical action and so on. All this fragmentation has been introduced by knowledge. We say we *know* that thought is different from feeling, both concretely and abstractly. We *know* that will is something else again. When I say "know," you get the feeling of knowing it, really. I *know* this is a table. It's not just abstract. I experience it concretely through the senses and also abstractly; they're fused. *That sense of certainty in knowledge comes from this fusion of the concrete and the abstract.* If I only had abstract knowledge, I'd say, "Well, yes, that's an abstraction; it may be so, possibly so."

There is some sort of illusion generated by this knowledge, and it's very persistent and pervasive. This knowledge does not belong to anybody – it is general consciousness, which manifests in each human being.

Participant Is that a collective subconscious?

Bohm We could call it that, but even to call it "subconscious." . . . You see, Freud introduced the idea of the conscious and the unconscious, or the subconscious. That suggests a division in saying that somewhere in another layer is the unconscious, and you've got to probe into that other layer. Now, what I'm going to propose is that if you introduce the division of conscious and unconscious, it's like dividing the two sides of a coin.

Participant In other words, it's just collective consciousness.

Bohm Yes, it's common consciousness, as the original root of the word meant. And perhaps people in primitive times had a sense of that, but since then we have come to know otherwise.

Participant We have increasingly come to know in fragments.

Bohm Yes, we have come to *know* that it cannot be as "simple" as a common consciousness; it's got to be this and this and this. And we *concretely* know it, not just abstractly. We experience it. We feel that since a lot of progress has happened between then and now, surely our knowledge must be better than theirs. If they knew it otherwise, then they have just made a mistake, right?

Participant Well, I certainly experience my individual consciousness, and I'm asking: On what are you basing this notion of a general consciousness?

Bohm I'm just proposing it now for your consideration.

Participant As an idea.

Bohm Yes, as an abstraction, but in order to counteract some of the other abstractions and concrete knowledge that you have, to propose this as another possible knowledge.

Participant Are you saying that the notion of the individual, the experience of the individual consciousness, is part of this general consciousness?

Bohm Yes, the general idea in the general consciousness is that you are a separate individual. That very idea produces the experience of a separate individual. You didn't invent that idea – you got it from all around you.

Participant We're very involved in finding an identity these days. It's very important to do that.

Bohm But it always has been, except when it was simple – when you identified with the whole community, or whatever. But still you had to have your own identity. Being exiled was the greatest tragedy, because you lost your identity then. If you have a different source of identity it won't bother you so much to be exiled, right? But that's all relative, that's all on the surface.

Participant So where does consciousness come from? If there's a consciousness that says, "Let's prepare for war," where do two-thirds of us get that notion? Do we hook into the mass consciousness of "Let's have a war" somehow?

Bohm Well, look, you can see how it happens. There was a film on the BBC about the way people were just before the First World War, and as the war was approaching, you could see people talking about it, saying "We've got to go into the army, we've got to be patriotic." People who would ordinarily be level-headed were just carried away into a wonderful feeling. I read elsewhere that people were elated; they were almost in ecstacy. They felt they were giving up the ordinary, meaningless dullness of everyday life for a great purpose.

That, evidently, is one of the things – we feel the need of a great purpose or meaning. Ordinary life doesn't give that, and in the collective consciousness there is this idea that life doesn't have much meaning in this fragmentary mode. So deep down we are waiting and waiting for this great purpose. It's even more than the collective – it's universal. The universal always has the supreme power, right? We'll see that in a moment.

Take Hitler shouting "Sieg Heil!" or whatever it is – all those other phrases. The people would shout back, and they had a sense of complete and perfect identity. They didn't know what they were being led into. Nor did the English people know before the First World War that they were going to be slaughtered in the millions in these trenches. They thought the war would be over in a month. It would be a glorious affair, and everybody would come back a hero. So you can see collective delusion forming just as individual delusion forms. They all experienced this. You couldn't walk up to them and say, "Look, this is just an idea." They would say, "No, you're an intellectual. You're just talking intellectually but we're having the real feeling. You're just cold and intellectualizing, just analyzing this glorious thing."

Participant There seems to be something very basic about the enjoyment of the feeling that we are one, that we belong together, that this is my identity. It seems to be a very basic need to move in that direction.

Bohm Well, that seems so, but again I'm questioning experience. I'm saying that it's not always what it seems to be.

Participant Well, it seems like an essential part of that is the sense of individual consciousness, or country or group consciousness. The essential similarity is that you're set apart from somebody else.

Bohm Together with certain ones and apart from other ones. That's the way it goes; it groups certain people together and others apart. That's the whole way thought works. The first step in thought is to put together what belongs together and keep apart what belongs apart, so you can't criticize that in itself. It's just merely a mistake in application of that principle. How do you come about this mistake in application? Because by the time you've done it, you apparently get proof – you experience your togetherness with your group and your division from another group. Then you say, "I'm not inventing it. I'm

not just classifying. My classification is based on real experience."
See the difficulty: knowledge says, "How else can I do it? If I can't count on experience, what do you want me to count on? I can't count on your pure intellectual abstractions."

Participant But isn't that the very nature of thought?

Bohm But that view won't help us. If that's the nature, then we're stuck. You're going to be forced to think, to do practical jobs, and you can't separate it; it slips over from practical life into the inward life.

Participant If I can watch it do this . . .

Bohm But now comes this question, *Who are "you" who are going to be different from it?* This is what we must look at. Are you different from it?

Participant Why do I have to be different? Why don't I watch this process going on?

Bohm Can you do it without being affected by it? A man who is angry can say, "Let me watch," but his watching will be affected by the anger and he will see evidence, for example, to justify why he should be angry, or else to prove he shouldn't be angry. With fear and pleasure it's the same. That is, a certain thing gives intense pleasure, and then you will tend to see whatever will help to sustain that pleasure. You see, the watcher is the same as the watched. This is the important point. Or the knower is the same as the known.

Participant But can't there be an observation of that without the watcher?

Bohm You can imagine it, but is there one? Knowledge can easily get to know about this and imagine it.

Participant I mean even without the watcher. There's no question that thought continues on. Isn't there a possibility of observing that without judging it – just seeing it?

Bohm It may be possible, but if we start out by assuming it is possible, that will be knowledge. You may know it's possible, but what about other people? Do they take your word? They may feel it's impossible. You see, we've got to get deeper.

I think we should explain a little more to show how difficult this problem is. Let's try to go into this relation of thinking and feeling, or thinking and will. You can see in the case of anger that words produce feelings. You know this abstractly, but you don't see it actually happening, as a rule. What happens is that words slowly pile up some state of the brain, and it suddenly explodes into a feeling. It happens so fast that the next thought comes along and says, "That wasn't due to thought; that was independent. That was just a feeling which tells me the state of my reality, and therefore I must deal with it. I must make my reality better." Similarly with fear.

Now, I want to compare this to a wave. You watch the waves slowly coming across the ocean towards the shore, and suddenly the wave breaks on the shore and crashes. If you didn't see the wave coming in, you would say that something is happening on the shoreline, right? Then you would say, "To change that, I must work on the shoreline." You would never get anywhere. Then you would say, "This problem seems to be impossible, hopeless." But we're saying that you don't notice the wave coming up, which crashes on the shore, goes back out, and starts the next wave going. This is the way thought and feeling work. Thought goes slowly; it makes a wave. As I said before, a large part of our thought is hardly noticed at all; it's become fairly automatic, and should be, a lot of it. But it's slowly building up certain conditions which suddenly explode. That explosion goes through the mind – it disturbs everything, it confuses everything, and thought then comes along and says, "What was that explosion? What shall I do about it?"

Now you see further confusion developing, because in the very act of doing that, thought has implicitly taken the disposition that it is something other than itself producing that explosion. If something is due to me, that's called *proprioception* – I know it's due to me. Now when certain nerves are damaged, the person doesn't know that, right? We know somebody who had her nerves damaged. She hit herself in the back of the head and said, "Who has hit me?" She didn't know she was moving her hand.

So with thought there's a failure of proprioception. Thought moves along building up certain conditions inside the brain which suddenly explode, and then thought comes along and says, "What was that explosion?", taking the attitude that it was something quite different from thought. Thought then says, "Therefore I must think what it was, and then decide if I can do something about it." Then you say, "Yes, that was *me* exploding. That was me way deep down inside, who was hurt so badly by what happened that anybody would have exploded." That makes it general and universal, right?

What *always* happens and what *everybody* would do is regarded as necessary. Therefore you say that it is necessary to explode that way. "Universal" implies necessity. Whether it's always the case in time or whether everybody does it, it's both. What you mean when you say that everybody does it, is that everybody would *always* do it. Anybody at anytime would do it, so it would always happen.

Now there may be, besides that, true feelings, which show something deeper, but unless you can tell which is which, you get very confused. One feeling may reveal the depths of the soul, and the other feeling may be nothing but thought exploding. But you make a mistake – you didn't see the thought building up, therefore the concrete experience you had was that something happened suddenly, with no cause that was visible, as if a new feeling had come in. Then you say, "This feeling is very disturbing. I must do something with it. I must first get to know what it is." That seems harmless enough, but it isn't. Knowledge is a very dangerous thing, or, as they say, a little knowledge. The point is that you say, "This is me exploding in anger." Or desire – desire is another explosion. The picture of desire that is given by both the ancient Greek culture and the ancient Hindu culture is that the arrow pierces the heart. Cupid, right? So suddenly there is that piercing thing that causes an explosion, the explosion of desire, which is not so different from the explosion of anger or fear.

So this *sudden explosion* is the general thing that's the cause of the trouble. It's as if somebody had piled up all the explosives and gotten the wires ready, and after hiding himself somewhere in the background he touched off this explosion. Then the person suddenly becomes conscious of this explosion and says, "That's *me* exploding. Anybody would have done it; I am justified." Or else saying, "I've done this so often and it was never any use. I'm going to try to stop this thing, these explosions from now on." It's like somebody saying, "This surf has been wearing away the beach. I'm going to try to stop it." But again, since he doesn't know the source, it has no meaning.

Participant Usually we say the source is outside. Behaviorists will say that it's a stimulus-response . . .

Bohm Yes, but without that thought that connected the stimulus with the response, it wouldn't happen. If you didn't have that period of thought that built up the anger, saying, "He treated me wrongly. I was sitting quietly and he did this. He's always doing this. He does it on purpose" (the various thoughts have been deposited over many years), it wouldn't happen. Between the stimulus and the response is a mechanism which thought has prepared.

After a while you may begin to notice what you're doing. What you can do, to begin with, is to use these words in connection with some event that happened, and watch what happens. You'll see the machine at work. You see, this is something you can observe beyond the experience that is created by the word. Which is, to use the words, saying, "I'm not interested in the *content* of the words but in their *effect*."

Participant   But I'm likely to see that only after the effect.

Bohm   Don't worry about it; I'm saying do it first. Use those words and perhaps you will see yourself doing it before, after a while.

Participant   I think there is one typical thing in what you just said. We try to think something out instead of actually doing it. We think it out and say it's impossible. Then you try to prove that it can't be done.

Bohm   If you say it's impossible to look, that conclusion of what's necessary and what's possible has a big effect. If something is necessary, then you will have the will to do it. Is that clear? You do not choose the content of your will, but the content of your will is immediately and inevitably whatever you regard as necessary and possible. If it's not possible, it's also not the content of your will. Like the fox and the sour grapes – as soon as he sees that it's not possible, he doesn't want the grapes. And not only that, he says the grapes are sour. He distorts to feel better!

Participant   I've experienced many times the anger, and then, looking back, I say things like, "He said that, and I got angry." It's a very common experience. I can't imagine – maybe I'm saying that I don't think it's possible – that I would notice the words before the anger. I can't even imagine why I would notice the words.

Bohm   Well, because you notice other things. Suppose you say, "I've been having the dangerous habit of lighting matches near gasoline tanks, because I light matches whenever I've felt like it. Then I began to notice that every time I lit a match . . ."

Participant   Are you saying that the words would become a cue? I would become aware that those kind of words . . .

Bohm   . . . are the same as matches that are lighting the gasoline that has already been poured out.

Participant   And I would notice that *before* I felt the emotion?

Bohm   Yes. Now, it's easy to see this happening in someone else. You can see somebody else whipping himself up with words. You've just got to watch films of Adolph Hitler talking to those masses, and you'll say, "What are they doing? He's whipping them up, they're allowing themselves to be whipped up, and there it is." You can see that emotions are being generated by words. But if you were in that group or if you were Adolph Hitler, you wouldn't feel that way.

Participant Are you suggesting that what's needed first is to see that something is necessary and possible, rather than to try and create the will directly?

Bohm More than that. You have to see that your will is entirely created out of the thought of what is necessary and possible. This has tremendous implications, because, for example, we in the western world have a long tradition of saying the will is free. You choose everything you will to do. Therefore people have to be punished very badly when they do the wrong thing. It's even part of the Christian tradition to say that Adam willfully chose evil, and therefore mankind had original sin, and that people have to be punished for it. This approach has been followed for thousands of years, but it doesn't do any good, obviously. It's like trying to turn back the waves at the shore.

Participant From what you are pointing out – particularly that knowledge catches itself in a trap from which it cannot get out – that is what the state of the world is like. Now, there seem to be two things which that entity which causes that trouble – which is us, the individual – does. First, it applies a mechanical process; it has mechanical reactions to whatever is happening. The other thing is that there is time involved. It never approaches things in any sort of instantaneous or immediate way, but always goes through the process of thinking, of abstractions.

Bohm No, that's the trouble, that it apparently is producing the immediate. You see, knowledge says, "I know about that already." But it fools itself. Knowledge says, "What you say is right, therefore I'm going to look for the immediate." This is the point I think we've got to watch out for. Certain thought always involves time and other thought is taking place slowly and unnoticed, and suddenly it produces an apparently immediate experience. Even a sense of insight or a flash of truth could be produced that way, right? Some people delude themselves; they think they've had great insights. The difficulty is that we have gone through a long process of saying that thought takes time, and people have said, "Okay, I won't trust thought. I'll trust my immediate intuition." The immediate intuition is the product of thought, but it happens with a sudden movement, right?

Participant So, is that what we should thoroughly examine?

Bohm Yes, that's what we have to examine: how thought and immediate intuition are connected. And also how the will is connected to that, and to desire, fear, and time. These are all related. All these are involved in the mess mankind is in. But this is the generative source which is common to all mankind – every race, every culture as far back as we know. When we study this we are beginning to get into the generative source of the problem. It is not merely thought, but it is the whole field of knowledge – its concrete as well as its abstract features. It's the whole field of knowing. I'm saying that the field of knowing is what is involved in the source. That knowing *may* be immediate and intuitive, though it also may *not* be, and therefore many people feel they have an intuitive apprehension of things, which is still, I'm trying to say, thought. But they experience it as knowing: they say, "I know this – I haven't thought it."

Now there may be a kind of knowing in which knowledge is infused with creative insights – you see something for the first time, or you get to know certain things. I think we have to be careful not to close everything, to make conclusions, because the thing that thought is doing constantly is to make conclusions and closing the issue, saying this is possible or this is not possible. And that very feeling of possibility and impossibility creates the will either to try, or to keep away. So you will find yourself experiencing the will not to bother because of what thought has done a minute ago or an hour ago. Then you say that will is spontaneous. Now, that's confused. It's very important not to go beyond the fact that is actually before us.

I'm trying to say with regard to anything beyond that fact, we simply say maybe so, maybe not. We just don't know, right? One of the points about knowledge that we were talking about is that in the beginning, knowledge said, "I only can do what I know." Knowledge says, "I've got to know, and I only know knowledge and that's all I can do." Therefore it seems to be stuck in a circle. That very thought would create the will just to go on with knowledge. So you will experience not knowledge, but the will to do this or that, and you say, "That's not knowledge, that's my immediate response." Knowledge says that's what it is. Knowledge is now apparently reflecting on something else, which is independent, and knowledge says, "That's my immediate response." So a very complex illusion is created. Now if knowledge gets to understand some of the structure of knowledge, then it may at least open the way to observe. But knowledge is now so fast, and producing apparent observations all the time, that it is very hard to observe it.

Participant You are saying that knowledge has, in a sense, split itself into two and taken that which it has learned – and to which it reacts most quickly – and called that part of itself . . .

Bohm . . . experience or intuition.

Participant It has elevated that and called it "true" or "truth."

Bohm It may call it truth, or experience, or intuition, or something like that.

Participant So by itself it has somehow separated itself.

Bohm Yes, it has produced this false separation, and this separation then comes out as the separation of the observer and the observed, or the thinker and the thought, the knower and the known. You see, the difficulty of saying "observe" is this: you say, "You must observe," but knowledge says, "I'm already observing." It doesn't even say it – it just produces the illusion of observation. But then when it comes to reflect on it, it says, "That's exactly what I've been doing."

The difficulty is that knowledge produces the appearance or illusion of observation and reflects on it, and calls that observation. Therefore, the injunction to observe has no effect, because knowledge says that's just what's been happening. That explains why people can go into this for endless years and nothing much happens.

Now we can either say we *know* all about this, or we can say that this is a *proposal* which is to be tested, so we don't introduce the same illusion. The trouble with knowledge is that it says, "I know" when it doesn't. It says, "I know with absolute certainty," when it doesn't. The knowledge with absolute certainty, or for always, is very dangerous. We'll have to discuss that, because that produces the experience of absolute certainty, and therefore there is nothing to be questioned. Then the whole questioning stops and you're just satisfied with illusion.

Participant In the physical world that certainty is useful. I mean, I am dead certain that the floor is going to hold me up, so I don't worry about it anymore.

Bohm That's right, though it may have a hole in it, and you may trip. But if it trips you, you don't keep on insisting that the floor is holding you up. But if something similar happens inwardly, we have the insistence that this "inward" is still reliable and certain – even absolutely certain. It is this absolute certainty of the inward – which tends to focus on the inward – that is the kind of knowledge which may produce darkness. That knowledge of what you are, who you are, what sort of person you've got to be, to whom you belong, what your desires are, what your fears are, what you can do, what you can't do . . .

Participant My self-identity.

Bohm Yes, and your collective identity as well.

Participant You're saying that that's the kind of knowledge which is endarkening.

Bohm Yes, it endarkens the brain.

Participant As distinct from, say, the knowledge of how to repair a jet engine.

Bohm Yes, or knowledge of where you're going and all sorts of knowledge of science and what not. One can say – remembering that knowledge is inseparable from the physical changes in the brain, especially this knowledge that produces these big explosions – that knowledge literally disrupts the brain physically.

Participant I'm having a bit of trouble with that, because the physical type of knowledge, just sheer pragmatic knowledge like stepping down the stairs too many times after you've already reached the last step, that, in a sense, is endarkenment, too. It's a mistake.

Bohm I'm glad you brought that up. I'm trying to make it clear what the mistake is. We will inevitably make mistakes. It is in the very nature of knowledge that it's incomplete, and when it's extended we will find that we make a mistake. Now the point is, knowledge is the whole function, not just one item. The function of knowledge is to admit your mistake, drop it, and learn, right?

You could say that where man took a wrong turn was, he got a certain kind of knowledge which was not only a mistake but which led him to make more mistakes to justify this mistake and hold on to it. That was where he began to go wrong, and where he's continually going wrong. We make a certain mistake about ourselves. Why is it so hard for a person to say, "I've made a mistake on an important point"? He doesn't say, "I've made a mistake." No, he says, "Somebody else made a mistake," or "It wasn't a mistake." So he makes a second mistake and a third mistake and a fourth; it goes on piling up. That's endarkenment. That is not only mental endarkenment, but it is a physical disruption of the brain.

11 DIALOGUE AS A NEW CREATIVE ORDER (1987)

In introducing the notion of a generative source for the turbulence in human consciousness (Chapter 10), Bohm was anticipating the more developed concept of generative order put forth with F. David Peat in Science, Order, and Creativity. *In its barest form, a generative order is a process whereby a limited number of simple components generate a diverse structure, as in the case of fractal geometry. A generative order may also be an implicate order, particularly if the whole is relevant to the creation of the parts, or if processes of enfoldment and unfoldment are present. Such an example is Bohm's causal interpretation as applied to quantum field theory (Chapters 4 and 6). In this model, particle-like structures (an explicate order) unfold from a latent field (a first-level implicate order). Yet active information is required for this unfoldment to occur, which comes from a super-implicate, or generative order (a second-level implicate order).*

Bohm proposes that a similar generative order can be discerned in the consciousness of society. All of a society's overt activities, artifacts, and individuals are its explicate order. Its first-level implicate order is the latent, relatively passive content of the entire culture – the pool of knowledge it has accumulated for millennia, as well as the somatic correlates of this knowledge. Its second-level implicate order – its generative order – is the values and meanings that inform the pool of knowledge with specificity and order, giving rise to dispositions, intentions, and actions that unfold into the explicate social order. This model, while not exhaustive of Bohm's view of the implicate nature of consciousness – which goes on to include the holomovement as a

Extract from Chapter 6 of D. Bohm, *Science, Order, and Creativity* (with F. David Peat), Routledge ([1987], 2000).

generative order – nonetheless provides a reference point for exploring Bohm's view of dialogue.

To effectively resolve the conflicts and contradictions in the structure of human experience, Bohm claims that values and meanings must be transformed at their concrete generative level. This view, expressed in various ways throughout this book, has primarily been discussed in the context of individual inquiry. Bohm, for example, in his 1958 letter to Yitzhak Woolfson, claims that it is "only through the individual and not through the collective that change can occur." And indeed, it remained Bohm's view that the individual, through his or her inherent link to implicate consciousness, does have the capacity to diminish the "pollution" in the generative order, albeit to a limited extent.

But Bohm's views of transformation at the concrete generative level were continually evolving, and the possibility of a collective approach to this issue captured his imagination in his later years. He came to feel that the efforts of scattered individuals would have only marginal effect on the generative social order, and that any enduring impact would require a collective approach, which carries the potential for exponential change.

Through dialogue, Bohm felt that a profound and generally unrealized creativity could be released in the generative order, thus shaping and informing the whole of society. To release this creativity, the "tacit infrastructure" of society – assumptions and values held rigidly but often outside the scope of conscious awareness – must be brought to light and understood in all its implications. In a dialogue context, this requires not only "suspending" one's own concrete values and assumptions (Chapters 7 and 8), but now extending this suspension to the values and assumptions expressed by others.

Bohm suggests that when a critical mass of participants in a dialogue is engaged in such mutual suspension, the mind is freed to move in new ways. Assumptions are held less rigidly, and with greater awareness. Meaning becomes a rich domain on its own terms, outside the restrictions of utilitarian application; it can follow the pathways opened via mutual suspension, and thus be shared in new and unexpected ways. And significantly, new orders can emerge, through questioning rigid definitions of "individual" and "collective," and encouraging the exploration of orders of consciousness between these traditional extremes.

Creativity, in almost every area of life, is blocked by a wide range of rigidly held assumptions that are taken for granted by society as a whole. Some of these have already been discussed,[1] but in addition, every society holds additional assumptions that are of such a shaky nature that they are not even admitted into discussion. There is therefore an unspoken requirement that everyone must subscribe to these assumptions, but that no one should ever mention that any such assumptions indeed exist. They are tacitly denied as operating within society, and even this denial is denied. The overall effect is to lead people to collude in "playing false" so they constantly distort all sorts of additional thoughts in order to protect these assumptions. Such bad faith enters deep into the overall generative order of society.

These rigidities and fixed assumptions, many of which must not be mentioned but must nevertheless be defended, may be compared with a kind of pollution that is constantly being poured into the stream of the generative order of society. It makes no sense to attempt to "clean up" parts of this pollution farther downstream while continuing to pollute the source itself. What is needed is either to stop the pollution at its source, or to introduce some factor into the stream that naturally "cleans up" pollution.

In the body a similar problem arises. As a person grows older, through infection, allergies, contaminants, misadventure, and the processes of aging, considerable "misinformation" or irrelevant information accumulates in the system. Indeed it is possible to look at a disease like cancer as arising from misinformation in the structure of DNA. Viruses also introduce misinformation, in the sense that DNA from the virus acts to replace some of the DNA in the host cell and therefore causes this cell to replicate foreign DNA rather than serving the needs of the body.

There are basically three ways of dealing with this problem of misinformation in the body. The first is to avoid the introduction of misinformation in the first place, for example, by keeping away from infection through good sanitation and a careful diet. Second, where misinformation exists, it may be possible to do something to remove it through various kinds of medical intervention. But more significantly, the third option involves the body itself, which possesses an immune system which is able to "clear up" misinformation in a natural way.

This is indeed the body's main mode of dealing with misinformation. This can be clearly seen from the fact that drugs are of little use in treating a disease like AIDS, which destroys this immune system itself. Furthermore, the whole practice of immunization relies on activating the immune system and so avoiding the onset of particular diseases.

The immune system itself is particularly complex and contains a very subtle kind of information that can respond to the whole "meaning" of what is happening to the order of the body. In this way it is able to distinguish misinformation from information needed for the body's healthy operation. It can be compared to a kind of "intelligence" that works within the body. Moreover there is evidence that this sort of "intelligence" can respond to the higher levels that are usually associated with thought and feeling. It is well known that depressing thoughts can inhibit the activity of the immune system, with the result that a person becomes more susceptible to infections. Indeed there is much evidence that a vigorous, creative state of mind and a strong "will to live" are conducive to general health and even to recovery from dangerous illnesses.[2] More generally, it could be said that good health is basically a manifestation of the overall creative intelligence, working in concert with the body, through various means that include exercise, diet, relaxation, and so on.

Returning to a consideration of society, clearly there is also a vast amount of misinformation in circulation which acts toward society's degeneration. The media and various modern means of communication have the effect of rapidly disseminating and magnifying this misinformation, just as they do with valid information. It should be clear that by "misinformation" is meant a form of *generative information* that is inappropriate, rather than simply incorrect statements of fact. In a similar way a small "mistake" in DNA can have disastrous consequences because it forms part of the generative order of the organism and may set the whole process in the wrong direction.

In society, the generative order is deeply affected by what has a very *general* significance. Indeed the generative order may be regarded as the *concrete activity of the general*. This takes the form of general principles, general aims, and generally accepted values, attitudes, and beliefs of all kinds that are associated with the family, work, religion, and country. In going from these general principles to the universal, it is clear that the effect on the generative order will become yet more powerful. When a given principle is regarded as universally valid, it means that it is taken as absolutely necessary. In other words, things cannot be otherwise, under any circumstances whatsoever. Absolute necessity means "never to yield." To have something in the generative order that can never give way, no matter what happens, is to put an absolute restriction on free play of the mind, and thus to introduce a corresponding block to creativity that is very difficult to move.

Of course, both the individual and society require a certain stability, and for this, thought must be able to hold itself fixed within certain appropriate limits and with a certain kind of *relative* necessity.

Over a limited period of time, certain values, assumptions, and principles may usefully be regarded as necessary. They are relatively constant, although they should always be open to change when evidence for the necessity of the latter is perceived. The major problem arises, however, when it is assumed, usually tacitly and without awareness and attention, that these values, assumptions, and principles have to be absolutely fixed, because they are taken as necessary for the survival and health of the society and for all that its members hold to be dear.

We have argued elsewhere[3] that science, which is in principle dedicated to the truth, tends to be caught up in necessity, which then leads to false play and a serious blockage of creativity. It is now clear that the assumptions of absolute necessity, with their predispositions to unyielding rigidity, are only part of a much broader spectrum of similar responses that pervade society as a whole. General principles, values, and assumptions, which are taken in this way to have absolute necessity, are thus seen as a major source of the destructive misinformation that is polluting the generative order of society.

As with the body, society attempts to deal with this sort of misinformation by trying to prevent it from entering its fabric, or attempting to "cure" it with some form of therapy. For example, on a rather superficial level, there are laws to prevent false information and information which may engender hatred, anger, and prejudice from being spread about various races, religions, and groups. Writers, dramatists, and filmmakers go some way to making people aware of prejudices and rigidly fixed attitudes. But in the long run, all these attempts are limited by the overwhelming, and yet often very subtle, pressures within society toward colluding to defend one's own group and its ideas. In addition, there is the whole problem of the intolerance and mistrust that have grown up between nations, religions, ideologies, and other groups which go all the way down to the family itself. To some extent psychotherapy and group therapies can help to clear up individual misinformation of this kind, which may go back to early childhood, or start in a later phase of life. But these approaches have very little effect in the larger sphere of society as a whole.

A particularly important piece of misinformation is the key assumption that creativity is necessary only in specialized fields. This assumption pervades the whole culture, but most people are generally not aware of it; there is always a tendency for misinformation to defend itself by leading people to collude in playing false, whenever such an assumption is questioned. Assuming the restricted nature of creativity is obviously of serious consequence for it clearly predetermines any program that is designed to clear up the misinformation within society and suggests that it cannot be creative.

All that seems to be left is to ask whether society contains some kind of "immune system" that could spontaneously and naturally clear up misinformation. If such a system exists, then it is certainly not obvious, nor does it appear to be in common operation within our society today.

DIALOGUE AND CULTURE

In this section it is proposed that a form of free dialogue may well be one of the most effective ways of investigating the crisis which faces society, and indeed the whole of human nature and consciousness today. Moreover, it may turn out that such a form of free exchange of ideas and information is of fundamental relevance for transforming culture and freeing it of destructive misinformation, so that creativity can be liberated. However, it must be stressed that what follows is not given in the spirit of a prescription that society is supposed to follow. Rather it is an invitation to the reader to begin to investigate and explore in the spirit of free play of ideas and without the restriction of the absolute necessity of any final goal or aim. For once necessity and absolute requirements or directions enter into the spirit of this exploration, then creativity is limited and all the problems that have plagued human civilization will surface yet again to overwhelm the investigation.

To begin, it should be noted that many of the ideas to be explored were first investigated by Patrick de Maré, who is a psychiatrist working in England.[4] De Maré has used his wide experience of dialogue in therapeutic groups to support his arguments. However, it is essential to emphasize that his ideas about dialogue are not concerned primarily with psychotherapy, but rather with the transformation of culture, along the general lines that have been indicated in this chapter.

We have previously shown[5] how rigid conditioning of the tacit infrastructure of scientific thought has led to a fragmentation in science and to an essential breakdown in communication between areas which are considered to be mutually irrelevant. Nevertheless a closer investigation of actual cases suggested that there is nothing inherent in science which makes such breaks in communication and fragmentation inevitable. Indeed wherever fragmentation and failures in communication arise, this clearly indicates that a kind of dialogue should be established.

The term *dialogue* is derived from a Greek word, with *dia* meaning "through," and *logos* signifying "the word." Here "the word" does not refer to mere sounds but to their meaning. So dialogue can be considered as a free flow of meaning between people in communication, in the sense of a stream that flows between banks.

A key difference between a dialogue and an ordinary discussion is that, within the latter, people usually hold relatively fixed positions and argue in favor of their views as they try to convince others to change. At best this may produce agreement or compromise, but it does not give rise to anything creative. Moreover, whenever anything of fundamental significance is involved, then positions tend to be rigidly non-negotiable and talk degenerates either into a confrontation in which there is no solution, or into a polite avoidance of the issues. Both these outcomes are extremely harmful, for they prevent the free play of thought in communication and therefore impede creativity.

In dialogue, however, a person may prefer a certain position but does not hold to it non-negotiably. He or she is ready to listen to others with sufficient sympathy and interest to understand the meaning of the other's position properly and is also ready to change his or her own point of view if there is good reason to do so. Clearly a spirit of goodwill or friendship is necessary for this to take place. It is not compatible with a spirit that is competitive, contentious, or aggressive. In the case of Einstein and Bohr,[6] these requirements were evidently met, at least initially. However, because each felt that a different notion of truth and reality was involved, which was not negotiable in any way at all, a real dialogue could never take place.

This brings us to an important root feature of science, which is also present in dialogue: to be ready to acknowledge any fact and any point of view as it actually is, whether one likes it or not. In many areas of life, people are, on the contrary, disposed to collude in order to avoid acknowledging facts and points of view that they find unpleasant or unduly disturbing. Science is, however, at least in principle, dedicated to seeing any fact as it is, and to being open to free communication with regard not only to the fact itself, but also to the point of view from which it is interpreted. Nevertheless, in practice, this is not often achieved. What happens in many cases is that there is a blockage of communication.

For example, a person does not acknowledge the point of view of the other as being a reasonable one to hold, although perhaps not correct. Generally this failure arises when the other's point of view poses a serious threat to all that a person holds dear and precious in life as a whole.

In dialogue it is necessary that people be able to face their disagreements without confrontation and be willing to explore points of view to which they do not personally subscribe. If they are able to engage in such a dialogue without evasion or anger, they will find that no fixed position is so important that it is worth holding at the expense of destroying the dialogue itself. This tends to give rise to a unity in plurality of the kind we have discussed elsewhere.[7] This is, of course, quite different from introducing a large number of compartmentalized positions that never dialogue with each other. Rather, a plurality of points of view corresponds to the earlier suggestion that science and society should consist not of monolithic structures but rather of a dynamic unity within plurality.

One of the major barriers to this sort of dialogue is the rigidity in the tacit infrastructure of the individual and society. The tacit infrastructure of society at large is contained in what is generally called culture. Within each society, however, there are many subcultures which are all somewhat different, and which are either in conflict with each other, or more or less ignore each other as having mutually irrelevant aims and values. Such subcultures, along with the overall culture, are generally rigidly restricted by their basic assumptions, most of which are tacit and not open to awareness and attention. Creativity is therefore, at best, an occasional occurrence, the results of which are quickly absorbed in a fairly mechanical way into the general tacit infrastructure.

At present, a truly creative dialogue, in the sense that has been indicated here, is not at all common, even in science. Rather the struggle of each idea to dominate is commonly emphasized in most activities in society. In this struggle, the success of a person's point of view may have important consequences for status, prestige, social position, and monetary reward. In such a conditioned exchange, the tacit infrastructure, both individually and culturally, responds very actively to block the free play that is needed for creativity.

The importance of the principle of dialogue should now be clear. It implies a very deep change in how the mind works. What is essential is that each participant is, as it were, suspending his or her point of view, while also holding other points of view in a suspended form and giving full attention to what they mean. In doing this, each participant has also to suspend the corresponding activity, not only of his or her own tacit infrastructure of ideas, but also of those of the others who are participating in the dialogue. Such a thoroughgoing suspension of tacit individual and cultural infrastructures, in the context of full attention to their contents, frees the mind to move in quite new ways. The tendency toward false play that is characteristic of the rigid infrastructures begins to die away. The mind is then able to respond to creative new perceptions going beyond the particular points of view that have been suspended.

In this way, something can happen in the dialogue that is analogous to the dissolution of barriers in the "stream" of the generative order, as discussed previously.[8] In the dialogue, these blockages, in the form of rigid but largely tacit cultural assumptions, can be brought out and examined by all who take part. Because each person will generally have a different individual background, and will perhaps come from a different subculture, assumptions that are part of a given participant's "unconscious" infrastructure may be quite obvious to another participant, who has no resistance to seeing them. In this way the participants can turn their attention more generally to becoming aware, as broadly as possible, of the overall tacit infrastructure of rigid cultural and subcultural assumptions and bringing it to light. As a result, it becomes possible for the dialogue to begin to play a part that is analogous to that played by the immune system of the body, in "recognizing" destructive misinformation and in clearing it up. This clearly constitutes a very important change in how the mind works.

There is, however, another extremely important way in which the operation of the mind can be transformed in such a dialogue. For when the rigid, tacit infrastructure is loosened, the mind begins to move in a *new order*. To see the nature of this order, consider first the order that has traditionally characterized cultures. Essentially this involves a strong fragmentation between individual consciousness – "what the individual knows all together" – and social consciousness – "what the society knows all together."

For the individual, consciousness tends to emphasize subjectivity in the sense of private aims, dreams, and aspirations that are shared to some extent with family and close friends, as well as a general search for personal pleasure and security. In society, however, consciousness tends to emphasize a kind of objectivity with common aims and goals, and there is an attempt to put conformity and the pursuit of the common welfare in the first place. One of the principal conflicts in life arises therefore in the attempt to bring these two fragments together harmoniously. For example, as a person grows up, he (or she) may find that his individual needs have little or no place in society. And in turn, as society begins to act on the individual consciousness in false and destructive ways, people become cynical. They begin to ignore the requirements of reality and the general good in favor of their own interests and those of their group.

Within this generally fragmentary order of consciousness, the social order of language is largely for the sake of communicating information. This is aimed, ultimately, at producing results that are envisaged as necessary, either to society or to the individual, or perhaps to both. Meaning plays a secondary part in such usage, in the sense, for example, that what are put first are the problems that are to be solved, while meaning is arranged so as to facilitate the solution of these problems. Of course, a society may try to find a common primary meaning in myths, such as that of the invincibility of the nation or its glorious destiny. But these lead to illusions, which are in the long run unsatisfactory, as well as dangerous and destructive. The individual is thus generally left with a desperate search for something that would give life real meaning. But this can seldom be found either in the rather crude mechanical, uncaring society, or in the isolated and consequently lonely life of the individual. For if there is not common meaning to be shared, a person can be lonely even in a crowd.

What is especially relevant to this whole conflict is a proper understanding of the nature of culture. It seems clear that in essence culture is meaning, as shared in society. And here "meaning" is not only *significance* but also *intention, purpose,* and *value*. It is clear, for example, that art, literature, science, and other such activities of a culture are all parts of the common heritage of *shared meaning*, in the sense described above. Such cultural meaning is evidently not *primarily* aimed at utility. Indeed, any society that restricts its knowledge merely to information that it regards as useful would hardly be said to have a culture, and within it, life would have very little meaning. Even in our present society, culture, when considered in this way, appears to have a rather small significance in comparison to other issues that are taken to be of vital importance by many sectors of the population.

The gulf between individual consciousness and social consciousness is similar to a number of other gulfs that have already been described, for example, between descriptive and constitutive orders, between simple regular orders of low degree and chaotic orders of infinite degree, and, of course, between the timeless and time orders.[9] But in all these cases, broad and rich new areas for creativity can be found by going to new orders that lie *between* such extremes. In the present case, therefore, what is needed is to find a broad domain of creative orders between the social and individual extremes. Dialogue therefore appears to be a key to the exploration of these new orders.

To see what is involved, note that as the above dialogue develops, not only do specific social and cultural assumptions "loosen up," but also much deeper and more general assumptions begin to be affected in a similar way. Among these, one of the most important is the assumption that between the individual consciousness and the social consciousness there is an absolute gulf. This implies that the individual must adjust to fit into the society, that society must be remade to suit the individual, or that some combination of both approaches must be carried out. If, however, the dialogue is sustained sufficiently, then all who participate will sooner or later be able to see, in actual fact, how a creative movement can take place in a new order between these extremes. This movement is present both externally and publicly, as well as inwardly, where it can be felt by all. As with alert attention to a flowing stream, the mind can then go into an analogous order. In this order, attention is no longer restricted to the two extreme forms of individual and social. Rather, attention is transformed so that it, along with the whole generative order of the mind, is in the rich creative domain "between" these two extremes.

The mind is then capable of new degrees of subtlety, moving from emphasis on the whole group of participants to emphasis on individuals, as the occasion demands. This is particularly significant for proper response to the strong emotional reactions that will inevitably arise, even in the friendliest group, whenever fundamental assumptions are disturbed. Because the mind is no longer rigidly committed to the individual or to the social extremes, the basic issues that arise in a disagreement between participants are to a considerable extent "defused." For the assumptions that are brought to the common attention are no longer implied to have absolute necessity. And as a result, the "emotional charge" that is inevitably associated with an assumption that is dear to one or more members of the group can be reduced to more manageable proportions, so that violent "explosions" are not likely to take place. Only a dialogue that can, at the same time, meet the challenge both of uncovering the intellectual content of a rigidly held basic assumption and of "defusing" the emotional charge that goes with it will make possible the proper exploration of the new order of mental operation that is being discussed here.

It is possible to have such dialogues in all sorts of circumstances, with many or just a few people involved. Indeed even an individual may have a kind of internal dialogue with himself or herself. What is essential here is the presence of the spirit of dialogue, which is, in short, the ability to hold many points of view in suspension, along with a primary interest in the creation of a common meaning. It is particularly important, however, to explore the possibilities of dialogue in the context of a group that is large enough to have within it a wide range of points of view, and to sustain a strong flow of meaning. This latter can come about because such a dialogue is capable of having the powerful nonverbal effect of consensus. In the ordinary situation, consensus can lead to collusion and to playing false, but in a true dialogue there is the possibility that a new form of consensual mind, which involves a rich creative order between the individual and the social, may be a more powerful instrument than is the individual mind. Such consensus does not involve the pressure of authority or conformity, for it arises out of a spirit of friendship dedicated to clarity and the ultimate perception of what is true. In this way the tacit infrastructure of society and that of its subcultures are not opposed, nor is there any attempt to alter them or to destroy them. Rather, fixed and rigid frames dissolve in the creative free flow of dialogue as a new kind of microculture emerges.

People who have taken part in such a dialogue will be able to carry its spirit beyond the particular group into all their activities and relationships and ultimately into the general society. In this way, they can begin to explore the possibility of extending the transformation of the mind that has been discussed earlier to a broader sociocultural context. Such an exploration would clearly be relevant for helping to bring about a creative and harmonious order in the world. It should be clear by now that the major barriers to such an order are not technical; rather they lie in the rigid and fragmentary nature of our basic assumptions. These keep us from changing in response to the actual situations and from being able to move together from commonly shared meanings.

Notes

1   See Bohm, D., and Peat, F. D., *Science, Order, and Creativity*, Routledge, London (2000).
2   See, for example, Cousins, N., *Anatomy of an Illness*, Bantam, New York (1981).
3   See Bohm and Peat, *Science, Order, and Creativity*, Ch. 1.
4   A brief, but fairly comprehensive presentation of de Maré's ideas can be found in *Group Analysis*, vol. XVII, no. 78, Sage, London (1985).
5   See Bohm and Peat, *Science, Order, and Creativity*, Chs 1 and 2.
6   See *ibid.*, Ch. 2.
7   See *ibid.*, Ch. 3.
8   See *ibid.*, Chs 4 and 5.
9   See *ibid.*, Ch. 3.

12 ON DIALOGUE AND ITS APPLICATION (1989)

While the material of Chapter 11 gives an overview of Bohm's dialogue per-
spective, and in particular links dialogue to the larger theoretical scope of
Bohm's work, this final chapter provides a "nuts and bolts" approach to
dialogue. How many people should be in a dialogue group? What somatic
and organizational structures are optimal for facilitating dialogue? How
often should groups meet, and for how long should they continue? How does
the traditional role of leadership factor into a dialogue? How are the difficul-
ties that inevitably arise in a dialogue – frustration, anxiety, fear, anger, role
identification – most effectively incorporated? These questions and many
others are addressed in a clear and straightforward manner – not as fixed
rules, but as recommendations based on considered observation and substan-
tial experimentation.

Bohm also outlines how suspension and proprioception can function at
the collective level, particularly the manner in which the physical body – both
group and individual – can be utilized as a mirror for the activity of values,
assumptions, and meanings. This leads into Bohm's "vision of dialogue," in
which a common, participatory consciousness can emerge from the group,
giving rise to a quality of impersonal fellowship which does not depend on
typical conventions of familiarity. Such a consciousness is seen by Bohm as
essential for true communion, which extends beyond the individual-collective
dichotomy to the unlimited ground – the holomovement – from which these
both arise.

One of the central themes that runs through all of Bohm's work is that the
nature of the universe – including our experience of it – is more porous, fluid,

Extract from Chapter 2 of D. Bohm, *On Dialogue* (ed. Lee Nichol), Routledge, London (1996).

and inherently whole than we generally recognize. We can of course ignore this dynamic wholeness, or objectify it in purely abstract terms. But if we wish to embody and give expression to the whole, we face the challenge of working through the fragmentation that currently defines our consciousness. For Bohm, dialogue was one way of bringing this fragmentation to light in a particularly direct and graphic manner, not an algorithm for creating novel forms of social entertainment. Propositions and theories – including those in this book – are essential for exposing us to new possibilities, but they can easily ensnare us and provide a new framework for delusion and false comfort.

At the end of the day, it may be that Bohm's vision, and his challenge, were deceptively simple. When books and speculations have served their purpose, we are left with the unvarnished truths of who we are and how we live in this world. Perhaps what matters then is that we take seriously our capacity, as Bohm suggests, to "soften up," to "open up the mind" – and to allow for the possibility of that strange energy we call love.

The way we start a dialogue group is usually by talking *about* dialogue – talking it over, discussing why we're doing it, what it means, and so forth. I don't think it is wise to start a group before people have gone into all that, at least somewhat. You can, but then you'll have to trust that the group will continue, and that these questions will come out later. So if you are thinking of meeting in a group, one thing which I suggest is to have a discussion or a seminar about dialogue for a while, and those who are interested can then go on to have the dialogue. And you mustn't worry too much whether you are or are not having dialogue – that's one of the blocks. It may be mixed. So we will discuss dialogue for a while – what is its nature?

I give a meaning to the word "dialogue" that is somewhat different from what is commonly used. The derivations of words often help to suggest a deeper meaning. "Dialogue" comes from the Greek word *dialogos*. *Logos* means "the word," or in our case we would think of the "meaning of the word." And *dia* means "through" – it doesn't mean "two." A dialogue can be among any number of people, not just two. Even one person can have a sense of dialogue within himself, if the spirit of the dialogue is present. The picture or image that this derivation suggests is of a *stream of meaning* flowing among and through us and between us. This will make possible a flow of meaning in the whole group, out of which may emerge some new understanding. It's something new, which may not have been in the starting point at all. It's something creative. And this shared meaning is the "glue" or "cement" that holds people and societies together.

Contrast this with the word "discussion," which has the same root as "percussion" and "concussion." It really means to break things up. It emphasizes the idea of analysis, where there may be many points of view, and where everybody is presenting a different one – analyzing and breaking up. That obviously has its value, but it is limited, and it will not get us very far beyond our various points of view. Discussion is almost like a ping-pong game, where people are batting the ideas back and forth and the object of the game is to win or to get points for yourself. Possibly you will take up somebody else's ideas to back up your own – you may agree with some and disagree with others – but the basic point is to win the game. That's very frequently the case in a discussion.

In a dialogue, however, nobody is trying to win. Everybody wins if anybody wins. There is a different sort of spirit to it. In a dialogue, there is no attempt to gain points, or to make your particular view prevail. Rather, whenever any mistake is discovered on the part of anybody, everybody gains. It's a situation called win–win, whereas the other game is win–lose – if I win, you lose. But a dialogue is something more of a common participation, in which we are not playing a game against each other, but *with* each other. In a dialogue, everybody wins.

Clearly, a lot of what is called "dialogue" is not dialogue in the way that I am using the word. For example, people at the United Nations have been having what are often considered to be dialogues, but these are very limited. They are more like discussions – or perhaps trade-offs or negotiations – than dialogues. The people who take part are not really open to questioning their fundamental assumptions. They are trading off minor points, like negotiating whether we have more or fewer nuclear weapons. But the whole question of two different systems is not being seriously discussed. It's taken for granted that you can't talk about *that* – that nothing will ever change that. Consequently their discussions are not serious, not deeply serious. A great deal of what we call "discussion" is not deeply serious, in the sense that there are all sorts of things which are held to be non-negotiable and not touchable, and people don't even want to talk about them. That is part of our trouble.

Now, why do we need dialogue? People have difficulty communicating even in small groups. But in a group of thirty or forty or more, many may find it very hard to communicate unless there is a set purpose, or unless somebody is leading it. Why is that? For one thing, everybody has different assumptions and opinions. They are basic assumptions – not merely superficial assumptions – such as assumptions about the meaning of life; about your own self-interest, your country's interest, or your religious interest; about what you really think is important.

And these assumptions are defended when they are challenged. People frequently can't resist defending them, and they tend to defend them with an emotional charge. We'll discuss that in more detail later, but I'll give an example now. We organized a dialogue in Israel a number of years ago. At one stage the people were discussing politics, and somebody said, just in passing, "Zionism is creating a great difficulty in good relations between Jews and Arabs. It is the principal barrier that's in the way." He said it very quietly. Then suddenly somebody else couldn't contain himself and jumped up. He was full of emotion. His blood pressure was high and his eyes were popping out. He said, "Without Zionism the country would fall to pieces!"

That fellow had one basic assumption, and the other person had another one. And those two assumptions were really in conflict. Then the question is, what can you do? You see, those are the kinds of assumptions that are causing all the trouble politically, all over the world. And the case I just described is relatively easier than some of the assumptions that we have to handle in politics. The point is that we have all sorts of assumptions, not only about politics or economics or religion, but also about what we think an individual should do, or what life is all about, and so forth.

We could also call these assumptions "opinions." An opinion is an assumption. The word "opinion" is used in several senses. When a doctor has an opinion, that's the best assumption that he can make based on the evidence. He may then say, "Okay, I'm not quite sure, so let's get a second opinion." In that case, if he is a good doctor he does not react to defend his assumption. If the second opinion turns out to be different from his, he doesn't jump up with an emotional charge, such as the fellow did on the question of Zionism, and say, "How can you say such things?" That doctor's opinion would be an example of a rational sort of opinion. But most are not of that nature – mostly they are defended with a strong reaction. In other words, a person identifies himself with them. They are tied up with his investment in self-interest.

The point is that dialogue has to go into all the pressures that are behind our assumptions. It goes into the process of thought *behind* the assumptions, not just the assumptions themselves.

DIALOGUE AND THOUGHT

It is important to see that the different opinions that you have are the result of past thought – all your experiences, what other people have said, and what not. That is all programmed into your memory. You may then identify with those opinions and react to defend them. But it doesn't make sense to do this. If the opinion is right, it doesn't need such a reaction. And if it is wrong, why should you defend it? If you are identified with it, however, you do defend it. It is as if you yourself are under attack when your opinion is challenged. Opinions thus tend to be experienced as "truths," even though they may only be your own assumptions and your own background. You got them from your teacher, your family, or by reading, or in yet some other way. Then for one reason or another you are identified with them, and you defend them.

Dialogue is really aimed at going into the whole thought process and changing the way the thought process occurs collectively. We haven't really paid much attention to thought as a process. We have *engaged* in thoughts, but we have only paid attention to the content, not to the process. Why does thought require attention? Everything requires attention, really. If we ran machines without paying attention to them, they would break down. Our thought, too, is a process, and it requires attention, otherwise it's going to go wrong.

I'll try to give some examples of the difficulty in thinking, in thought. One of these difficulties is *fragmentation*, which originates in thought – it is thought which divides everything up. Every division we make is a result of how we think. In actuality, the whole world is shades merging into one. But we select certain things and separate them from others – for convenience, at first. Later we give this separation great importance. We set up separate nations, which is entirely the result of our thinking, and then we begin to give them supreme importance. We also divide religions by thought – separate religions are entirely a result of how we think. And in the family, the divisions are in thought. The whole way the family is set up is due to the way we think about it.

Fragmentation is one of the difficulties of thought, but there is a deeper root, which is that thought is very active, but the process of thought thinks that it is doing nothing – that it is just telling you the way things are. Almost everything around us has been determined by thought – all the buildings, factories, farms, roads, schools, nations, science, technology, religion – whatever you care to mention. The whole ecological problem is due to thought, because we have thought that the world is there for us to exploit, that it is infinite, and so no matter what we did, the pollution would all get dissolved away.

When we see a "problem," whether pollution, carbon dioxide, or whatever, we then say, "We have got to solve that problem." But we are constantly *producing* that sort of problem – not just that particular problem, but that sort of problem – by the way we go on with our thought. If we keep on thinking that the world is there solely for our convenience, then we are going to exploit it in some other way, and we are going to make another problem somewhere. We may clear up the pollution, but may then create some other difficulty, such as economic chaos, if we don't do it right. We might set up genetic engineering, but if ordinary technology can produce such vast difficulties, imagine the kind of thing genetic engineering could get us into – if we go on with the same way of thinking. People will be doing genetic engineering for whatever suits their fancy and the way they think.

The point is that thought produces results, but thought says it didn't do it. And that *is* a problem. The trouble is that some of those results that thought produces are considered to be very important and valuable. Thought produced the nation, and it says that the nation has an extremely high value, a supreme value, which overrides almost everything else. The same may be said about religion. Therefore, freedom of thought is interfered with, because if the nation has high value it is necessary to continue to think that the nation has high value. Therefore you've got to create a pressure to think that way. You've got to have an impulse, and make sure everybody has got the impulse, to go on thinking that way about his nation, his religion, his family, or whatever it is that he gives high value. He's got to defend it.

You cannot defend something without first *thinking the defense*. There are those thoughts which might question the thing you want to defend, and you've got to push them aside. That may readily involve self-deception – you will simply push aside a lot of things you would rather not accept by saying they are wrong, by distorting the issue, and so on. Thought defends its basic assumptions against evidence that they may be wrong.

In order to deal with this, we have got to look at thought, because the problem is originating in thought. Usually when you have a problem, you say, "I must think about it to solve it." But what I'm trying to say is that *thought is the problem*. What, therefore, are we going to do? We could consider two kinds of thought – individual and collective. Individually I can think of various things, but a great deal of thought is what we do together. In fact, most of it comes from the collective background; language is collective. Most of our basic assumptions come from our society, including all our assumptions about how society works, about what sort of person we are supposed to be, and about relationships, institutions, and so on. Therefore we need to pay attention to thought, both individually and collectively.

In a dialogue, people coming from different backgrounds typically have different basic assumptions and opinions. In almost any group you will probably find a great many different assumptions and opinions of which we are not aware at the moment. It is a matter of culture. In the overall culture there are vast numbers of opinions and assumptions which help make up that culture. And there are also sub-cultures that are somewhat different from one another according to ethnic groups, or to economic situations, or to race, religion, or thousands of other things. People will come to such a gathering from somewhat different cultures or sub-cultures, with different assumptions and opinions. And they may not realize it, but they have some tendency to defend their assumptions and opinions reactively against evidence that they are not right, or simply a similar tendency to defend them against somebody who has another opinion.

If we defend opinions in this way, we are not going to be able to have a dialogue. And we are often *unconsciously* defending our opinions. We don't usually do it on purpose. At times we may be conscious that we are defending them, but mostly we are not. We just feel that something is so true that we can't avoid trying to convince this stupid person how wrong he is to disagree with us.

Now, that seems the most natural thing in the world – it seems that that's inevitable. Yet if you think of it, we can't really organize a good society if we go on that basis. That's the way democracy is supposed to work, but it hasn't. If everybody has a different opinion, it will be merely a struggle of opinions. And the one who is the strongest will win. It may not necessarily be the right one; it may be that none of them are right. Therefore, we won't be doing the right thing when we try to get together.

This problem arises whenever people meet for dialogue, or legislators try to get together, or businessmen try to get together, or whatever. If we all had to do a job together, we would likely find that each one of us would have different opinions and assumptions, and thus we would find it hard to do the job. The temperature could go way up. In fact, there are people facing this problem in large corporations. The top executives may all have different opinions, hence they can't get together. So the company doesn't work efficiently, it starts to lose money and goes under.

There are some people who are trying to form groups where top business executives can talk together. If politicians would do that, it would be very good. Religious people would be the hardest to get together. The assumptions of the different religions are so firmly embedded that I don't know of any case of two religions, or even subgroups of any given religion, where they ever got together once they had split. The Christian church, for instance, has been talking about trying to get together for ages and it stays about the same all the time. They talk and they appear to get a little bit closer, and then it never happens. They talk about unity and oneness and love and all that, but the other assumptions are more powerful; they are programmed into us. Some religious people are trying to get together; they are really sincere – they are as serious as they can be – but it seems that they cannot do it.

Scientists also get into the same situation. Each one may hold to a different view of the truth, so they can't get together. Or they may have different self-interests. A scientist who is working for a company that produces pollution may have a certain self-interest in proving that the pollution is not dangerous. And somebody else might have self-interest in proving that it is dangerous. And perhaps then somewhere there is an unbiased scientist who tries to judge it all.

Science is supposed to be dedicated to truth and fact, and religion is supposed to be dedicated to another kind of truth, and to love. But people's self-interest and assumptions take over. Now, we're not trying to judge these people. Something is happening, which is that assumptions or opinions are like computer programs in people's minds. Those programs take over against the best of intentions – they produce their own intentions.

We could say, then, that a group of about twenty to forty people is almost a microcosm of the whole society – like the groups we have just looked at, it has a lot of different opinions and assumptions. It is possible to have a dialogue with one person or with two, three, or four, or you can have the attitude of the dialogue by yourself, as you weigh all the opinions without deciding. But a group that is too small doesn't work very well. If five or six people get together, they can usually adjust to each other so that they don't say the things that upset each other – they get a "cozy adjustment." People can easily be very polite to each other and avoid the issues that may cause trouble. And if there is a confrontation between two or more people in such a small group, it seems very hard to stop it; it gets stuck. In a larger group, we may well start out politely. After a while, though, people can seldom continue to avoid all the issues that would be troublesome. The politeness falls away pretty soon. In a group of less than about twenty it may not, because people get to know each other and know the rough edges that they have to avoid. They can take it all into account; it's not too much. But in a group of forty or fifty it is too much.

So when you raise the number to about twenty, something different begins to happen. And forty people is about as many as you can conveniently arrange in a circle – or you might put two circles concentrically. In that size group, you begin to get what may be called a "microculture." You have enough people coming in from different subcultures so that they are a sort of microcosm of the whole culture. And then the question of culture – the collectively shared meaning – begins to come in.

That is crucial, because the collectively shared meaning is very powerful. The collective thought is more powerful than the individual thought. As we said, the individual thought is mostly the result of collective thought and of interaction with other people. The language is entirely collective, and most of the thoughts in it are. Everybody does his own thing to those thoughts – he makes a contribution. But very few change them very much.

The power of the group goes up much faster than the number of people. I've said elsewhere that it could be compared to a laser. Ordinary light is called "incoherent," which means that it is going in all sorts of directions, and the light waves are not in phase with each other so they don't build up. But a laser produces a very intense beam which is coherent. The light waves build up strength because they are all going in the same direction. This beam can do all sorts of things that ordinary light cannot.

Now, you could say that our ordinary thought in society is incoherent. It is going in all sorts of directions, with thoughts conflicting and canceling each other out. But if people were to think together in a coherent way, it would have tremendous power. That's the suggestion. If we have a dialogue situation – a group which has sustained dialogue for quite a while in which people get to know each other, and so on – then we might have such a coherent movement of thought, a coherent movement of communication. It would be coherent not only at the level we recognize, but at the *tacit level*, at the level for which we have only a vague feeling. That would be more important.

"Tacit" means that which is unspoken, which cannot be described – like the knowledge required to ride a bicycle. It is the *actual* knowledge, and it may be coherent or not. I am proposing that thought is actually a subtle tacit process. The concrete process of thinking is very tacit. The meaning is basically tacit. And what we can say explicitly is only a very small part of it. I think we all realize that we do almost everything by this sort of tacit knowledge. Thought is emerging from the tacit ground, and any fundamental change in thought will come from the tacit ground. So if we are communicating at the tacit level, then maybe thought is changing.

The tacit process is common. It is shared. The sharing is not merely the explicit communication and the body language and all that, which are part of it, but there is also a deeper tacit process which is common. I think the whole human race knew this for a million years; and then in five thousand years of civilization we have lost it, because our societies got too big to carry it out. But now we have to get started again, because it has become urgent that we communicate. We have to share our consciousness and be able to think together, in order to do intelligently whatever is necessary. If we begin to confront what's going on in a dialogue group, we sort of have the nucleus of what's going on in all society. When you are by yourself you miss quite a bit of that; even one-on-one you don't really get it.

ENGAGING IN DIALOGUE

A basic notion for a dialogue would be for people to sit in a circle. Such a geometric arrangement doesn't favor anybody; it allows for direct communication. In principle, the dialogue should work without any leader and without any agenda. Of course, we are used to leaders and agendas, so if we were to start a meeting without a leader – start talking and have no agenda, no purpose – I think we would find a great deal of anxiety in not knowing what to do. Thus, one of the things would be to work through that anxiety, to face it. In fact, we know by experience that if people do this for an hour or two they do get through it and start to talk more freely.

It may be useful to have a facilitator to get the group going, who keeps a watch on it for a while and sort of explains what's happening from time to time, and that kind of thing. But his function is to work himself out of a job. Now, that may take time. It may be that people must meet regularly and sustain the dialogue. That form might be to meet week after week, or bi-weekly or whatever, and sustain it a long time – a year or two or more. In that period, all those things we mentioned would come out. And people would begin to learn really to depend less and less on the facilitator – at least that's the idea behind it. Now, the whole of society has been organized to believe that we can't function without leaders. But maybe we can. That's the suggestion. Of course, it's an experiment. We can't guarantee that it is going to happen. But that is what takes place in any new venture – you consider all the evidence, you consider what's the best idea, what to say about it, what your theories about it are, and then you go ahead and try it.

At the beginning of a dialogue we would not expect that personal problems or questions would enter into it. If people sustained the dialogue week after week, or month after month, then maybe they could. Everything can enter, but the people have to get to know each other and trust each other and establish that relationship of sharing. It would be too much to expect to start with that. And in fact, a personal problem may not be all that important anyway, although if someone has one, the group could consider it. There is no reason why they couldn't. However, I don't think we would begin with that, at least not often. *The group is not mainly for the sake of personal problems; it's mainly a cultural question.* But the personal could come into the group, because personal problems and culture get mixed up.

It is important to understand that a dialogue group is not a therapy group of some kind. We are not trying to cure anybody here, though it may happen as a byproduct. But that's not our purpose. Dr. Patrick de Maré, a friend of mine who has gone into this, calls it "socio-therapy," not individual therapy. The group is a microcosm of society, so if the group – or anyone – is "cured," it is the beginning of the larger cure. You can look at it that way if you like. That's limited, but still it's a way to look at it. Nor is this a so-called "encounter group," which is aimed at a particular type of therapy where people's emotions, and so forth, can come up. We are not particularly aiming for that, but we are not saying that emotions should never come up, because in certain cases, if people confront each other emotionally it will bring out their assumptions. In the dialogue people should talk directly to one another, one to one, across the circle. Then the time would come, if we got to know each other a bit and could trust each other, when you could speak very directly to the whole group, or to anybody in it.

Some time ago there was an anthropologist who lived for a long while with a North American tribe. It was a small group of about this size. The hunter-gatherers have typically lived in groups of twenty to forty. Agricultural group units are much larger. Now, from time to time that tribe met like this, in a circle. They just talked and talked and talked, apparently to no purpose. They made no decisions. There was no leader. And everybody could participate. There may have been wise men or wise women who were listened to a bit more – the older ones – but everybody could talk. The meeting went on, until it finally seemed to stop for no reason at all and the group dispersed. Yet after that, everybody seemed to know what to do, because they understood each other so well. Then they could get together in smaller groups and do something or decide things.

In the dialogue group we are not going to decide what to do about anything. This is crucial. Otherwise we are not free. We must have an empty space where we are not obliged to do anything, nor to come to any conclusions, nor to say anything or not say anything. It's open and free. It's an empty space. The word "leisure" has that meaning of a kind of empty space. "Occupied" is the opposite of leisure; it's full. So we have here a kind of empty space where anything may come in – and after we finish, we just empty it. We are not trying to accumulate anything. That's one of the points about a dialogue. As Krishnamurti used to say, "The cup has to be empty to hold something."

We see that it is not an arbitrary imposition to state that we have no fixed purpose – no absolute purpose, anyway. We may set up relative purposes for investigation, but we are not wedded to a particular purpose, and are not saying that the whole group must conform to that purpose indefinitely. All of us might want the human race to survive, but even that is not our purpose. Our purpose is really to communicate coherently in truth, if you want to call that a purpose.

You could say that generally our culture goes in for large groups of people for two reasons. One is for entertainment and fun. The other is to get a useful job done. Now I'm going to propose that in a dialogue we are not going to have any agenda, we are not going to try to accomplish any useful thing. As soon as we try to accomplish a useful purpose or goal, we will have an assumption behind it as to what is useful, and that assumption is going to limit us. Different people will think different things are useful. And that's going to cause trouble. We may say, "Do we want to save the world?" or "Do we want to run a school?" or "Do we want to make money?" or whatever it may be. That's also going to be one of the problems in corporate dialogues. Will they ever give up the notion that they are there primarily to make a profit? If they could, this would be a real transformation of mankind. I think that many business executives in certain companies are feeling unhappy and really want to do something – not merely to save the company. Just as we are, they are unhappy about the whole world. It's not the case that all of them are money-grubbing or exclusively profit-oriented.

When a dialogue group is new, in general people talk around the point for a while. In all human relations nowadays, people generally have a way of not directly facing anything. They talk around things, avoiding the difficulties. This practice will probably continue within a dialogue group. If you keep the group going for a while though, that tendency begins to break down. At a dialogue one evening a fellow spoke up, saying, "Okay, we're all talking about philosophy. Can I read this nice bit of philosophy I brought?" And some people said "no," so he didn't read it. It seemed a bit of a shock, but it worked out.

It all has to be worked out. People will come to a group with different interests and assumptions. In the beginning they may have negotiation, which is a very preliminary stage of dialogue. In other words, if people have different approaches, they have to negotiate somehow. However, that is not the end of dialogue; it is the beginning. Negotiation involves finding a common way of proceeding. Now, if you only negotiate, you don't get very far, although some questions do have to be negotiated.

A great deal of what nowadays is typically considered to be dialogue tends to focus on negotiation; but as we said, that is a preliminary stage. People are generally not ready to go into the deeper issues when they first have what they consider to be a dialogue. They negotiate, and that's about as far as they get. Negotiation is trading off, adjusting to each other and saying, "Okay, I see your point. I see that that is important to you. Let's find a way that would satisfy both of us. I will give in a little on this, and you give in a little on that. And then we will work something out." Now that's not really a close relationship, but it begins to make it possible to get going.

So the suggestion is that people could start dialogue groups in various places. The point would not be to identify with the group, but rather, what is important is this whole process. You might say, "This is a wonderful group," but it's actually the process that counts.

I think that when we are able to sustain a dialogue of this sort you will find that there will be a change in the people who are taking part. They themselves would then behave differently, even outside the dialogue. Eventually they would spread it. It's like the Biblical analogy of the seeds – some are dropped in stony ground and some of them fall in the right place and they produce tremendous fruit. The thing is that you cannot tell where or how it can start. The idea here, the communication here, the kind of thought we're having here, is a kind of seed which may help this to come about. But we mustn't be surprised if many of these groups are abortive and don't get going. That doesn't mean it can't happen.

The point is not to establish a fixed dialogue group forever, but rather one that lasts long enough to make a change. If you keep holding it for too long it may become caught up in habits again. But you have to keep it up for a while, or else it won't work. It may be valuable to keep the dialogue going for a year or two, as we said, and it is important to sustain it regularly. If you sustain it, all these problems will arise; it cannot avoid bringing out the deep assumptions of the people who are participating. The frustration will arise, the sense of chaos, the sense that it's not worth it. The emotional charge will come. The fellow with the assumptions about Zionism probably wanted to be very polite. But suddenly somebody said something that outraged him, and he couldn't control himself. It's going to happen that the deep assumptions will come to the surface, if we stick with it. But if you understand that you do nevertheless have to stick with it, then something new will come.

Now, dialogue is not going to be always entertaining, nor is it doing anything visibly useful. So you may tend to drop it as it gets difficult. But I suggest that it is very important to go on with it – to stay with it through the frustration. When you think something is important you will do that. For example, nobody would climb Mount Everest unless for some reason he thought it was important, as that could also be very frustrating and not always entertaining. And the same is true if you have to make money, or do all sorts of things. If you feel that they are necessary, you do them.

I'm saying that it is necessary to share meaning. A society is a link of relationships among people and institutions, so that we can live together. But it only works if we have a culture – which implies that we share meaning; i.e., significance, purpose, and value. Otherwise it falls apart. Our society is incoherent, and doesn't do that very well; it hasn't for a long time, if it ever did. The different assumptions that people have are tacitly affecting the whole meaning of what we are doing.

SUSPENDING ASSUMPTIONS

We have been saying that people in any group will bring to it their assumptions, and as the group continues meeting, those assumptions will come up. Then what is called for is to *suspend* those assumptions, so that you neither carry them out nor suppress them. You don't believe them, nor do you disbelieve them; you don't judge them as good or bad. Normally when you are angry you start to react outwardly, and you may just say something nasty. Now suppose I try to suspend that reaction. Not only will I now not insult that person outwardly, but I will suspend the insult that I make *inside* of me. Even if I don't insult somebody outwardly, I am insulting him inside. So I will suspend that, too. I hold it back, I reflect it back. You may also think of it as suspended in front of you so that you can look at it – sort of reflected back as if you were in front of a mirror. In this way I can see things that I wouldn't have seen if I had simply carried out that anger, or if I had suppressed it and said, "I'm not angry" or "I shouldn't be angry."

So the whole group now becomes a mirror for each person. The effect you have on the other person is a mirror, and also the effect the other person has on you. Seeing this whole process is very helpful in bringing out what's going on, because you can see that everybody's in the same boat.

What's required then is that we notice the connection between the thoughts going on in the dialogue, the feelings in the body, and the emotions. If you watch, you'll see from the body language, as well as from the verbal language, that everybody's in much the same boat – they're just on opposite sides. The group may even polarize so that two very powerful groups are against each other. But one of the things we're aiming for is that this *should* come out. We're not trying to suppress it.

Therefore, you simply see what the assumptions and reactions mean – not only your own, but the other people's as well. We are not trying to change anybody's opinion. When this meeting is over, some-body may or may not change his opinion. This is part of what I consider dialogue – for people to realize what is on each other's minds without coming to any conclusions or judgments. Assumptions will come up. And if you hear somebody else who has an assumption that seems outrageous to you, the natural response might be to get angry, or get excited, or to react in some other way. But suppose you suspend that activity. You may not even have known that you had an assumption. It was only because he came up with the opposite one that you find out that you have one. You may uncover other assumptions, but we are all suspending them and looking at them and seeing what they mean.

You have to notice your own reactions of hostility, or whatever, and you can see by the way people are behaving what their reactions are. You may find, as with anger, that it could go so far that the meeting could blow up. If temperatures do rise, then those who are not com-pletely caught up in their particular opinions should come in to defuse the situation a bit so that people could look at it. It mustn't go so far that you can't look at it. The point is to keep it at a level where the opinions come out, but where you can look at them. Then you may have to see that the other person's hostility provokes your own. That's all part of the observation, the suspension. You become more familiar with how thought works.

THE IMPULSE OF NECESSITY

We've been discussing dialogue and thought, and the importance of giving attention to the whole process – not merely to the content of all the different opinions and views – and to how we hold it all together. Also we're all watching the process of how it affects us, our feelings and states of the body, and how other people are affected. This is really something of crucial importance, to be listening and watching, observing, to give attention to the actual process of thought and the order in which it happens, and to watch for its incoherence, where it's not working properly, and so on. *We are not trying to change anything, but just being aware of it. And you can notice the similarity of the difficulties within a group to the conflicts and incoherent thoughts within an individual.*

I think that as we do this we will find that certain kinds of thoughts play a greater role than other kinds. One of the kinds that is most important is the thought of *necessity*. What is necessary cannot be otherwise; it's just got to be that way. It is interesting that the word *necessary* has a Latin root, *necesse*, meaning "don't yield." It really means "what cannot be turned aside." Ordinarily as we go through life, problems come up and they can be turned aside, or if they can't be turned aside then *we* turn aside, and that is the way we resolve things. But then there may arise a necessity, as I said, which cannot be turned aside; but we may have our own necessity which also cannot be turned aside. Then we feel frustrated. Each necessity is absolute, and we have a conflict of absolute necessities. Typically, it may come up that your own opinion cannot be turned aside, nor can the other person's, and you feel the other person's opinion working within you, opposing you. So each person is in a state of conflict.

Necessity creates powerful impulses. Once you feel that something is necessary, it creates an impulse to do it or not to do it, whatever it may be. It may be very strong and you feel compelled, propelled. Necessity is one of the most powerful forces – it overrides all the instincts eventually. If people feel something is necessary, they'll even go against the instinct of self-preservation and all sorts of things. In the dialogue, both individually and collectively – this is important – the conflicts come up around this notion of necessity. All the serious arguments, whether in the family or in the dialogue, are about different views of what is *absolutely necessary*. Unless it takes that form, then you can always negotiate it and decide what has first priority, and adjust it. But if two things are absolutely necessary you cannot use the usual way of negotiation. That is the weak point about negotiation. When two different nations come up and each one says, "I'm sovereign, and what I say has to go, it's absolutely necessary," then there is no answer unless they can change that.

The question is what to do if there is a clash of two absolute necessities. The first thing that happens is that we get this emotional charge and we can build up powerful feelings of anger, hate, frustration, as I described before. As long as that absolute necessity remains, nothing can change it, because in a way each person says that they have a valid reason to stick to what they've got, and they have a valid reason to hate the other person for getting in the way of what is absolutely necessary: "He rather obstinately and stupidly refuses to see this," and so on. One may say that it's regrettable that we have to kill all these people, but it is absolutely necessary, in the interests of the country, the religion, or whatever it may be. So you see the power of that notion.

So in the dialogue we are expecting the notions of absolute necessity to come up, to clash with each other. People avoid that, because they know that there's going to be trouble and they skirt those questions. But if we sustain the dialogue it's going to come up. The question is what happens then.

We discussed previously that something can happen, if people will stay with it, which will change their whole attitude. At a certain moment we may have the insight that each one of us is doing the same thing – sticking to the absolute necessity of his idea – and that nothing can happen if we do that. If so, it may raise the question, "Is it absolutely necessary? So much is being destroyed just because we have this notion of it being absolutely necessary." Now if you can question it and say, "Is it absolutely necessary?" then at some point it may loosen up. People may say, "Well, maybe it's not absolutely necessary." Then the whole thing becomes easier, and it becomes possible to let that conflict go and to explore new notions of what is necessary, creatively. The dialogue can then enter a creative new area. I think this is crucial.

What about these notions of necessity which we have to set up or discover? If an artist just puts on his paint in arbitrary places, you would say there wasn't anything to it; if he just follows somebody else's order of necessity, he's mediocre. He's got to create his own order of necessity. Different parts of the form he is making must have an inner necessity or else the thing has not really much of a value. This artistic necessity is creative. The artist has his freedom in this creative act. Therefore, freedom makes possible a *creative perception of new orders of necessity*. If you can't do that, you're not really free. You may say you're doing whatever you like and that's your impulse, but I think we've seen that your impulses can come from your thoughts. For example, the thought of what is necessary will make an impulse, and people who are in international conflict will say, "Our impulse is to go to war and get rid of these people who are in our way," as if that were freedom. But it isn't. They're being driven by that thought. So doing what you like is seldom freedom, because what you like is determined by what you think and that is often a pattern which is fixed. Therefore we have a creative necessity which we discover – you can discover individually or we can do so collectively in the group – of how to operate in a group in a new way. Any group which has problems really has got to solve them creatively if they're serious problems. It can't just be by trade-offs and negotiations of the old ways.

I think this is one of the key points, then – to realize when you come to an assumption, that there is an assumption of absolute necessity which you're getting into, and that's why everything is sticking.

PROPRIOCEPTION OF THOUGHT

You can see the whole scope of this question of dialogue giving attention to thought may look rather elementary or simple in the beginning, but it actually gets to the root of our problems and opens the way to creative transformation.

We come back to the realization that the thing which has gone wrong with thought is basically, as I said before, that it does things and then says or implies that it didn't do them – that they took place independently, and that they constitute "problems." Whereas what you really have to do is to stop thinking that way so that you stop creating that problem. The "problem" is insoluble as long as you keep on producing it all the time by your thought. Thought has to be in some sense aware of its consequences, and presently thought is not sufficiently aware of its consequences. That ties up with something similar in neurophysiology called *proprioception*, which really means "self-perception" – the body can perceive its own movement. When you move the body you know the relation between intention and action. The impulse to move and the movement are seen to be connected. If you don't have that, the body is not viable.

We know of a woman who apparently had a stroke in the middle of the night. She woke up and she was hitting herself. People came in and turned on the light and that's what they found. What happened was that her motor nerves were working, but her sensory nerves were no longer working. So she probably touched herself, but she didn't know that she'd touched herself, and therefore she assumed that somebody else was touching her and interpreted that as an attack. The more she defended, the worse the attack got. The proprioception had broken down. She no longer saw the relation between the intention to move and the result. When the light was turned on, proprioception was established in a new way, by sight.

The question is: can thought be proprioceptive? You have the intention to think, which you're not usually aware of. You think because you have an intention to think. It comes from the idea that it is necessary to think, that there's a problem. If you watch, you'll see an intention to think, an impulse to think. Then comes the thought, and the thought may give rise to a feeling, which might give rise to another intention to think, and so on. You're not aware of that, so the thought appears as if it were coming by itself, and the feeling appears to be coming by itself, and so on. That gives the wrong meaning, as in the case of the woman we talked about just now. You may get a feeling that you don't like from a thought, and then a second later say, "I've got to get rid of that feeling," but your thought is still there working, especially if it's a thought that you take to be absolutely necessary.

In fact, the problems we have been discussing are basically all due to this lack of proprioception. *The point of suspension is to help make proprioception possible, to create a mirror so that you can see the results of your thought.* You have it inside yourself because your body acts as a mirror and you can see tensions arising in the body. Also other people are a mirror, the group is a mirror. You have to see your intention. You get an impulse to say something and you see it there, the result, at almost the same time.

If everybody is giving attention, then there will arise a new kind of thought between people, or even in the individual, which is proprioceptive, and which doesn't get into the kind of tangle that thought gets into ordinarily, which is not proprioceptive. We could say that practically all the problems of the human race are due to the fact that thought is not proprioceptive. Thought is constantly creating problems that way and then trying to solve them. But as it tries to solve them it makes it worse because it doesn't notice that it's creating them, and the more it thinks, the more problems it creates – because it's not proprioceptive of what it's doing. If your body were that way you would very quickly come to grief and you wouldn't last very long. And it may be said that if our culture were that way, our civilization would not last all that long, either. So this is another way in which dialogue will help collectively to bring about a different kind of consciousness.

COLLECTIVE PARTICIPATION

All of this is part of collective thought – people thinking together. At some stage we would share our opinions without hostility, and we would then be able to *think together*, whereas when we defend an opinion we can't. An example of people thinking together would be that somebody would get an idea, somebody else would take it up, somebody else would add to it. The thought would flow, rather than there being a lot of different people, each trying to persuade or convince the others.

In the beginning people won't trust each other. But I think that if they see the importance of the dialogue, they will work with it. And as they start to know each other, they begin to trust each other. It may take time. At first you will just come into the group bringing all the problems of the culture and the society. Any group like this is a microcosm of society – it has all sorts of opinions, people not trusting each other, and so on. So you begin to work from there. People talk at first in a perhaps rather trivial way, and then later less trivially. Initially they talk about superficial issues, because they're afraid of doing more, and then gradually they learn to trust each other.

The object of a dialogue is not to analyze things, or to win an argument, or to exchange opinions. Rather, it is to suspend your opinions and to look at the opinions – to listen to everybody's opinions, to suspend them, and to see what all that means. If we can see what all of our opinions mean, then we are sharing a *common content*, even if we don't agree entirely. It may turn out that the opinions are not really very important – they are all assumptions. And if we can see them all, we may then move more creatively in a different direction. We can just simply share the appreciation of the meanings; and out of this whole thing, truth emerges unannounced – not that we have chosen it.

If each of us in this room is suspending, then we are all doing the same thing. We are all looking at everything together. The content of our consciousness is essentially the same. Accordingly, a different kind of consciousness is possible among us, a participatory consciousness – as indeed consciousness always is, but one that is frankly acknowledged to be participatory and can go that way freely. Everything can move between us. Each person is participating, is partaking of the whole meaning of the group and also taking part in it. We can call that a true dialogue.

Something more important will happen if we can do this, if we can manage it. Everybody will be sharing all the assumptions in the group. If everybody sees the meaning together, of all the assumptions, then the content of consciousness is essentially the same. Whereas if we all have different assumptions and defend them, each person is then going to have a different content, because we won't really take in the other person's assumptions. We'll be fighting them, or pushing them away, trying to convince or persuade the other person.

Conviction and persuasion are not called for in a dialogue. The word "convince" means to win, and the word "persuade" is similar. It's based on the same root as are "suave" and "sweet." People sometimes try to persuade by sweet talk or to convince by strong talk. Both come to the same thing, though, and neither of them is relevant. There's no point in being persuaded or convinced. That's not really coherent or rational. If something is right, you don't need to be persuaded. If somebody has to persuade you, then there is probably some doubt about it.

If we could all share a common meaning, we would be participating together. We would be partaking of the common meaning, just as people partake of food together. We would be taking part and communicating and creating a common meaning. That would be participation, which means both "to partake of" and "to take part in." It would mean that in this participation a common mind would arise, which nonetheless would not exclude the individual. The individual might hold a separate opinion, but that opinion would then be absorbed into the group, too.

Thus, everybody is quite free. It's not like a mob where the collective mind takes over – not at all. It is something *between* the individual and the collective. It can move between them. It's a harmony of the individual and the collective, in which the whole constantly moves toward coherence. So there is both a collective mind and an individual mind, and like a stream, the flow moves between them. The opinions, therefore, don't matter much. Eventually we may be somewhere between all these opinions, and we start to move beyond them in another direction – a tangential direction – into something new and creative.

A NEW CULTURE

A society is a link of relationships that are set by people in order to work and live together – rules, laws, institutions, and various things. It is done by thinking and agreeing that we are going to have them, and then we do it. And behind that is a culture, which is shared meaning. Even to say that we want to set up a government, people must agree to a common meaning of what kind of government they want, what's good government, what's right, and so on. Different cultures will produce different functions of government. And if some people don't agree, then we have political struggle. When it goes further, it breaks down into civil war.

I am saying society is based on shared meanings, which constitute the culture. If we don't share coherent meaning, we do not make much of a society. And at present, the society at large has a very incoherent set of meanings. In fact, this set of "shared meanings" is so incoherent that it is hard to say that they have any real meaning at all. There is a certain amount of significance, but it is very limited. The culture in general is incoherent, and we will thus bring with us into the group – or microcosm or microculture – a corresponding incoherence.

If all the meanings can come in together, however, we may be able to work toward coherence. As a result of this process, we may naturally and easily drop a lot of our meanings. But we don't have to begin by accepting or rejecting them. The important thing is that we will never come to truth unless the overall meaning is coherent. All the meanings of the past and the present are together. We first have to apprehend them, and just let them be; and this will bring about a certain order.

If we can work this through, we will then have a coherent meaning in the group, and hence the beginning of a new kind of culture – a culture of a kind which, as far as I can tell, has never really existed. If it ever did, it must have been very long ago – maybe in some groups in the primitive Stone Age conditions. I am saying that a genuine culture could arise in which opinions and assumptions are not defended incoherently. And that kind of culture is necessary for the society to work, and ultimately for the society to survive.

Such a group might be the germ or the microcosm of the larger culture, which would then spread in many ways – not only by creating new groups, but also by people communicating the notion of what it means.

Also, one can see that it is possible that this spirit of the dialogue can work even in smaller groups, or one-on-one, or within the individual. If the individual can hold all of the meanings together in his own mind, he has the attitude of the dialogue. He could carry that out and perhaps communicate it, both verbally and non-verbally, to other people. In principle, this could spread. Many people are interested in dialogue now. We find it growing. The time seems to be ripe for this notion, and it could perhaps spread in many different areas.

I think that something like this is necessary for society to function properly and for society to survive. Otherwise it will all fall apart. This shared meaning is really the cement that holds society together, and you could say that the present society has some very poor-quality cement. If you make a building with very low-quality cement, it cracks and falls apart. We really need the right cement, the right glue, and that is shared meaning.

DIFFICULTIES IN DIALOGUE

We have talked about the positive side of dialogue. However, this attempt at dialogue can be very frustrating. I say this not only theoretically, but also from experience. We've mentioned some of the difficulties: it's frustrating to have all these opinions; there may be anxiety. Besides that, you will find other problems in trying to have a dialogue in a group of any size. Some people want to assert themselves; that's their way of going about things. They talk easily and they become dominant. They may have an image of themselves as dominant, and they get a certain amount of security out of it, a lift out of it. Other people, however, do not have such great self-esteem in this area; they tend to hold back, especially when they see somebody who is dominant. They are afraid that they'll make fools of themselves, or something of the kind.

There are various roles that people adopt. Some people adopt the dominant role, some adopt the role of the weak, powerless person who can be dominated. They sort of work together, with each other. Those "roles," which are really based on assumptions and opinions, will also interfere with the operation of dialogue. So a person has built some assumptions about himself, whether it's one way or the other. Also, since his childhood people have told him that that's what he is, that he is this way or that way. He has had bad experiences or good experiences, and it all built up. These are some of the problems which will arise when we try to have a dialogue.

A further difficulty is you find that very often there is an impulse or pressure, a compulsion almost, to get in there quickly and get your point of view across, particularly if you are one of the "talkers." Even if you're not, you have that pressure, but you're holding back because you're frightened. Therefore, there is no time for people to absorb what has been said, or to ponder it. People feel under pressure to get in, and people feel left out. The whole communication breaks down for this very elementary reason. This is nothing deep at all, but still we have got to address it. Very often when you don't give space in a group, everybody jumps in right away with whatever he has in his mind. But at the same time, you shouldn't be mulling it over in your mind – picking on one point and turning it over – while the conversation goes on to something else. If you stop to think about one point, by the time you have thought about it the group has moved on, and what you were going to say is now irrelevant. As you were thinking, "What does all that mean and what shall I say about it?" it became too late, because the topic has changed. So there is sort of a subtle situation in between, where you are not jumping in too fast, nor holding back too much. There may be silent periods, and so on.

So while we don't have "rules" for the dialogue, we may learn certain principles as we go along which help us – such as that we must give space for each person to talk. We don't put this as a rule; rather we say that we can see the sense of it, and we are learning to do it. So we see the necessity or value of certain procedures that help.

Also, if someone wants the group to accomplish his idea or purpose, it would probably start a conflict. The dialogue is aimed for those people who can commonly agree that this is the way to go about it. If people don't agree that this is the way to go about it, then there is no reason to be in it. Frequently you find that as the dialogue goes on and the group continues, some people leave and others come in. There are those who feel, "Well, this is not for me."

Now, how are you going to deal with the frustrations within the group? As we said before, things may make you angry or frustrated or may frighten you. Your assumptions may be revealed and challenged, and you may find the opinions of others to be outrageous. Also, people may be frightened and anxious if there is no leader and no topic and nothing "to do." So you have to get through all of that.

These are the problems that are going to arise – that have arisen in all the groups that I've seen. And you can expect that they are almost inevitable, and may ask, "Then what is the point in going on with all of this?" So we must explore that.

THE VISION OF DIALOGUE

Let me give what I call a "vision of dialogue." You don't have to accept it, but it may be a way to look at it. Let's suppose we stick with this, and we face the emotional charge – all this irritation, all this frustration – which actually can develop into hate if very powerful assumptions are there. We could say that hate is a neurophysiological, chemical disturbance of a very powerful kind, which is now endemic in the world. Wherever you look, you see people hating each other. So suppose you stick with this. You may get an insight, a shared insight, that we're all in the same position – *everybody has an assumption, everybody is sticking to his assumption, everybody is disturbed neurochemically.* The fundamental level in people is the same; the superficial differences are not so important.

It's possible to see that there's a kind of "level of contact" in the group. The thought process is an extension of the body process, and all the body language is showing it, and so on. People are really in rather close contact – hate is an extremely close bond. I remember somebody saying that when people are in really close contact, talking about something which is very important to them, their whole bodies are involved – their hearts, their adrenalin, all the neurochemicals, everything. They are in far closer contact with each other than with some parts of their own bodies, such as their toes. So, in some sense there is established in that contact "one body." And also, if we can all listen to each other's opinions, and suspend them without judging them, and your opinion is on the same basis as anyone else's, then we all have "one mind" because we have the *same content* – all the opinions, all the assumptions. At that moment the difference is secondary. Then you have in some sense one body, one mind. It does not overwhelm the individual. There is no conflict in the fact that the individual does not agree. It's not all that important whether you agree or not. There is no pressure to agree or disagree.

The point is that we would establish, on another level, a kind of bond, which is called impersonal fellowship. You don't have to know each other. In England, for example, the football crowds prefer not to have seats in their football stands, but just to stand bunched against each other. In those crowds very few people know each other, but they still feel something – that contact – which is missing in their ordinary personal relations. And in war many people feel that there's a kind of comradeship which they miss in peacetime. It's the same sort of thing – that close connection, that fellowship, that mutual participation. I think people find this lacking in our society, which glorifies the separate individual. The communists were trying to establish something else, but they completely failed in a very miserable way. Now a lot of them have adopted the same values as we have. But people are not entirely happy with that. They feel isolated. Even those who "succeed" feel isolated, feel there's another side they are missing.

I am saying that this is a reason for dialogue. We really do need to have it. This reason should be strong enough to get us through all the frustration we talked about. People generally seem ready to accept frustration with anything that they regard as important. Doing your job or making money, for example, is often frustrating; it produces anxiety, Yet people will say, "That is important! We have to stick with it." They feel that way about all sorts of things. I'm saying that if we regard dialogue as important, as necessary, we will say about it as well, "We will stick to it." But if we don't think it is necessary, we might say, "Okay, what's the point? This is too much trouble. Let's give it up. It's not producing anything." You see, you have to explore anything new for a while. In science, or anywhere, you usuallly have to go through a period where you are not getting anywhere while you are exploring. It can, nevertheless, be very discouraging.

If we can all suspend carrying out our impulses, suspend our assumptions, and look at them all, then we are all in the same state of consciousness. And therefore we have established the thing that many people say they want – a common consciousness. It may not be very pleasant, but we have got it. People tend to think of common consciousness as "shared bliss." That may come; but if it does, I'm saying that the road to it is through this. We have to share the consciousness that we *actually* have. We can't just impose another one. But if people can share the frustration and share their different contradictory assumptions and share their mutual anger and stay with it – if everybody is angry together, and looking at it together – then you have a common consciousness.

If people could stay with power, violence, hate, or whatever it is, all the way to the end, then it would sort of collapse, because ultimately they would see that we are all the same. And consequently they would have participation and fellowship. People who have gone through that can become good friends. The whole thing goes differently. They become more open and trusting to each other. *They have already gone through the thing that they are aftaid of*, so the intelligence can then work.

There's a story I would like to relate in this connection. I knew a man in London who had been a child psychiatrist. He told me that somebody once brought to him a girl about seven years old who was very disturbed. She refused to talk to anybody. They brought her hoping that he would help to get her talking. So he tried for about an hour and got nowhere. Finally, getting exasperated, he said, "Why don't you talk to me?" She answered, "Because I hate you." He thought that he had to bring time into this somewhere to defuse it. So he said, "How long will you hate me?" She said, "I'll hate you forever." He was then a bit worried, so he brought time in again. He asked, "How long will you hate me forever?" Then she burst out laughing and the whole thing was broken. The energy which had been there was now available. The absurdity of the thing was shown to her – that the thing was incoherent. She was saying that she was going to hate him forever, and she could see that that wouldn't really be so; and if that's not so, then the idea that she has got to go on with the hatred is not necessary either.

When you have anger, it has a reason, or a cause. You say that you are angry because of this, this, or that. It builds up to rage and hate, at which point it no longer has a particular reason anymore – it just sustains itself. That energy of hate is sort of locked up, and then it's looking for an occasion to discharge. The same holds with panic. You are usually aware of a reason for your fear, but by the time you get to panic it goes on by itself. However, the sort of energy that goes around at that level may also in a vague way be the kind of energy we are talking about for creativity – namely, *an energy without a reason.*

But there is a great deal of violence in the opinions that we are defending. They are not merely opinions, they are not merely assumptions; they are assumptions with which we are *identified* – which we are therefore defending, because it is as if we are defending ourselves. The natural self-defense impulse, which we got in the jungle, has been transferred from the jungle animals to these opinions. In other words, we say that there are some dangerous opinions out there – just as there might be dangerous tigers. And there are some very precious animals inside us that have to be defended. So an impulse that made sense physically in the jungle has been transferred to our opinions in our modern life. And in a dialogue, we get to be aware of that in a collective way.

As long as we have this defensive attitude – blocking and holding assumptions, sticking to them and saying, "I've got to be right," and that sort of thing – then intelligence is very limited, because intelligence requires that you don't defend an assumption. There is no reason to hold to an assumption if there is evidence that it is not right. The proper structure of an assumption or of an opinion is that it is open to evidence that it may not be right.

That does not mean that we are going to impose the opinions of the group. In this way the collective can often be troublesome. The group may act like a conscience, inducing powerful guilt feelings in its members, because we are all so built that we tend to regard what everybody agrees on as true. Everybody may or may not have a different opinion – it is not that important. It isn't necessary that everybody be convinced to have the same view. This sharing of mind, of consciousness, is more important than the content of the opinions. And you may see that these opinions are limited anyway. You may find that the answer is not in the opinions at all, but somewhere else. Truth does not emerge from opinions; it must emerge from something else – perhaps from a more free movement of the tacit mind. So we have to get meanings coherent if we are going to perceive truth, or to take part in truth. That is why I say the dialogue is so important. If our meanings are incoherent, how are we going to participate in truth?

I think this new approach could open the way to changing the whole world situation – ecologically, and in other ways. For instance, the ecological movement, the "green movement," is now in danger of fragmenting and splitting, because many of those groups have different opinions about how to deal with the problems. So they can wind up fighting each other as much as they fight for the ecology. Consequently, it seems particularly urgent that the green movement get into dialogue.

People concerned with the ecology are clearly aware of some of our planetary problems, but I think that many of them may not be as aware of their assumptions and tacit thought processes. I think it is important to call attention to this explicitly in a clear way, so that it becomes clear what the basic problem is. These kinds of activities go together. Cleaning up the rivers and planting trees and saving the whales should go together with dialogue, and with seeing the general problem of thought. They all belong together, because any one of those activities by itself is not enough. If we all just talk about thought and think about thought for a long while, the whole planet may be destroyed in the meantime. But I think that dialogue will work in this tacit level of mental process, where the most significant things take place.

There are situations where people have differing assumptions and opinions, where one faction is interested and the other isn't. Still, somehow, we have got to have a dialogue. Even if one faction won't participate, we who are willing can participate in a dialogue between our thought and their thought. We can at least dialogue among ourselves as far as we can, or you may by yourself. That is the attitude of dialogue. And the further this attitude could spread, the more I think it would help to bring order. If we really could do something creative, it might still affect other people on a tacit level. It would really communicate at the tacit level, both with words and beyond words. But if we keep on repeating the same old story, then it won't happen.

This notion of dialogue and common consciousness suggests that there is some way out of our collective difficulties. And we have to begin at the grass roots, as it were, not to begin at the top of the heap with the United Nations or with the President. I know that there are people in the US State Department who are familiar with this idea of dialogue, which shows how these ideas do percolate and may even reach the highest levels. This indicates that things can communicate very fast in this modem world – though that may look very insignificant at first. In three to five steps it might reach all sorts of levels. Just as the destructive things communicate, so this idea of dialogue could communicate, too.

As we ourselves stay with the frustrations of dialogue, the meaning of what we are doing may be much more than will appear at first sight. In fact, we could say that instead of being part of the problem, we become part of the solution. In other words, our very movement has the quality of the solution; it is part of it. However small it is, it has the quality of the solution and not the quality of the problem. However big the larger movement is, it has the quality of the problem, not of the solution. Accordingly, the major point is to start something which has the quality of the solution. As I have said, we don't know how fast or slowly it would spread. We don't know how fast a movement in the mind – in the thought process and beyond the thought process, this sharing together – will spread.

People sometimes say, "All we really need is love." Of course, that's true – if there were universal love, all would go well. But we don't appear to have it. So we have to find a way that works. Even though there may be frustration and anger and rage and hate and fear, we have to find something which can take all that in.

To illustrate the point, here is a story about the two leading physicists of this century, Albert Einstein and Niels Bohr. Einstein remembered that when he first met Bohr, he felt close to him. He wrote of a feeling of love for him. They talked physics in a very animated way, and so on. But they finally came upon a point where they had two different assumptions, or opinions, about what was the way to truth. Bohr's judgments were based on his view of quantum theory, and Einstein's on his view of relativity. They talked it over again and again in a very patient way, with all goodwill. It went on for years, and neither of them yielded. Each one just repeated what he had been saying before. So finally they found that they weren't getting anywhere, and they gradually drifted apart. They didn't see each other for a long time after that.

Then one year, both of them were at the Institute for Advanced Study at Princeton, but they still didn't meet each other. A mathematician named Herman Weyl said, "It would be nice if they got together. It's a pity that they don't." So he arranged a party to which Einstein and Bohr and their respective students were invited. Einstein and his associates stayed at one end of the room, and Bohr and his associates stayed at the other end. They couldn't get together because they had nothing to talk about. They couldn't share any meaning, because each one felt his meaning was true. How can you share if you are sure you have truth and the other fellow is sure he has truth, and the truths don't agree? How can you share?

Therefore, you have to watch out for the notion of truth. Dialogue may not be concerned directly with truth – it may arrive at truth, but it is concerned with *meaning*. If the meaning is incoherent you will never arrive at truth. You may think, "My meaning is coherent and somebody else's isn't," but then we'll never have meaning shared. You will have the "truth" for yourself or for your own group, whatever consolation that is. But we will continue to have conflict.

If it is necessary to share meaning and share truth, then we have to do something different. Bohr and Einstein probably should have had a dialogue. I'm not saying that they could have had one, but in a dialogue they might have listened properly to each other's opinion. And perhaps they both would have suspended their opinions, and moved out beyond relativity and beyond quantum theory into something new. They might have done that in principle, but I don't think that this notion of dialogue had occurred to scientists then.

Science is predicated on the concept that science is arriving at truth – at a *unique* truth. The idea of dialogue is thereby in some way foreign to the current structure of science, as it is with religion. In a way, science has become the religion of the modern age. It plays the role which religion used to play of giving us truth; hence different scientists cannot come together any more than different religions can, once they have different notions of truth. As one scientist, Max Planck, said, "New ideas don't win, really. What happens is that the old scientists die and new ones come along with new ideas." But clearly that's not the right way to do it. This is not to say that science couldn't work another way. If scientists could engage in a dialogue, that would be a radical revolution in science, in the very nature of science. Actually, scientists are in principle committed to the concepts involved in dialogue. They say, "We must listen. We shouldn't exclude anything."

However, they find that they can't do that. This is not only because scientists share what everybody else shares – assumptions and opinions – but also because the very notion which has been defining science today is that we are going to *get* truth. Few scientists question the assumption that thought is capable of coming to know "everything." But that may not be a valid assumption, because thought is abstraction, which inherently implies limitation. The *whole* is too much. There is no way by which thought can get hold of the whole, because thought only abstracts; it limits and defines. And the past from which thought draws contains only a certain limited amount. The present is not contained in thought; thus, an analysis cannot actually cover the moment of analysis.

There are also the relativists, who say that we are never going to get at an absolute truth. But they are caught in a paradox of their own. They are assuming that relativism is the absolute truth. So it is clear that people who believe that they are arriving at any kind of absolute truth can't make a dialogue, not even among themselves. Even different relativists don't agree.

So we can see that there is no "road" to truth. What we are trying to say is that in this dialogue we share all the roads and we finally see that none of them matters. We see the meaning of all the roads, and therefore we come to the "no road." Underneath, all the roads are the same because of the very fact that they are "roads" – they are rigid.

We've said that in a dialogue there will be frustrations, but you might become better friends if you can get through all that. Not that we demand affection. We don't demand friendship; we don't demand anything, though friendship may come. *If you see other people's thought, it becomes your own thought, and you treat it as your own thought. And when an emotional charge comes up, you share all the emotional charges, too, if they affect you; you hold them together with all the thoughts.* Often, when there is an emotional charge somebody can come in to defuse the issue a bit so that it doesn't run away – as the child psychiatrist defused it with his asking, "How long will you hate me forever?" Or some other sort of humor may defuse the issue, or something else – some appropriate remark which you can't foresee.

Sometimes you may find that you are about to raise a question, but someone else brings it up. In such a case, that thought is probably latent in the group as a whole, implicit. And one person may say it, or somebody else may say it. Then another person may pick it up and carry it along. If the group is really working, that would be thinking together – a common participation in thinking – as if it were all one process. That one thought is being formed together. Then, if somebody comes up with another assumption, we all listen to that, we share that meaning. Now that would be the "vision of dialogue."

SENSITIVITY IN DIALOGUE

What we have been discussing has not been common in human society, although it is really what is necessary if the society is to cohere. If people would do this in government or in business or internationally, our society would work differently. But then, that requires *sensitivity* – a certain way of knowing how to come in and how not to come in, of watching all the subtle cues and the senses and your response to them – what's happening inside of you, what's happening in the group. People may show what is happening to them in the stance of their body – by their "body language" – as well as by what they say. They are not trying to do this purposefully, but you will find that it develops. That's part of the communication. It will be non-verbal as well as verbal. You're not *trying* to do it at all. You may not even be aware that it is happening.

Sensitivity is being able to sense that something is happening, to sense the way you respond, the way other people respond, to sense the subtle differences and similarities. To sense all this is the foundation of perception. The senses provide you with information, but you have to be sensitive to it or you won't see it. If you know a person very well, you may pass him on the street and say, "I saw him." If you are asked what the person was wearing, however, you may not know, because you didn't really look. You were not sensitive to all that, because you saw that person through the *screen of thought*. And that was not sensitivity.

So sensitivity involves the senses, and also something beyond. The senses are sensitive to certain things to which they respond, but that's not enough. The senses will tell you what is happening, and then the consciousness must build a form, or create some sense of what it *means*, which holds it together. Therefore, meaning is part of it. You are sensitive to the meaning, or to the lack of meaning. It's perception of meaning, if you want to put it that way. In other words, it is a more subtle perception. The meaning is what holds it together. As I said, it is the "cement." Meaning is not static – it is flowing. And if we have the meaning being shared, then it is flowing among us; it holds the group together. Then everybody is sensitive to all the nuances going around, not merely to what is happening in his own mind. From that forms a meaning which is shared. And in that way we can talk together coherently and think together. Whereas generally people hold to their assumptions, so they are not thinking together. Each one is on his own.

What blocks sensitivity is the defense of your assumptions and opinions. But if you are defending your opinions, you don't judge yourself and say, "I shouldn't be defending." Rather, the fact is that you are defending, and you then need to be sensitive to *that* – to all the feelings in that, all the subtle nuances. We are not aiming for the type of group that condemns and judges, and so forth – we can all realize that that would get in the way. So this group is not going to judge or condemn. It is simply going to look at all the opinions and assumptions and let them surface. And I think that there could then be a change.

Krishnamurti said that "to be" is to be related. But relationship can be very painful. He said that you have to think/feel out all your mental processes and work them through, and then that will open the way to something else. And I think that is what can happen in the dialogue group. Certain painful things can happen for some people; you have to work it all out.

We once had a dialogue in Sweden, in which the group seemed to divide itself into two factions. There were a lot of "New Age" people, and from the beginning they began to talk about the virtues of love and the fact that the place was full of love all around, that it was all love everywhere. Part of the group remained silent for a while, but in the next hour they started to talk. They intimated that the love talk was all sentimental nonsense and didn't mean anything. Then one fellow got so excited that he couldn't stand it, and he walked out. He eventually came back, and they finally got together again. Polarization had taken place, which is a typical difficulty that can arise. Someone noticed the polarization happening and said with a bit of good humor, "There are two groups here – the love group and the hate group." That broke the tension a little, and the two sides could then begin to talk. They didn't necessarily convince each other, but each was able to see the meaning of the other side's position, and the two polarized groups were able to talk to each other.

Now, *that* was a more important point than whether they convinced each other. They might find that they both have to give up their positions so that something else can come about. It was not important whether one favored love or one favored hate or another favored being suspicious and careful and somewhat cynical, or whatever. Really, underneath they were similar, because they both had rigid positions. Loosening that position, then, was the key change.

On the whole, you could say that if you are defending your opinions, you are not serious. Likewise, if you are trying to avoid something unpleasant inside of yourself, that is also not being serious. A great deal of our whole life is not serious. And society teaches you that. It teaches you *not* to be very serious – that there are all sorts of incoherent things, and there is nothing that can be done about it, and that you will only stir yourself up uselessly by being serious. But in a dialogue you have to be serious. It is not a dialogue if you are not – not in the way I'm using the word.

There is a story about Freud when he had cancer of the mouth. Somebody came up to him and wanted to talk to him about a point in psychology. The person said, "Perhaps I'd better not talk to you, because you've got this cancer which is very serious. You may not want to talk about this." Freud's answer was, "This cancer may be fatal, but it's not serious." And actually, of course, it was just a lot of cells growing. I think a great deal of what goes on in society could be described that way – that it may well be fatal, but it's not serious.

LIMITED DIALOGUE

Sometimes people feel a sense of dialogue within their families. But a family is generally a hierarchy, organized on the principle of authority which is contrary to dialogue. The family is a very authoritative struc-ture, based on obligation, and that sort of thing. It has its value, but it is a structure within which it might be difficult to get dialogue going. It would be good if you could. Perhaps that could happen in some families.

In general it is difficult, though, because there is no place in the dialogue for the principle of authority and hierarchy. We want to be free of hierarchy and authority as we move. You must have some authority to "run" things; that's why we say that if you have a "purpose," then you are bound to bring in some authority somewhere. But in dialogue, inso-far as we have no purpose and no agenda and we don't have to do anything, we don't really need to have an authority or a hierarchy. Rather, we need a place where there is no authority, no hierarchy, where there is no special purpose – sort of an empty place, where we can let anything be talked about.

As we said, you can also have a dialogue in a more limited way – perhaps with a purpose or a goal in mind. It would be best to accept the principle of letting it be open, because when you limit it, you are accept-ing assumptions on the basis of which you limit it, assumptions that may actually be getting in the way of free communication. So you are not looking at those assumptions.

However, if people are not ready to be completely open in their communication, they should do what they can. I know some university professors who are interested in applying the principles of dialogue to corporate problems. One of them recently had a meeting with the executive officers of a corporation that makes office furniture. They wanted to have this sort of meeting, because they knew that they were not functioning efficiently and that they couldn't agree. The higher officers had all sorts of assumptions that blocked everything. So they asked the professor to come in. He started a dialogue which they found very interesting, and now they want to have a whole series of them.

Naturally, that sort of dialogue will be limited – the people involved do have a definite purpose, which is limiting – but even so, it has considerable value. The principle is at least to get people to come to know each other's assumptions, so they can listen to their assumptions and know what they are. Very often people get into problems where they don't really know what the other person's assumption is, and they react according to what they think it is. That person then gets very puzzled and wonders: what is he doing? He reacts, and it all gets very muddled. So it is valuable if they can at least get to realize each other's assumptions.

The professor told me about two interesting cases. One involved a company which had trouble with people in the higher executive branches who were not very happy and were not getting on with each other. The company's usual way of solving it was to offer them a higher salary, sort of a sweetener, and a lot of mediocre people were given the very highest possible positions. It went on and on, and pretty soon there were so many people with high salaries that the company couldn't afford it; they were failing. They said, "What can we do? Well, we've got to have somebody who's tough, who will tell these people, 'You have to accept another position.'" The negotiator that they used explained the new policies by saying, "The company just can't afford it." But he was avoiding the issue. He was not straightforwardly saying, "This whole approach is wrong." Now, if the company is to work efficiently, there must be a mutual agreement that they are not going to give a person a higher position just to alleviate a psychological problem between people. That's not a right way to proceed. Everybody should understand that that is not the right way of working, otherwise the company won't succeed. Therefore, a dialogue was needed so that they could really begin talking with each other in order to come to see clearly the salient points: "That's the way we are thinking, that's where the problems are coming from, and that's the way we have to go." So within the framework of assuming that the company has to survive, there was a limited kind of dialogue – not the kind we ultimately want to have here, but still it was good in some way.

Now, I am suggesting that the human race has got to do that. We could say that the human race is failing for the same sort of reason that the company was failing.

The second case involved the negotiating group itself, the university people whose specialty it is to go into companies and help solve these problems. They were organizing a meeting among themselves with the same purpose – just so they could talk. They had a series of meetings where it happened that two of their people could never quite meet on any issue. One of them constantly had the assumption that the right thing to do was to bring out the trouble – to confront somebody with it. And the other person had the opposite assumption, which was that you shouldn't do that. He wanted other people to draw him out. He felt that he couldn't say something unless other people created the space for him to talk, and drew him out. The first fellow wouldn't do that, he did the opposite. So they couldn't meet. The whole thing went on for a long time in confusion, with the one person waiting to be drawn out, and the other person not understanding that this was the case. Finally they got to talking, and each one actually brought up childhood experiences which were behind his assumptions, and then it opened up.

The fellow who was working as facilitator during this time did very little. In fact, several of the people appealed to the facilitator and said, "Why don't you talk?" The facilitator may come in from time to time and comment on what is going on, or on what it all means. In a more general group he should eventually be able to be just a participant. Probably in the company group this wouldn't work, though; he couldn't become just a participant – such a group has too limited an objective.

This second example might be an illustration of when the personal may have to come into the general, because in certain cases there are blocks due to particular assumptions that the person got hold of in childhood, or in some other way. And in this example, they were finally able to uncover those assumptions. They weren't trying to heal each other, or to do therapy; nevertheless, it had a therapeutic effect. But that's a secondary thing.

Some people feel that that type of corporate dialogue is only furthering a corrupt system. However, there is a germ of something different. I think that if you go into society, you will find that almost everything is involved in this corrupt game. So it doesn't accomplish anything to dismiss it all. The executives have got to make the company work; and in fact, if all these companies would work more efficiently we would all be a lot better off. It's partly because they are in such a mess that we are in trouble, that society is inefficient, that the whole thing is falling apart. If the government and the companies could all work efficiently, we wouldn't be so wasteful, even though that by itself wouldn't solve all the problems.

For the society to be working right, all those things have got to work efficiently and coherently. If we look at what is going on in the world today, in this or in any country, we can say that it is not working coherently. Most companies are not really working coherently. And slowly the thing is sinking. I think that if you can get this notion across in whatever situation – the germ of the notion of dialogue – if you can get people to look at it, it's a step. You could say that heads of state are not likely to have the kind of dialogue that we are talking about. But if they will have any kind at all, if they'll begin to accept this principle, it's a step. It may make a change; for instance, the kind of waste of energy which is going on in the production of armaments could be cut down. If we could stop the tremendous amount that's being spent on armaments – let's say a trillion dollars a year – that could be used for ecological regeneration and all sorts of constructive things. And possibly some of that might happen. Those political figures who are more aware of the ecological problem might, for instance, make the President more aware of it, if they would really talk. Not that we can expect the politicians to solve the problems we face. But I'm saying that if there's a slight movement toward something more open, the rate of destruction will slow down. If we go on at this rate, we may have very little time to do anything.

We can't do anything at the level of presidents or prime ministers. They have their own opinions. But the various ideas filter, as we've said. Somehow the notion of something a little bit like dialogue has filtered to that level, and it may have an effect; that's all I am saying. I think that in the US government there are some people who are more this way, and some people who are more the other way. We don't know how it is going to come out, but there is a certain movement toward something more open. I don't say that it is going to solve the whole thing; I am saying that if it slows down the destruction, that's important, because unless the destruction is slowed down to give time for something new to emerge, it will be too late.

There may be no pat political answer to the world's problems. However, the important point is not the answer – just as in a dialogue, the important point is not the particular opinions – but rather the softening up, the opening up of the mind, and looking at all the opinions. If there is some sort of spread of that attitude, I think it can slow down the destruction.

So we've said that it is crucial to be able to share our judgments, to share our assumptions, to listen to each other's assumptions. In the case of Einstein and Bohr it didn't lead to violence that they did not; but in general, if somebody doesn't listen to your basic assumptions you feel it as an act of violence, and then you are inclined to be violent yourself. Therefore, this is crucial both individually and collectively. Dialogue is the collective way of opening up judgments and assumptions.

BEYOND DIALOGUE

We should keep in mind, nonetheless, that the dialogue – and in fact, all that we've been talking about – is not only directed at solving the ills of society, although we do have to solve those ills. We would be much better off if we didn't have them. If we survive and we want to have a worthwhile life, we have to deal with those problems. But ultimately that's not the entire story. That's only the beginning. I'm suggesting that there is the possibility for a transformation of the nature of consciousness, both individually and collectively, and that whether this can be solved culturally and socially depends on dialogue. That's what we're exploring.

And it's very important that it happens together, because if one individual changes it will have very little general effect. But if it happens collectively, it means a lot more. If some of us come to the "truth," so-called, while a lot of people are left out, it's not going to solve the problem. We would have another conflict – just as there is conflict between different parts of the Christian faith or the Muhammadan faith or various others, even though they all believe in the same God, the same prophet or the same savior.

Love will go away if we can't communicate and share meaning. The love between Einstein and Bohr gradually evaporated because they could not communicate. However, if we can really communicate, then we will have fellowship, participation, friendship, and love, growing and growing. That would be the way. The question is really: do you see the necessity of this process? That's the key question. If you see that it is absolutely necessary, then you have to do something.

And perhaps in dialogue, when we have this very high energy of coherence, it might bring us beyond just being a group that could solve social problems. Possibly it could make a new change in the individual, and a change in the relation to the cosmic. Such an energy has been called "communion." It is a kind of participation. The early Christians had a Greek word, *koinonia*, the root of which means "to participate" – the idea of partaking of the whole and taking part in it; not merely the whole group, but the *whole.*

BIBLIOGRAPHY

Bohm, David, *Quantum Theory*, Prentice-Hall, Englewood Cliffs, New Jersey (1951).

— *Causality and Chance in Modern Physics*, University of Pennsylvania Press, Philadelphia (1957) and Routledge, London (1957).

— *The Special Theory of Relativity*, Routledge, London ([1965], 1996).

— *Fragmentation and Wholeness*, The Van Leer Jerusalem Foundation, Jerusalem (1976).

— *Wholeness and the Implicate Order*, Routledge, London (1980).

— *The Ending of Time* (with J. Krishnamurti), Harper and Row, San Francisco (1985).

— *Unfolding Meaning: A Weekend Of Dialogue* (ed. Donald Factor), Routledge, London ([1985], 1987).

— *Science, Order, and Creativity* (with F. David Peat), Routledge, London ([1987], 2000).

— *Changing Consciousness: Exploring the Hidden Source of the Social, Political, and Environmental Crises Facing Our World* (with Mark Edwards), Harper, San Francisco (1991).

— *The Undivided Universe: An Ontological Interpretation of Quantum Mechanics* (with Basil Hiley), Routledge, London (1993).

— *Thought as a System*, Routledge, London (1994).

— *On Dialogue* (ed. Lee Nichol), Routledge, London (1996).

— *On Creativity* (ed. Lee Nichol), Routledge, London (1998).

— *The Bohm–Biederman Correspondence: Volume 1 – Creativity and Science* (ed. Paavo Pylkkänen), Routledge, London (1999).

— *The Limits of Thought* (with J. Krishnamurti) (ed. Ray McCoy), Routledge, London (1999).

Davies, P. C. W. and Brown, J. R. (eds), *The Ghost in the Atom*, Cambridge University Press, Cambridge ([1986], 1999).

Griffin, D. (ed.), *Physics and the Ultimate Significance of Time: Bohm, Prigogine and Process Philosophy*, State University of New York Press, Albany (1986).

Hiley, B. J. and Peat, F. D. (eds), *Quantum Implications: Essays in Honour of David Bohm*, Routledge, London (1987).

Peat, F. D., *Infinite Potential: The Life and Times of David Bohm*, Addison Wesley, Reading, Mass. (1997).

Pylkkänen, P. (ed.), *The Search for Meaning*, Thorsons, Wellingborough (1989).

Rosen, S. M., *Science, Paradox, and the Moebius Principle*, State University of New York Press, Albany (1994).

Weber, R., *Dialogues With Scientists and Sages: The Search for Unity*, Routledge and Kegan Paul, London (1986).

Wilber, K. (ed.), *The Holographic Paradigm and Other Paradoxes: Exploring the Leading Edge of Science*, Shambhala, Boulder (1982).

INDEX

Aristotle 217; ideas of movement 128
art: socially informed 293
astronomy: abstracted knowledge
  69–70; Big Bang 100; motion 26–7,
  28, 30–1
atomic particles: "bottom-level" reality 169, 170; de Broglie's model
  142–3, 144; dimensional reality
  96–7; division into particles 81, 82;
  enfoldment 91–2; holomovement
  134–6; locality and non-locality
  191; many particle system 184;
  and meaning 180; movements 78;
  non-local, non-causal relations
  83–4; ontological interpretation
  183; predictions and relations 25;
  qualitative infinity of nature
  17–19; quantum process 83,
  184–7; short term properties
  34–5; standard Copenhagen
  interpretation 183; structure of 9,
  17–18; and waves 184, 193–7
autonomy: being 36; causal theory
  21; defining 136; extent of 23;
  kinds of motion 29; reciprocal
  relationships 24–5; stable
  existence of things 20

background: causes and conditions
  9–10; substructures 10
bees 165
being: abstract character of 32–6;

and autonomy 36; Heraclitus
  and becoming 32; and light 156;
  qualitative infinity of natures
  33–6
Bell, John 191
Biederman, Charles 192
biology: life and implicate order
  101–2; misinformation 291–2;
  motion 27
Bohm-Biederman Correspondence
  (Pylkkänen) 6
Bohr, Niels: ambiguity and meaning
  169–71; and de Broglie 142–3;
  dialogue with Einstein 295,
  329–30, 338; experimental
  situation 190
body: as mirror in dialogue 315, 319
brain function: dysfunctional
  feedback loop 200–1; holographic
  model 79, 105–6; material
  substance of 174; meaning 178–9;
  neurophysiology of self-image
  243; process-structure 223–4;
  proprioception 236; thalamus 200;
  totality 212
Broglie, Louis de 140, 141, 142

cancer 291
Cartesian grid 84
cause: active information 190–1;
  non-locality 4; ontological
  interpretation 183; Piaget's theory

267, 281–6; mechanical 239–43
see desires; ego
energy: active information 189–90;
cosmic sea of 99–101, 102, 139;
dual aspect with matter 174–5;
and light 154; zero-point 97–9
enfoldment 139, 193; aspects of
holomovement 131–6; co-present
elements of 110–11; defining
78–9, 129–30; glycerine and dye
experiment 87–91, 132–3, 135;
higher level 147–8; holograms 86;
matter and energy 174–5; particles
and structure 91–2; quantum
theory 78–9; super-implicate order
140–1
experience: attribution 237–8; "I"
and "me" 247–9; immediate
269–72; and knowledge 285; self-
observation 235; structure of 290

fear: process-structure 225
free will/freedom: and choice 254;
individuality 253, 256–7; inner
will 255; and knowledge 253,
254–7; realizing one's potential
260; totality 258–60
Freud, Sigmund 334

Galileo Galilei: frame of reference 41;
motion 128
generative order: defining 289;
dialogue 297; and society 292–4;
transformations 290; values and
assumptions 292–3
Gibson, J. G. 52, 63; structuring perception 56–7; tactile perception 51
glycerine and dye experiment 87–91,
132–3, 135, 192–4
God: super-super-implicate order 147
good and evil: individual vision
204–8; value judgement 247
gravitational field 98
Green movement: internal dialogue
328

Hebb 60
Heisenberg, Werner: microscope
experiment 127–8, 135;
uncertainty principle 72
Held, R. 52

Heraclitus: Becoming 32
Hiley, Basil J. 1, 183–4
Hitler, Adolf 204; fictional memory
loss 265
HIV/AIDS 291–2
holograms 85–6, 192–3; method of
125–7; model of brain function
105–6; quantum waves 195–7;
super-implicate order 144–5;
theoretical implications of
127–8; three-dimensional
reality 97
holomovement 78, 87, 254, 289–90;
aspects of 131–6; consciousness
79–80; defining 193; dye in glycerine 132–3, 135; flowing 150–1;
implicate order 93–4; law of
holonomy 136–7; life 102–3;
manifest reality 148–9;
timelessness 259
Hopkins, Gerard Manley v
human beings: confusion 215–17

identity: and change 249–52; and
dialogue 301; individual
consciousness 279–80; and
knowledge 287–8
imagination: abstract knowledge
268–9; children's development
46–7, 67; higher-level abstraction
70; scientific investigation 141–4
implicate order 2, 78, 139; and
consciousness 103–4; defining
128–30; epistemology 148; and
explicate order 86; and generative
order 289; intrinsic 91; and life
101–2; mathematics 144–5;
movement 110–11; multidimensional 94–7; music 106–7, 108; not
manifest 93; quantum potential
184; soma-significance 158;
structure 87–91, 111; super-
140–1; super-super 147; vision
107–8; wave function 193–7
individuals: confusion and truth
204–8; distraction 237, 244; and
free will 253, 256–7; "I" and
"me" 247–9; identity and change
249–52; illusion-generating
structures 232–4; mechanical
emotions 239–43; self-image 243;

self-observation 235–6, 236–8;
value judgement 245–8 see also
ego
*Infinite Potential – The Life and
Times of David Bohm* (Peat) 1
information: active 159, 184, 186–90;
as significance/meaning 161–9;
172–9
instruments: holograms 125–8;
shaping theory 124–5; tools of
consciousness 115
intention: cause and effect 181; and
meaning 166–8, 298; and soma-
significance 172–3
Iran 266
Israel 201, 230; Zionism 304, 314

Jesus Christ 154

Kant, Immanuel: *a priori* knowledge
61–2, 63; understanding 148
Kierkegaard, Søren 152
knowledge: abstract and concrete
261–2, 264, 266–70, 276, 278;
conditioning 267–8; consciousness
277–80; and creativity 286;
cultural pool of 289; and emotions
267, 281–6; "endarkenment" 261,
288; and free will 253, 254–7;
function of 253, 271–2, 273–4;
"general" 263–5; higher level
abstractions 66–7, 69–70;
immediate experience 269–72,
285; inward and outward 271,
275–7, 287; Kant's *a priori*
61–2, 63; and meaning 262;
misinformation 261, 265, 273;
past and present 257–9; process
of abstraction 67; scientific
accumulation 73–4; skill 264–5;
and structure of perceptions
74–6; tacit 166; and thought
272–3, 281–2; totality of understanding 208–13; and truth 272,
274, 287
Krishnamurti, Jiddu 148, 199; being
related 333; distraction 244; free
will and individuality 256–7;
knowledge as difficulty 264;
memories 226; totality of
understanding 209

language 192; collective nature of
307, 309; communication 298
laws, natural: "holonomy" 136–7;
limitations of 13–15; mechanistic
philosophy 9, 11–12; motion 10;
qualitative infinity of nature 15
see also cause; Newton, Isaac
leadership: dialogue 301
Leibniz, Gottfried: monads 114
Lewis, G. N. 152, 155
life: common ground 119–20;
holomovement 102–3; implicate
order 101–2; origin of 27
light: and being 156; and energy 154;
frozen 140, 152; and mysticism
152–7; and vision 107–8
literature: socially informed 293
love: and communication 338–9; and
dialogue 302

magnetism: soma-significance 159,
160–2
maps, geographical: Mercator's
projection 155
maps, mental 39–40; children's
development of 46–7, 68
mathematics: children's development 47–8, 67; fractal geometry
289; of quantum theory 170,
185–6, 193; super-implicate order
144–5
matter: the body 115–17; and consciousness 104–5; conservative
nature of 253–4; dual aspect with
energy 174–5; life and 101–2;
structure of 87–91; transformations 18–19, 33, 34 see also atoms
Maxwell, James Clerk 185
meaning: actualizing 168–9; arouses
energy 180; Bohr and ambiguity
169–71; consciousness 171–2;
creativity and 180–2; and dialogue
290; dialogue and group culture
321–3; Einstein's reflections 162;
and intention 166–8, 298; and
knowledge 262; and meaninglessness 179–80; shared 330;
signa-somatic activity 179;
society and reality 176–82;
soma-significance 159–60,
171–2